Detection Methods for Irradiated Foods
Current Status

Detection Methods for Irradiated Foods
Current Status

Edited by

Cecil H. McMurray
The Department of Agriculture for Northern Ireland

Eileen M. Stewart
The Department of Agriculture for Northern Ireland and The Queen's University of Belfast

Richard Gray
The Department of Agriculture for Northern Ireland

Jack Pearce
The Department of Agriculture for Northern Ireland and The Queen's University of Belfast

THE ROYAL SOCIETY OF CHEMISTRY
Information Services

The Proceedings of an International Meeting on Analytical Detection Methods for Irradiation Treatment of Foods, held in Belfast 20–24 June 1994

Special Publication No. 171

ISBN 0-85404-770-0

A catalogue record for this book is available from the British Library

© The Royal Society of Chemistry 1996

For the chapters starting on pages 93, 121, 229, 269, 299, 326, 345, 349 and 377
© International Atomic Energy Agency 1996

All rights reserved.

Apart from any fair dealing for the purpose of research or private study, or criticism or review as permitted under the terms of the UK Copyright, Designs and Patents Act, 1988, this publication may not be reproduced, stored or transmitted, in any form or by any means, without the prior permission in writing of The Royal Society of Chemistry, or in the case of reprographic reproduction only in accordance with the terms of the licences issued by the Copyright Licensing Agency in the UK, or in accordance with the terms of the licences issued by the appropriate Reproduction Rights Organization outside the UK. Enquiries concerning reproduction outside the terms stated here should be sent to The Royal Society of Chemistry at the address printed on this page.

Published by The Royal Society of Chemistry,
Thomas Graham House, Science Park, Milton Road,
Cambridge CB4 4WF, UK

Printed in Great Britain by Bookcraft (Bath) Ltd.

Foreword

This book in the main comprises the scientific record of an International Meeting on Analytical Detection Methods for Irradiation Treatment of Foods held at The Queen's University of Belfast in Northern Ireland between 20-24 June 1994.

The meeting was attended by 57 delegates from 19 countries, of whom 21 were participants in a Research Co-ordination Programme on Analytical Detection Methods for Irradiation Treatment of Foods (ADMIT) sponsored by the Joint Division of two United Nations organisations, the Food and Agriculture Organisation and the International Atomic Energy Agency.

The International Meeting was used as an opportunity to hold the third and final Research Co-ordination Meeting of ADMIT, which duly took place during two sessions immediately preceding and following the main proceedings; the majority of ADMIT members also presented papers during the main proceedings.

As well as encouraging the basic development of detection tests, one of the functions of ADMIT was to critically assess the various test methods resulting from the research programme and to determine their suitability for general use by public health laboratories and others concerned with trade in irradiated foods, both national and international.

The Belfast ADMIT meeting re-confirmed the "General Principles for the Development of Detection Methods" which are ideally required or desirable. It was determined that it is not necessary for each method to exhibit all the characteristics listed below in order to be useful. However, the list serves as a standard for the "ideal" method against which proposed methods could be evaluated.

Two sets of criteria have been elaborated:

Technical criteria which are to be met if a qualitative or quantitative test is to be successfully elaborated, and

Practical criteria which are desirable if a method is to be applied widely by authorities concerned with the control of labelling of irradiated food moving in trade.

1   *Technical Criteria*

(a)   **Discrimination** - the parameter measured in the irradiated food should be absent in the non-irradiated food; alternatively the parameter should be well characterised in the non-irradiated food and changes induced by irradiation should be distinct and separable.

(b)   **Specificity** - other food processing methods and storage should not induce comparable changes to irradiation.

(c)   **Applicability** - the test should apply throughout the dose range relevant to the irradiation of the food tested.

(d) **Stability** - the parameter should be useful for at least the storage life of the irradiated food.

(e) **Robustness** - the measurement should be insensitive to the following effects, or its response should be known with sufficient confidence, *e.g.*:
- dose rate
- temperature at any stage of treatment or storage
- other storage variables (O_2, moisture *etc.*)
- further processing
- admixture with other foods

(f) **Independence** - the method should not require samples of the non-irradiated food from the particular batch tested.

(g) **Reproducibility** and **repeatability**

(h) **Accuracy** and proper statistical **validation**

(i) **Sensitivity** - the method should be capable of detecting doses below the commercially applicable dose.

(j) **Dose dependence** - the method should be capable of generating a dose response curve. This criterion concerns the measurement of dose applied to the food.

2 *Practical Criteria*

(a) **Simplicity** - the method should not demand high levels of technical skill, data interpretation or specialised equipment.

(b) **Low cost**

(c) **Small sample size**

(d) **Speed** of measurement.

(e) The method should apply to a **wide range** of food and food types.

(f) **Non-destructive** measurement of the parameter.

(g) The method should be capable of **easy standardisation** and **cross calibration**.

(h) Confidence that the method is **resistant to fraud**. It would be desirable if, *e.g.*, the parameters were inherent to food rather than to the associated packaging, mineral dusts *etc.*

Standards of Health and Safety Laboratory Practices should be adhered to.

Current Status of Analytical Detection Methods for Irradiation Treatment of Foods

	Meat		Fish		Egg		Shellfish		MRM	Seeds & Herbs	Grains	Cheese	Vegetables	Fruit	
	Bone	Flesh	Bone	Flesh	Shell	Contents	Cuticle	Meat						Fresh	Dried
ESR	B	G	B	G	D	G	D	G	D	C	F, R	F, R	D, R	D, R	D
TL	F, S	G	H, S	G	F, S	G	E, S	B	G	B	H	G	D, R	D, R	H
PSL	G	G	G	G	H	G	H	G	G	D	H	G	E, R	E, R	E, R
Hydrocarbons	G	B	G	E	G	B	G	E	H	E	H	B	H, R	B, R	H, R
Cyclobutanones	G	B	G	H	G	B	G	E	E	H	H	F	H, R	E	H
σ-Tyrosine	G	E	G	H	G	H	G	E	H	G	H	H	G	E	H
Gas Evolution	G	E, R	G	H	G	G	G	F, R	H	E, R	H	G	G	G	G
Peroxides	G	E	G	H	G	H	G	F	H	G	H	H	G	G	G
Immunological															
(1) Cyclobutanones	G	E	G	E	G	E	G	E	E	H	H	E	H	E	H
(2) DNA	G	H	G	H	G	H	G	H	H	H	F	H	H	H	H
(3) Protein	G	H	G	H	G	E	G	H	H	G	H	H	H	H, R	H, R
DNA - Comet	C, R	C	G	E	G	E	G	F	F	H	H	H	F	F	H
- MtDNA	G	D	G	E	G	G	G	E	F	H	H	H	F	F	H
Microbiological	G	E	G	H	G	H, R	G	E	H	B	H	G	G	G	H
Viscometry	G	G	G	G	G	G	G	G	G	B, R, S	G	G	G	G	G
NIR	G	G	G	G	G	G	G	G	G	E, R	H	G	G	G	G
Impedance	G	G	G	G	G	G	G	G	G	G	H	G	C, R	G	G
Germination	G	G	G	G	G	G	G	G	G	G	E	G	E	E	G

A = Agreed International Protocol, B = Protocol Under Evaluation, C = Intercomparison in Progress, D = Research Advanced, E = Research Underway
F = Preliminary Data, G = Not Applicable, H = Applicable but Not Tested, R = Applicable to Certain Types, S = Screening Method Only

MRM = Mechanically Recoverd Meat

The Belfast ADMIT meeting went on to evaluate the current status of each of the proposed methods for a defined set of foods and food products. The table provides an agreed summary guide to the status of each of the detection methods which are considered in detail in this book. It will be seen that tests for a wide range of potentially irradiated food items are in an advanced state of development. Certain tests have already been accepted by some national authorities as constituting standard tests for use by control authorities. These methods, and the principal foods for which they are applicable, are as follows:

Electron Spin Resonance:	Meat (bone), fish (bone), some fruits
Thermoluminescence:	Seeds and herbs (including spices), some fruits and vegetables, shellfish
Hydrocarbons:	Meat (flesh), liquid egg, cheese, some fruits
2-Alkylcyclobutanones:	Meat (flesh), liquid egg
Microbiological:	Seeds and herbs (including spices)
Viscometry:	Some spices
Impedance:	Some vegetables

It should be noted that more than one method may be applicable to a particular food group, thus allowing the possibility of multiple testing to increase confidence in results. Many more tests are in an advanced state of development by national laboratories, and work will undoubtedly continue to improve and refine these techniques to such an extent that they will be capable of adoption internationally.

CECIL H McMURRAY

Conference Chairman
and
Chairman of the 3rd ADMIT Meeting

Preface

Food irradiation is a food processing technology which has a number of important attributes. First of all, it can improve the safety of food through reduction of pathogenic microorganisms and secondly, by reducing those organisms which cause food spoilage the quality of food can be improved and the shelf-life extended. Like any other food processing technique the process must be used properly. The correct dose of irradiation (whether gamma rays, electron beams or X-rays) along with other processing variables, such as temperature and atmosphere must be applied. Close attention must be paid to these variables in order to achieve a high organoleptic quality which will ensure that the final product is acceptable to the consumer.

Extensive studies on the safety and wholesomeness of irradiated products have demonstrated the safety of the food produced. In spite of the many advantages of irradiation, consumers in many countries have remained sceptical of the applicability of this food processing technique. However, when it has been possible to carry out properly conducted consumer trials the advantages of irradiated food have become evident. Often these trials can be quantified in terms of improved freshness, appearance and flavour of irradiated food compared with non-irradiated controls. Because improved safety does not readily lend itself to direct consumer evaluation, the point is often not appreciated that many irradiated products are considerably safer on account of the reduction in *Salmonella spp.*, *Campylobacter spp.*, *E. coli* and other pathogenic microorganisms which can be present in untreated foods. There are particular advantages in reducing the pathogenic load of food for young, elderly and immuno-compromised people who are often most at risk from food poisoning.

One of the key issues which came to the forefront in determining consumer acceptance of irradiated foods was the demand by consumers and their representative organisations for methods which would discriminate between irradiated and non-irradiated products. This requirement led to the demand that irradiated food be labelled. In order to validate the correctness of labelling, analytical techniques were needed which would allow irradiated and non-irradiated foods to be differentiated. Thus, a practical basis was sought to allow consumers to exercise a free choice as to which food they wished to purchase.

This book is the proceedings of a symposium which was held in Belfast, Northern Ireland between 20-24 June 1994, and organised by a local committee under the auspices of the ADMIT (Analytical Detection Methods for Irradiation Treatment of Foods) Co-ordinated Programme sponsored by the Food and Agriculture Organisation of the United Nations and the International Atomic Energy Agency through their Joint FAO/IAEA Division of Nuclear Techniques in Food and Agriculture. The meeting was constituted as a full international conference open to a wide range of participants. A key component of the meeting was the third and final Research Co-ordination Meeting of ADMIT. Organising the conference in this way provided an ideal forum for enabling the presentation of all the techniques which had been suggested as being suitable for the detection of irradiated food and permitted a detailed evaluation of the effectiveness of each of the proposed methods. This book contains the text of the scientific papers presented at the conference.

The ADMIT programme had previously held two co-ordinating meetings in Jachranka, Poland (1990) and Budapest, Hungary (1992). Each of these meetings provided an opportunity for scientists active in the field to compare the progress on method development and most importantly, to devise an extensive set of criteria which enabled candidate techniques to be evaluated.

Readers of the detailed text which follows should be aware that at the time when the hunt for suitable detection tests began in earnest in the mid 1980s, previous research had failed to result in the development of methods which were sufficiently reliable and sensitive to be of practical use. Indeed, there were many people who thought the task inherently impossible given the very small changes which were known to be induced in food at the irradiation dose levels used (up to 10 kGy). However, now, as a result of an extensive research effort, the results of which are summarised in this volume, a range of methods using a wide variety of chemical, physical, and biological techniques is now available. In addition, these methods are capable of being used in a variety of ways. Some have the capability of acting as screening methods, while others can provide definitive discrimination which, if required, can be confidently used in defending claims in a court of law.

Some detection tests, such as luminescence, depend on changes in extrinsic components (*e.g.* adhering minerals) while other tests depend on changes to intrinsic components (*e.g.* production of 2-alkylcyclobutanones from food lipids).

One mandatory requirement for a suitable technique is that it must be able to discriminate at the irradiation dose levels used in commercial practice. Detailed perusal of the contents of the papers presented in this book will clearly demonstrate that the choice of method will also depend on the food. Some methods, such as electron spin resonance spectroscopy, will not only permit detection of primary irradiated products but will also allow detection of these irradiated products in secondary and tertiary foods (*e.g.* the detection of irradiated mechanically recovered chicken meat in a cooked hamburger) such is the sensitivity and specificity of the method.

One important facet of the work in developing methods has been the use of interlaboratory blind trials, several of which are reported here. These trials have had a number of important consequences. First of all, they have allowed the validity of a method to be established and secondly, permitted the transfer of the analytical methodology to other laboratories on a world-wide basis. Finally, the proof that methods do work has permitted some of them to be adopted as official methods in a number of countries and now also by the European Union (European Committee for Standardisation (CEN)). Such acceptance is due to the robustness, reproducibility and ease of use by which discriminating analysis can be carried out with confidence.

No doubt further research will continue to refine and provide for future innovation. However, there is no doubt that the contents of this book provide a landmark in the development of successful methods for discriminating between irradiated and non-irradiated foods should these be required to enforce labelling requirements as a means of satisfying the demands of some concerned consumers.

The success of the meeting and this record of the proceedings is due to the hard work of a large number of researchers internationally. All who have worked in this field are especially indebted to the work of two people who provided the inspiration and willpower

Preface

to move the science of the detection of irradiated foods forward. We refer specifically to two colleagues; Dr Hilary Stevenson and Dr David Deeble who tragically died on 5 October 1994 and 22 November 1993 respectively.

As a mark of tribute to their work we wish to respectfully dedicate this volume to them. Obituaries for each are included as a memory to their contribution to the field. The editors are particularly indebted to Hilary for the contribution she made in assisting with organising the Belfast meeting.

Two other names must be acknowledged, namely Dr Leslie Ladomery who during the lifetime of the ADMIT programme acted as its Scientific Secretary, and Paisan Loaharanu for his unstinting support over the years in co-ordinating and promoting work on food irradiation on a world-wide basis.

This book, while written by different authors, is a complete synthesis of the current status in the detection of irradiated food and should be a valuable source of reference to those who either wish to use the techniques or alternatively, seek to further advance the field.

CECIL H McMURRAY
EILEEN M STEWART
RICHARD GRAY
JACK PEARCE

Contents

General Introduction

Physical Mechanisms of Irradiation Technologies and their Characteristic Effects
J. McKeown and N.H. Drewell — 3

The Contribution of Analytical Detection Methods to the Enforcement of Good Irradiation Practice
D.A.E. Ehlermann — 14

Physical Methods
Electron Spin Resonance (ESR) Spectroscopy

EPR Spectroscopy for the Detection of Foods Treated with Ionising Radiation
W. Stachowicz, G. Burlinska, J. Michalik, A. Dziedzic-Goclawska and K. Ostrowski — 23

Detection of Irradiation Treatment in Crustacea by Electron Spin Resonance (ESR) Spectroscopy
E.M. Stewart, M.H. Stevenson and R. Gray — 33

Time Course Study of the EPR Spectra of Seeds of Soft Fruit Irradiated in Wet and Dry States
S.M. Glidewell, N. Deighton, A.E. Morrice and B.A. Goodman — 45

The Use of ESR Spectroscopy for the Detection of Irradiated Mechanically Recovered Meat (MRM) in Tertiary Food Products
M.H. Stevenson, E. Marchioni, R. Gray, E.M. Stewart, M. Bergaentzle and F. Kuntz — 53

ESR Dosimetry of Irradiated Chicken Legs and Chicken Eggs
S. Onori and M. Pantaloni — 62

ESR Detection of Free Radicals in Gamma Irradiated Spices and Other Foodstuffs
J.R. Pilbrow, G.J. Troup, D.R. Hutton and C.R. Hunter — 70

Identification of γ-Irradiated Spices by Electron Spin Resonance (ESR) Spectroscopy
S. Uchiyama, M. Murayama, Y. Kawamura and Y. Saito — 85

ESR Identification of Irradiated Foodstuffs: LARQUA Research
J. Raffi — 93

Interlaboratory Tests to Identify Irradiation Treatment of Various Foods via Gas Chromatographic Detection of Hydrocarbons, ESR Spectroscopy and TL Analysis
G.A. Schreiber, N. Helle, G. Schulzki, B. Linke, A. Spiegelberg, M. Mager and K.W. Bögl
98

Interlaboratory Trials of the EPR Method for the Detection of Irradiated Spices, Nutshell and Eggshell
M.F. Desrosiers, D.M. Yaczko, A. Basi and W.L. McLaughlin
108

Physical Methods
Thermoluminescence

Thermoluminescence and Photostimulated Luminescence Techniques to Identify Irradiated Foods
G.A. Schreiber
121

Recent Advances in Thermoluminescence and Photostimulated Luminescence Detection Methods for Irradiated Foods
D.C.W. Sanderson, L.A. Carmichael and J.D. Naylor
124

Luminescence Detection of Shellfish
D.C.W. Sanderson, L.A. Carmichael, J.Q. Spencer and J.D. Naylor
139

Comparison of Thermoluminescence Detection Methods for Irradiated Spices
Y. Kawamura, M. Murayama, S. Uchiyama and Y. Saito
149

Thermoluminescence Identification of Irradiated Foodstuffs: LARQUA Research
G. Lesgards, A. Fakirian and J. Raffi
158

Detection of Irradiated Spices by Thermoluminescence Analysis
K.M. Hammerton and C. Banos
168

The Detection of Irradiation of Foodstuffs by a Thermoluminescence Method, and the Limits of its Practical Applicability
J. Kispéter and L.I. Kiss
172

Thermoluminescence Detection of Irradiated Herbs and Spices: An Australasian Trial
P.B. Roberts and K.M. Hammerton
178

Other Physical Methods

Attempts to Elaborate Detection Methods for some Irradiated Food and Dry Ingredients — 185
S. Barabássy, M. Sharif, J. Farkas, J. Felföldi, Á. Koncz, Z. Formanek and K. Kaffka

Detection of Irradiated Potatoes by Impedance Measurement — 202
T. Hayashi, S. Todoriki, K. Otobe and J. Sugiyama

Applicability of Viscosity Measurement to the Detection of Irradiated Peppers — 215
T. Hayashi, S. Todoriki and K. Kohyama

Collaborative Study of Viscosity Measurement of Black and White Peppers — 229
T. Hayashi

Chemical Methods
Lipids

Progress in the Detection of Irradiated Foods by Measurement of Lipid-Derived Volatiles — 241
W.W. Nawar, Z. Zhu, H. Wan, E. DeGroote, Y. Chen and T. Aciukewicz

Identification of Irradiated Seafood — 249
K.M. Morehouse

Irradiation Detection in Complex Lipid Matrices by Means of On-Line Coupled (LC)LC-GC — 259
G. Schulzki, A. Spiegelberg, K.W. Bögl and G.A. Schreiber

Validation of the Cyclobutanone Protocol for Detection of Irradiated Lipid Containing Foods by Interlaboratory Trials — 269
M.H. Stevenson

The Use of 2-Substituted Cyclobutanones in the Development of an Enzyme-Linked Immunosorbent Assay (ELISA) for the Detection of Irradiated Foods — 285
L. Hamilton, C.T. Elliott, D.R. Boyd, W.J. McCaughey and M.H. Stevenson

Application of DCI to the Lipid Method — 293
J. Raffi, G. Lesgards, I. Pouliquen, P. Giamarchi and A. Fakirian

Other Chemical Methods

The Status of Detection Methods Based on Radiolytic Products 299
P.B. Roberts

Determination of o-Tyrosine in Shrimps, Fish, Mussels, Frog Legs and Egg-White 303
W. Meier, H. Hediger and A. Artho

Identification of Irradiated Foods by an Immunochemical Method 310
T. Kume and T. Matsuda

Use of the Peroxide Method for Identifying Irradiated Food 317
S.C. Qi and J.L. Wu

A Rapid and Simple Screening Test to Identify Irradiated Food Using Multiple Gas Sensors 326
H. Delincée

Gas Evolution as a Rapid Screening Method for Detection of Irradiated Foods 331
P.B. Roberts, D.M. Chambers and G.W. Brailsford

Determination of Hydrogen in Ice and in Irradiated Frozen Chicken 335
C.H.S. Hitchcock

DNA Methods

Introduction to DNA Methods for Identification of Irradiated Foods 345
H. Delincée

Application of the DNA "Comet Assay" to Detect Irradiation Treatment of Foods 349
H. Delincée

Detection of Irradiated Fresh, Chilled, and Frozen Foods by the Mitochondrial DNA Method 355
E. Marchioni, M. Bergaentzle, F. Kuntz, S. Todoriki and C. Hasselmann

Immunological Detection of Modified DNA Bases in Irradiated Food 367
J.H.H. Williams, A.L. Tyreman, D.J. Deeble, M. Jones, C.J. Smith, J.F. Christiansen and P.C. Beaumont

Biological Methods

Biological Methods for the Detection of Irradiated Foods 377
K.M. Hammerton

Development of Half-Embryo Test and Germination Test for Detection 383
of Irradiated Fruits and Grains
Y. Kawamura, M. Murayama, S. Uchiyama and Y. Saito

Detection of Irradiated Spices with a Microbiological Method - 392
DEFT/APC Method
K.M. Hammerton and C. Banos

Poster Presentations

Control of Food Irradiation in Denmark 399
T. Leth

Influencing Factors on ESR Dose Assessment in Irradiated Chicken Legs 401
S. Baccaro, P. Fuochi, S. Onori and M. Pantaloni

The Nature and Origin of the EPR Spectra from Irradiated Bone and 402
Hydroxyapatite
S.M. Glidewell and B.A. Goodman

Preliminary Studies on the Detection of Irradiated Prawns using 406
2-Alkylcyclobutanones
B.T. McMurray, W.C. McRoberts, J.T.G. Hamilton and M.H. Stevenson

Identification of Irradiated Fruit from the Pectin-Derived EPR Signal 408
N. Deighton, S.M. Glidewell, B.A. Goodman, G.P. McMillan and M.C.M. Perombelon

Practical Method for Detecting Irradiated Chicken and Turkey 410
D. Schwartz, L. Lakritz and K. Kohout

Application of a Microbiological Screening Method for the Indication of 412
Irradiation of Poultry Meat
G. Wirtanen, S. Salo, M. Karwoski and A-M. Sjoberg

Participants 415

Subject Index 423

Acknowledgements

We would like to express our gratitude to the following organisations for their sponsorship of the conference:

Joint FAO/IAEA Division of Nuclear Techniques in Food and Agriculture
Department of Agriculture for Northern Ireland
Queen's University of Belfast
Northern Ireland Tourist Board
Belfast City Council

Thanks are due to the people who chaired the various sessions and to all who contributed papers and discussions at the meeting. We also acknowledge William Graham, Lynne Hamilton, Brian McMurray and Solveig Moore whose co-operation and support contributed to the smooth running of the meeting. We are also grateful to Kathleen Dowds and her staff for secretarial support during the conference. Finally, we would like to thank the staff of the Royal Society of Chemistry for help in preparing this book.

OBITUARY

MARY HILL (HILARY) STEVENSON OBE, BSc, BAgr, MSc, PhD, CChem, FRSC, FIFST

21 January 1947 - 5 October 1994

Dr Hilary Stevenson was a native of the district of Ballyrashane, Coleraine, Northern Ireland and had a farming background. She attended Coleraine High School and then entered The Queen's University of Belfast from which she graduated with a First Class Honours Degree in both Chemistry (1969) and Agricultural Chemistry (1970). After graduating Hilary embarked on a career with the Department of Agriculture for Northern Ireland as a Lecturer at Loughry College of Agriculture and Food Technology. During her time at Loughry she was seconded to the University of Strathclyde where she obtained an MSc in Food Science and Food Microbiology. In 1974 she transferred within the Department of Agriculture for Northern Ireland to take up a position in the Agricultural Chemistry Research Division.

In 1974 Hilary was also appointed to the staff of the Faculty of Agriculture and Food Science at The Queen's University of Belfast as a lecturer and at the time of her death she was Reader in Food Science at the University. Throughout her career Hilary had wide-ranging teaching responsibilities in the Agriculture and Food Science courses. Her contributions, particularly in relation to the teaching of Food Chemistry and Human Nutrition, and on many other matters, were appreciated and valued both by colleagues and students alike. Her support, advice and encouragement as a supervisor was also greatly appreciated by her numerous post-graduate students during the course of their MSc and PhD studies. She was always a source of wise counsel.

Hilary's initial research interests were in poultry nutrition and she undertook studies in the mineral metabolism of the laying hen, the protein nutrition of broilers and nutrition and metabolism in the goose. She published extensively in these areas and obtained her PhD for studies on the mineral metabolism of poultry in 1981.

In the mid 1980s Hilary's research interests diversified and she will be remembered as one of the most influential scientists in the world in recent decades in developing the science of food irradiation. She played an important role in devising new methods for the detection of irradiated food, a goal which was sought by many governments and consumers. Her enthusiasm and commitment was an example to many others both within and outside her active area of research interest. In spite of a prolonged illness, she continued to make a substantial input to her field and, indeed, had a major role in bringing this international conference on *Analytical Detection Methods for Irradiation Treatment of Foods* to Belfast in June 1994. She was actively involved not only in the organisation of the meeting, but demonstrated to the conference that the work which she led was outstanding on a world-wide basis.

Hilary directed a number of international programmes and was research co-ordinator for a programme on behalf of the European Union's Community Bureau of Reference (BCR). She also actively participated in conferences organised by the International Atomic Energy Agency and the Food and Agriculture Organisation of the United Nations. She collaborated with many other workers across Europe, the United States of America and South Africa.

In the course of her work Hilary also made many contributions to scientific literature and had many invitations to speak on her subject. She contributed to and edited with a colleague the book *Food Irradiation and the Chemist* and at the time of her death was working on this book *Detection Methods for Irradiated Food - Current Status*. It was for her contribution to the field of irradiation detection methods that Hilary was awarded the Order of the British Empire (OBE) in the 1993 Birthday Honours.

Hilary was a Fellow of the Institute of Food Science and Technology (IFST). She played a full and active part in Branch affairs and had served as Chairperson of the Northern Ireland Branch Committee of IFST. She was also a member of the Nutrition Society and a Fellow of the Royal Society of Chemistry. As well as her research activities in poultry science, Hilary also contributed notably to the success of journal *British Poultry Science* through her involvement in refereeing. She also served on the Council of Management and Business Committee of British Poultry Science Limited. Her role in these areas was greatly valued.

Hilary's input to both poultry science and food science was significant and would have been much greater but for her untimely death. As a person who made wide-ranging contributions she will be sadly missed and, in particular, she will be remembered for her professionalism, integrity, humanity and fortitude.

This deep sense of loss is felt by her colleagues, associates and friends and most acutely by her husband Noel who was always a great support to Hilary, especially during her illness.

Hilary Stevenson will be remembered as a person who radiated warmth and friendliness towards everyone she met, and through the many people who had the privilege of knowing her, the memory of a truly exceptional person will live on.

OBITUARY

DAVID JOHN DEEBLE BSc, PhD

16 March 1949 - 22 November 1993

David Deeble graduated from the University of Newcastle-upon-Tyne in 1969 with a First Class Honours Degree in Chemistry. He continued his academic studies at the University and whilst studying with George Scholes for the degree of PhD, David began to study the effects of radiation-induced damage on nucleic acids. After completing his PhD studies, he worked for two years with the Northumbrian Water Authority as a Trade Effluent Officer and Chemist before returning to his former University to pursue teaching and research. In 1982, David moved to the Max-Planck Institut für Strahlenchemie, Mulheim to work with Professor Clemens von Sonntag. During his time in Germany he met his future wife, Alina. In October 1985 David returned to the UK to work at the North East Wales Institute (NEWI).

During his time at NEWI, David's research interests continued to focus upon the effects of radiation on nucleic acids and in that field of chemistry he really was one of the world's leading experts. In addition he developed a broad interest in the effects of free radicals on carbohydrate systems, in particular hyaluronic acid, and the role of such reactions in arthritis. David was a key researcher in the Ministry of Agriculture,. Fisheries and Food (MAFF) funded project at NEWI looking at assays for detecting irradiated food.

As a teacher of students, David was both well liked and respected in view of his wide circle of knowledge. In particular he cared about his students and their progress.

Just prior to his death, David had started to give 'schools-type' lectures in chemistry. On these occasions he was an assured presenter always ready to adapt the content of the lecture to suit the needs of the audience. In many ways this typified his approach - always willing to talk to people about science but always at their level of understanding.

We were devastated by the sad news of his death. We try to draw comfort from the following:- (i) we knew him; (ii) we could call him our friend and (iii) he would have called us his friend. They say only the good die young, in our book 44 is still very young.

Paul C Beaumont
John H H Williams

General Introduction

PHYSICAL MECHANISMS OF IRRADIATION TECHNOLOGIES AND THEIR CHARACTERISTIC EFFECTS

J. McKeown and N. H. Drewell

AECL Accelerators
436B Hazeldean Road
Kanata, Ontario K2L 1T9
Canada

1 INTRODUCTION

The search for a reliable marker of irradiated food is widespread. Presently, leading methods of identification are (1) electron spin resonance (ESR) spectroscopy to detect and quantify radiation treated meat containing bone,[1-4] and (2) gas chromatography to determine lipid volatile hydrocarbons in irradiated fatty foods. Other methods based on chemical (*e.g.* o-tyrosine in poultry[5]) and microbiological (Acinetobacter and Moraxella[6]) principles show potential, but there are significant practical difficulties.

Most reported work in this field uses ^{60}Co as the source of radiation. The limited availability of ^{60}Co will likely require that a significant part of future world capacity will have to be supplied by electron accelerators. Therefore, electron accelerators must be included in any search for a radiation marker. A key parameter that distinguishes a ^{60}Co facility from most electron-beam facilities is the dose rate applied to the product. Any dose rate dependence of existing and future markers must, therefore, be established. In this paper, advantage is taken by the authors of the accessibility to a range of S-band and L-band linear accelerators in research and commercial operating environments to explore specific characteristics that could lead to a key identifying parameter.

2 THE INTERACTION PROCESS

Chemical and biological effects of radiation are the results of electron collisions. The electrons can be introduced to the treated product directly from an accelerator source or indirectly by photons generated either by a radioactive source or from accelerator-produced electrons that pass close to a heavy nucleus. The spatial and temporal distribution of the electron flux generated in the product is characteristic of the source, and classification could depend on free radical densities and life-times.

Differences in the directivity of the sources and in penetration of the primary emission, determine the required exposure time for a given dose. As World Health Organization regulations for food permit a maximum energy of 10 MeV for electrons and 5 MeV for Bremsstrahlung irradiation, Figure 1 shows that for a single-sided irradiation with a

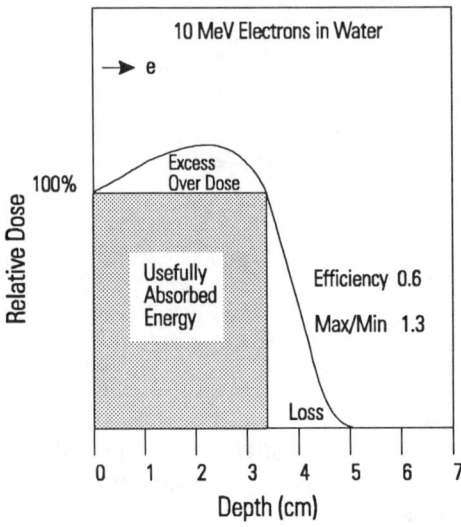 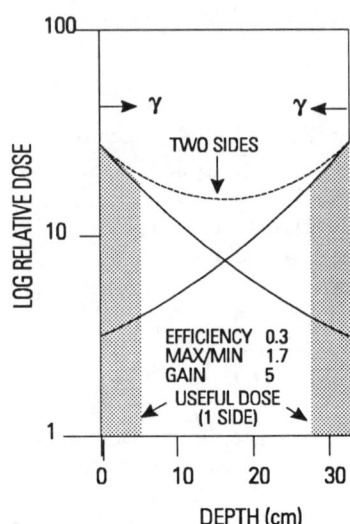

Figure 1 Variation in dose for monoenergetic electrons incident normally on a water surface

Figure 2 Variation in dose in water for two-sided irradiation with a Bremsstrahlung beam from 5 MeV electrons incident on a tantalum target

parallel beam of 10 MeV electrons, all of the usefully absorbed energy is deposited in the first 3.4 cm of water. This definition relates to a convention that equates input dose to exit dose. Consequently, about 60% of the input dose is considered useful and the max/min ratio is 1.3. Using 5 MeV electrons and a Bremsstrahlung converter in front of the target, most of the useful energy from the resultant photon beam would pass straight through the target as shown in Figure 2, if action were not taken to improve efficiency. Double-sided irradiation allows up to 32 cm of water to be irradiated with a max/min ratio of 1.7 and an efficiency of 30%.

3 DOSE DISTRIBUTIONS FROM DIFFERENT TECHNOLOGIES

The above data for parallel beams are specific to accelerator-based radiation. Photons from a radioactive gamma source are, of course, emitted isotropically. In a typical cobalt irradiator, the source pencils are mounted in a rack of area of approximately 10 m^2. Nevertheless, when this directionality and source density are compounded with the more penetrating aspect of gamma rays, the range of dose intensity available from cobalt and electron technologies exceeds five orders of magnitude. Accelerator-based photons are somewhere in between.

The temporal behaviour of dose intensity from cobalt irradiators and S-band linear accelerators (linacs) covers a similar range. This is a consequence of distinct phenomena, because the cobalt irradiator emits continuously, whereas the linac generally emits for only a small fraction of the time.

If the instantaneous dose rate of an rf linac is R_D, then this must be integrated over the time duration of the rf micropulse t_b (25 to 75 ps). The target is peppered with these micropulses during the time duration, t_m, of the modulator pulse (1 to 200 μs). The duty factor of the machine is typically 0.1 to 5.0%, hence the dose is summed over all modulator pulses at the pulse repetition rate of f_r during the time interval t_c that the product is under the beam. Practical considerations usually dictate that the highly directional electron beam must be diffused across the target with frequency f_s, and $G(t, f_s, f_r)$ describes the appropriate scanning function. The charge distribution of the beam in the direction of travel of the conveyor is represented by $f(x)$, and the irradiated product traverses the beam of width x^1 on the conveyor in the time t_c, hence the total dose D_T can be represented by:

$$D_T = \frac{1}{x^1 \cdot t_c} \int_0^{x^1} f(x) \int_0^{t_c} G(t\ f_s\ f_r) \sum_1^{t_c \cdot f_r} \sum_1^{t_m/t_b} \int_0^{t_b} R_D\ dt\ dt\ dx$$

The parameters t_c, x^1, f_s, f_r, t_m and t_b are specific to the accelerator-based facilities available to the authors and except for t_b, could be made free parameters as required by the experiments.

Details of the equation are illustrated in Figure 3, which shows pictorially, the dose rates available over various averaging times with different technologies. Details of the generic temporal behaviour of irradiation products created by the IMPELA® accelerator are shown, for illustration purposes, in Figures 3b and 3c for time scales near t_b and t_m.

The top three of the four traces in Figure 3b represent concentrations of products whose instantaneous growth rates are proportional to the dose rate, but which decay at differing rates. Of these, only trace A would normally be of interest as a marker, because it does not decay on this time scale, and therefore its long-term growth is proportional to the dose. Trace B decays appreciably during the rf cycle, and C decays quickly enough so that its profile essentially mimics that of the dose-rate profile itself. The bottom trace (D) shows the build up of a species that would be the chemical product of two short-lived populations illustrated by trace C. Barring saturation effects, its instantaneous growth rate is proportional to the square of the dose rate, but in any case is non-linear with dose rate, such as are radical-radical interactions. When the dose and dose rate are sufficiently high, this interaction could lead to observable dose rate effects. Again, as a potential marker, it is shown growing without decay on this time scale.

Traces A, B and D are continued in Figure 3c, on the somewhat longer time scale of the modulator pulse. Potential markers A and D continue to build up. Species B does not build up on this time scale and essentially follows the modulator pulse profile.

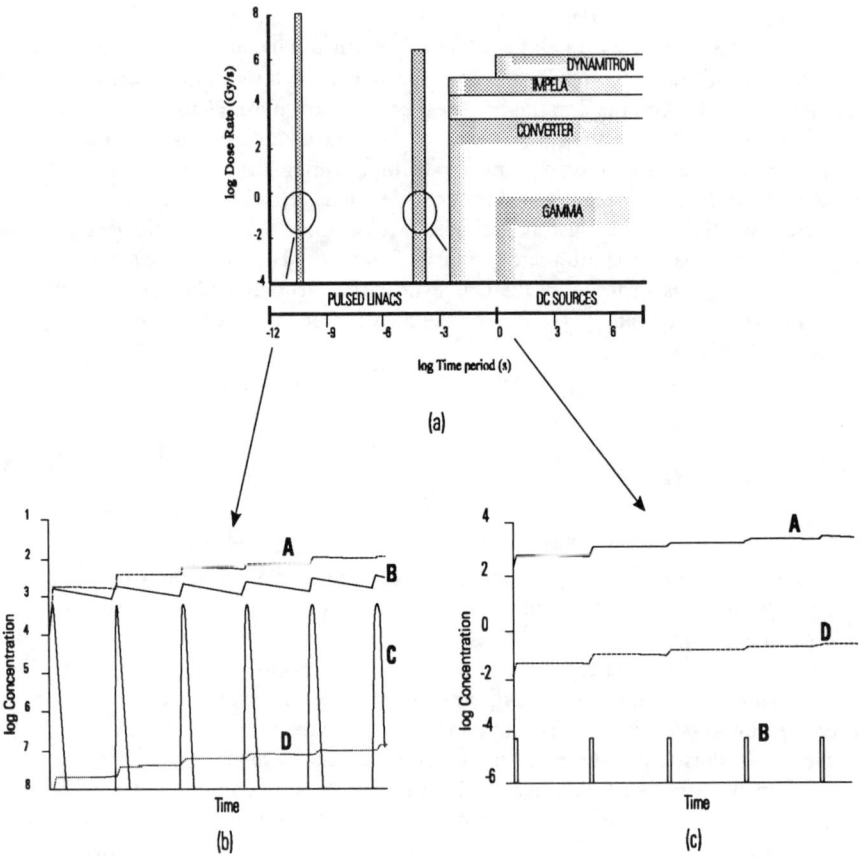

Figure 3 (a) *Variation in dose rate with time for different technologies*
(b) *An exploded view of radiolytic response in the rf pulse time scale*
(c) *An exploded view of radiolytic response in the modulator pulse time scale*

The effects of dose rate, especially via linear accelerators, have been extensively studied by McLaughlin *et al.*[7] The motivation in those experiments was to investigate the relative response of different plastic and dyed plastic dosimeters as a function of dose rate when irradiated with gamma radiation and electron beams. Figure 4 reproduced from McLaughlin's paper shows the pronounced effects. The present objective is to explore the possibility of similar effects in irradiated food that are uniquely characteristic of the dose rate.

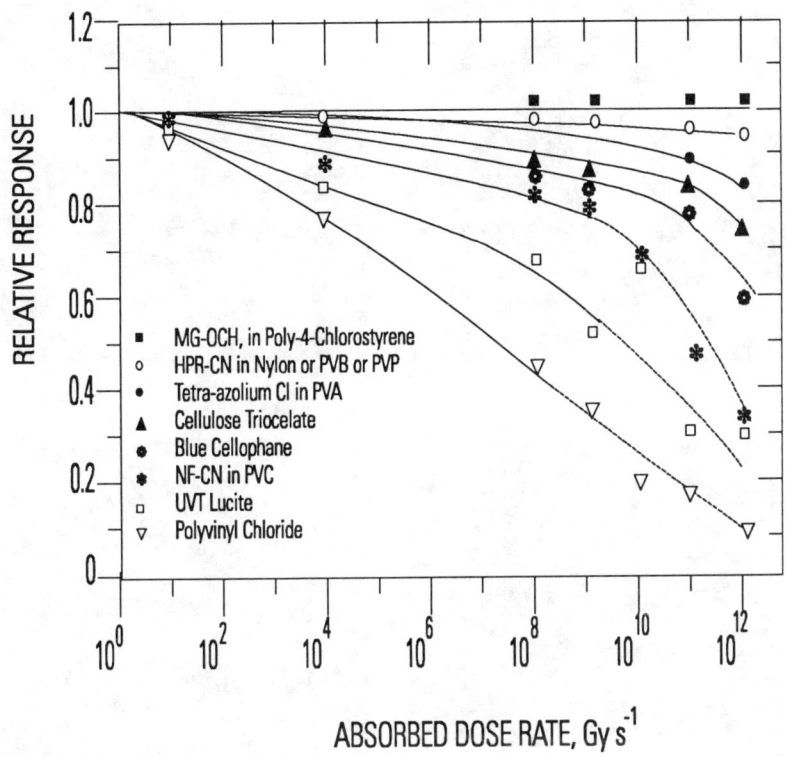

Figure 4 *Changes in relative response for different plastic and dyed plastic dosimeters as a function of absorbed dose rate for a total dose of 25 kGy (McLaughlin, et al., 1979b[7] Reproduced by permission)*

4 IRRADIATION FACILITIES

The characteristic parameters of facilities involved in the search are listed in Table 1. The Gamma Cell 220 and the I-10/1 accelerator are situated in the Whiteshell Laboratory, Manitoba. The IMPELA prototype and PHELA accelerators are located at the Chalk River Laboratory, Ontario. The first commercial IMPELA is installed at an irradiation facility owned by E-BEAM Services, Cranbury, New Jersey and the second commercial IMPELA at an irradiation facility owned by Iotron Industries Canada Inc., Vancouver, British Columbia. Photographs of the same facilities are shown in Figure 5. Two experiments were performed in an attempt to find a unique marker from the effects of radiation on selected targets using these facilities.

Gamma Cell - Whiteshell

I - 10/1 - Whiteshell

IMPELA - New Jersey

PHELA - Chalk River

IMPELA - Chalk River

IMPELA - Vancouver

Figure 5 *Photographs of irradiation facilities using different technologies*

Table 1 *Characteristic Parameters of the AECL Facilities. Optimization and Calibration have Resulted in Small Differences in Operational Parameters.*

Facility	Pulse width (µs)	PRF (Hz)	Energy (MeV)	Power (kW)	Average Dose Rate (Gy s^{-1})
Gamma Cell 220					1.6
Photon Converter	200	250	5	3	10.0
I-10/1	4	300	10	1	3.1E02
PHELA	6	250	12	4	1.2E03
IMPELA prototype	200	250	10	1-50	1.6E04
IMPELA New Jersey	208	240	10	10-50	1.6E04
IMPELA Vancouver	197	254	10	10-50	1.6E04

5 RESULTS

5.1 Microorganism Sensitivity

A primary objective of the first series of experiments was to compare the relative efficacies of electron-beam and gamma ray irradiation for the inactivation of microorganisms. Prepared specimens of different species of microorganisms were irradiated[8] with gamma and electron-beams from the facilities described above at the Whiteshell Laboratory. The irradiation was done under wet and dry conditions. Figure 6 shows a comparison between the I-10/1 and ^{60}Co gamma rays for *Bacillus pumulis*. For dried film, the sensitivity of the microorganisms is indistinguishable, but for wet conditions, electron-beam would seem to be more efficient

Oxygen depletion in and around the microorganism was investigated by Saunders *et al.*[9] by comparing electron-beam from the short pulsed I-10/1 and gamma sterilisation for selected spores and vegetative bacteria in a wet medium irradiated in air. The results are shown in Table 2. The pulsed electron-beam delivered an average dose rate of 130 Gy s^{-1} while the average gamma dose rate was 2 Gy s^{-1}. The dose required for a 1 log 10 reduction (D_{10}) in the viability of these microorganisms was either unaffected or lowered when using high dose rate electron-beam treatment rather than gamma treatment.

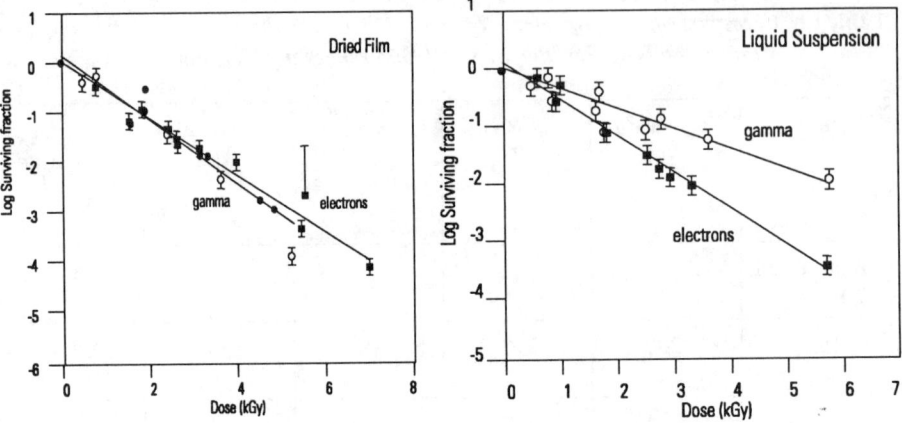

Figure 6 *The surviving fraction of Bacillus pumulis as a function of dose for gamma rays and the I-10/1 linac under wet and dry conditions*

All of these results illustrate that, on average, electron-beam sterilization is at least as efficient as gamma treatment, and bring into question the role of oxygen diffusion as a limiting step for inactivating microorganisms by radiation. The fact that high dose rate

Table 2 *Effect of Dose Rate on the Survival of Selected Spores and Bacteria*

Type of Spore or Bacteria	D_{10} Value[b] (kGy)		Sterilization Dose[c] (kGy)	
	Gamma Cell	I-10/1	Gamma Cell	I-10/1
Clostridium sporogenes:[a]				
B-9	2.01	1.92	16.1	15.4
B-10	2.57	1.99	20.6	15.9
Listeria monocytogenes:				
Scott A	0.24	0.23	1.9	1.8
81-861	0.45	0.20	3.6	1.6
Escherichia coli	0.43	0.17	3.4	1.4

[a]*Clostridium sporogenes:* batch B-9 and B-10 obtained from the American Type Culture Collection (ATCC-13325); *Listeria monocytogenes:* strains Scott A and 81-861 obtained from Health and Welfare Canada; *Escherichia coli:* obtained from the American Type Culture Collection (ATCC-25922)
[b]D_{10} dose inactivates 90% of the organisms present
[c]Based on a sterility assurance level (SAL) of 10^{-6} and an initial bioburden of 100 organisms/unit

General Introduction

electron-beam treatment is at least as efficient as gamma irradiation, suggests that either enough oxygen is present within microorganisms to allow lethal oxidation reactions to occur, or that, as the dose rate is increased, the decrease in the number of lethal oxidation reactions is more than compensated for by other reactions that do not involve oxygen.

The relatively small differences provide little promise of a distinguishing marker between the short pulse linac and the cobalt irradiator.

A related series of experiments[8] was conducted at the Chalk River Laboratory using an IMPELA accelerator. This machine has a lower frequency (1300 MHz) and the ability to vary the width of the modulator pulse. Similar experiments were carried out previously by Tallentiere *et al.*[10] Figure 7 shows the surviving fraction of *Bacillus stearothermophilus* as a function of dose for both wet and dry film using 40 μs and 200 μs pulses. Pulse length seems to have no effect and any difference in sensitivity between electrons and gamma rays is small.

5.2 ESR Dosimetry

As ESR spectroscopy is now well established as a marker for irradiated food, experiments[11] were performed at Chalk River with free radicals trapped in commercial alanine/paraffin dosimeters and in-house sucrose/silicone dosimeter pellets to check the dose rate dependence of the ESR signal. Figure 8a shows the calibration curve for the ESR signal from alanine as a function of absorbed dose in a gamma field, as provided by Nordion International with their calibrated source, certified by the U.S. National Institute of Standards and Technology (NIST). This calibration curve was used to determine the dose generated by an electron beam from IMPELA at Chalk River as the power level was increased by increasing the pulse width. A second determination of the dose was obtained using Far West Technology dosimeters as transfer dosimeters from the gamma

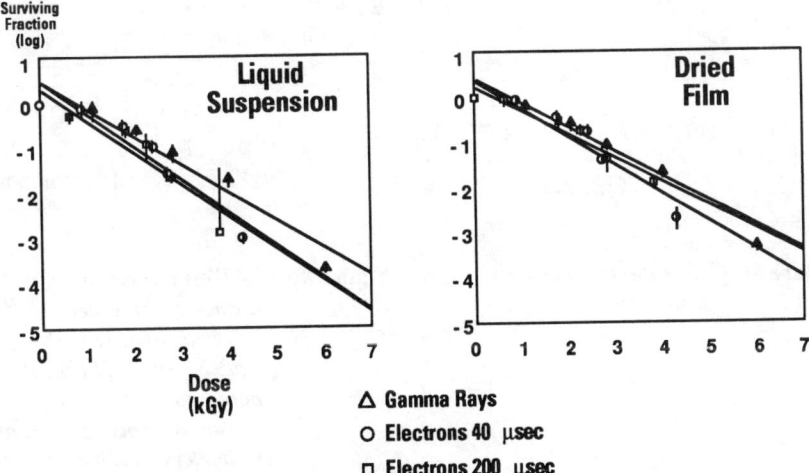

Figure 7 *The surviving fraction of Bacillus stearothermophilus as a function of dose for wet and dry film using the variable pulse width IMPELA beam*

cell at Whiteshell, which was previously calibrated against a NPL standard. Since the dose registration mechanism is entirely different in the two dosimeters, it is highly likely that the linear curve in Figure 8b demonstrates a dose rate independence for both ESR spectrometry and Far West dosimeters. It also follows that the interlaboratory calibrations are consistent to within 5%.

6 CONCLUSION

There is strong evidence that the ESR signals are independent of dose rate over a range from a few Gy s^{-1} to 10^8 Gy s^{-1}. Thus the ESR signal may be used as a reference signal when determining the dose rate response of other potential markers. Preliminary data suggests that the viability of tested bacteria does not depend strongly on dose rate.

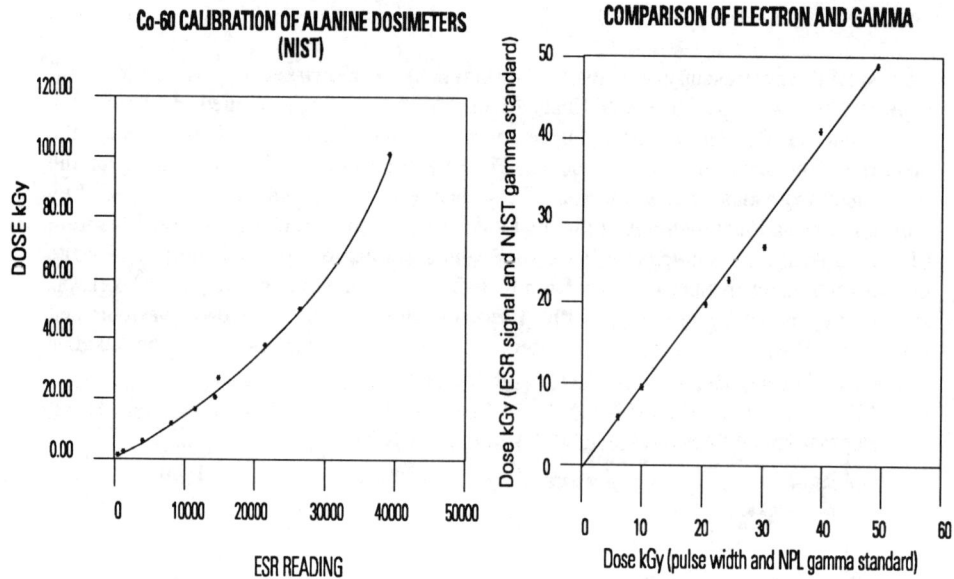

Figure 8a *Calibration of alanine dosimeters*

Figure 8b *Far West dosimeters calibrated in the gamma cell give the expected dose as the pulse width of IMPELA is increased. The dose as measured using alanine dosimeters calibrated from NIST Gamma standard is plotted on the ordinate*

Acknowledgments

The support of our colleagues at Chalk River Laboratories and Whiteshell laboratory is gratefully acknowledged. In particular, D. Smyth, J. Borsa, L. Lucht, W. L. McLaughlin and C. Saunders provided vital information from their work published elsewhere.

References

1. N. Chauqui-Offermanns, *J. Radiat. Steril.*, 1992, **1**, 29.
2. J. S. Lea, N. J.F. Dodd and A. J. Swallow, *Int. J. Food Sci. Technol.*, 1988, **23**, 625.
3. M. H. Stevenson and R. Gray, *J. Sci. Food Agric.*, 1989, **48**, 261.
4. R. Gray, M. H. Stevenson and D. J. Kilpatrick, *Radiat. Phys. Chem.*, 1990, **35**, 284.
5. L. R. Karam and M. J. Simic, "Health Impact, Identification, and Dosimetry of Irradiated Foods," WHO, Copenhagen, 1988, p. 297.
6. P. Colin, C. Lahellec, G. Bennejan, M. J. Laisney and M. T. Toquin, *Viandes et Produits Carnes*, 1989, **10**, 17.
7. W. L. McLaughlin, A. C. Lucas, B. Kaspar, and A. Miller, *Radiat. Phys. Chem.*, 1979b, **14**, 467. See also "Dosimetry for Radiation Processing," Published by Taylor and Francis, p. 187.
8. L. Lucht, J. Barnard, J. Borsa, D. Smyth and C. Saunders, *J. Radiat. Steril.*, **2**, 1994.
9. C. Saunders, L. Lucht and T. McDougall, *J. Med. Dev. Diag. Indust.*, **May**, 1993.
10. A. Tallentiere, D. J. W. Barber, J. Booz and H. G. Ebert, "Interrelationship Between Pulse Length and Dose Rate within the Pulse in the Enhancement of Anoxic Damage in Wet *Bacillus megaterium*," Symposium on Microdosimetry, Belgium, **May**, 1978 p. 1061.
11. C. L. Greenstock, A. Trivedi, D. Smyth and G. Van Dyk, "Pulsed Electron Beam Measurements on the IMPELA-10/50 Machine Using ESR Dosimetry," 38th Health Physics Society Annual Meeting, Atlanta, **July**, 1993.

THE CONTRIBUTION OF ANALYTICAL DETECTION METHODS TO THE ENFORCEMENT OF GOOD IRRADIATION PRACTICE

D. A. E. Ehlermann

Institute of Process Engineering
Federal Research Centre for Nutrition
Karlsruhe
Germany

1 INTRODUCTION

Good Manufacturing Practice (GMP) is a generally accepted principle combining several features in order to achieve the optimal quality. In other words, GMP is observed if every possible measure is taken to contribute to the best possible quality of the end product. In this sense, Good Irradiation Practice (GIP) may be understood as one of the many elements of GMP dealing especially with aspects of processing food using ionising radiation. On the contrary the operator of an irradiation facility may consider GIP as the central idea including that GMP - in this case excluding the radiation aspects - is observed in addition to GIP standards. Regardless of this theoretical dispute, it is obvious that food irradiation requires classical GMP to be applied without exception and that good practice is also applied in radiation processing.

The main interest in analytical methods for the identification of radiation processed food arose from the need to enforce the ban on this process, to verify correct labelling, or to ensure that it is used only for the very limited number of applications which are permitted. In this field, identification methods already introduced into the official food inspection systems have contributed considerably towards making evident several cases of the fraudulent application of radiation processing. At present, as radiation processing of food is becoming more and more accepted, the number of national clearances is increasing and the European Community is preparing for a Directive to harmonise the food laws of the member states with regard to food irradiation. Therefore, it should be considered how the analytical detection methods which have been developed could contribute to enforcing good manufacturing practices, once the main goal ceases to be the suppression of this process.

Before coming to the possible contributions of analytical detection methods in the enforcement of GIP it is advisable to discuss and elucidate some components of GIP and the objectives of regulations on food irradiation.

2 THE SELF-CONTROLLING NATURE OF THE PROCESS

Processing of food by ionising radiation is, in principle, a completely self-controlling technique: any treatment outside reasonable limits is not effective in achieving the goals and aims of the process. There is no scientific or technical reason to put additional or tighter limits on the process. For economic reasons the operator of an irradiation facility has no interest in not fulfilling the contracted treatment and, in doing so, becoming liable for tainting the goods. Of course, radiation processing is not self-controlling in the physical sense that a stable minimum for the optimum treatment exists and some inherent forces for a too high or too low treatment readjust the process parameters for the optimum treatment. The dose range around the optimum treatment within which the economic correction forces are not pronounced is the technical dose range for the process. A commercial irradiation facility would always, for reasons of technology and facility design, deliver a certain range of treatment instead of a uniform dose throughout the goods.

The critical quantity which governs radiation processing is radiation absorbed dose or 'dose'. These *reasonable* dose limits are:

> on the low-dose side, the technically effective minimum dose necessary to achieve the process aims (*e.g.* complete sprout inhibition, ripening delay, reduction of spoilage causing microorganisms, elimination of pathogenic microorganisms and organisms, or sterility)

> on the high-dose side, the maximum tolerable dose for the product in order to prevent impairment of quality (*e.g.* off-flavours, colour changes, physiological damage).

These limits are termed 'reasonable' as their values are derived from laboratory studies and adjusted for the need of industrial practice, and take into consideration any health aspects and, also of legal requirements. To balance between such limits requires the skill of the operators of irradiation facilities. It is of economic interest to the operator to process any item at a dose as low as possible, close to the technical minimum. An unnecessarily high dose implies absorption of excess energy for which no returns are obtained; treatment outside the technical limits may cause losses of the product for which the operator of the irradiation facility is liable.

It is as a consequence, as long as regulations on food irradiation allow for a technically effective dose range, that the self-controlling nature of radiation processing would be sufficient to control the process in detail. There is no health hazard to the consumer from a treatment of any product outside of these reasonable limits; in a case where the quality of the product is impaired the consumer may complain as for food processed by traditional methods. This economic correcting effect could be very strong, *e.g.* when potatoes sprout because of too low a treatment or are glassy and too sweet (like under-chilled potato) because of overtreatment. Consequently, there is no justification for special restrictions. It should not be overlooked that the dose limits called 'reasonable' vary over a wide range from product to product or from intended effect to effect; however, the selection of doses for a given treatment would always be governed by the technical possibilities (*i.e.* the dose homogeneity an irradiation facility is designed for) and the self-controlling nature of the process.

3 THE OVER-REGULATION IN CLEARANCES

Aiming to prevent misuse or fraudulent use of radiation processing, authorities tend to impose rather tight regulations with regard to dose limits. In some cases these limits do not allow for the range of technically useful doses. The whole approach to regulating a process in such detail with specific dose limits for a given situation, application and product is inherently an over-regulation compared to clearances for classes or groups of similar products or applications; however, the latter approach requires wider dose ranges to be permitted. An excessively strict regulation requires that, for any minute changes, a new application for a revision of the regulated dose ranges is needed.

4 THE MINIMAL REGULATION

What is really necessary and indispensable in a regulation is to set the general frame for dose ranges in radiation processing. It is the function of authorities to protect the health of the public. In this respect, the 'overall average dose', currently 10 kGy, at which any radiation processed food is judged wholesome by Joint Expert Committee on the Wholesomeness of Irradiated Foods (JECFI) and is also the recommended dose by Codex Alimentarius, is the reference quantity for regulations. This setting of an average value implies that half of the total mass under consideration receives a dose above this average; the Code of Practice of the Codex Alimentarius Standard sets a limit in a way that at least 97.5% of the mass under consideration must receive a dose below 15 kGy. However, it should be kept in mind that regulations of this kind are difficult to control and to enforce on individual items. The quantity 'overall average dose' as defined by JECFI and Codex Alimentarius cannot be determined under practical circumstances. Further regulations would be needed for only a few exceptional cases, *e.g.* to permit explicit radiation processing of spices with doses up to 30 kGy as in the USA. While not prescribing the details of the process, such a generous regulation relies on the self-controlling nature of the process. The adherence to technological and reasonable dose limits can be controlled best at the irradiation facility and in the log-book there.

5 GMP AND GIP

Pre- and post-irradiation storage and handling must be chosen in order to comply with GMP and to preserve the quality of the product. The main and essential component of GIP is the control of the dose applied. GIP aims at the technically achievable lowest possible uniformity ratio, *i.e.* a narrow range between maximum and minimum doses. It also aims at achieving the technically effective minimum dose throughout the product. It takes into account that during the irradiation process GMP applies (*e.g.* temperature control). Also, any maximum dose limit imposed must be respected for legal reasons or in order to avoid impairment of quality. In the following considerations, most attention is paid to those elements where analytical detection methods may contribute. This is dose control.

6 GENERAL CONTRIBUTIONS OF ANALYTICAL DETECTION METHODS

Originally intended to enforce the ban on food irradiation, the majority of identification methods have been refined to detect radiation treatments at dose levels considerably lower than those which are technically effective. Therefore, the argument that radiation processing of food must not be permitted for the reason that analytical detection of the treatment is impossible is no longer valid. Consequently, the labelling of any product may be verified if the radiation treatment was well above the detection limit and, hence, a technically useful and justified treatment by ionising radiation had been applied, *i.e.* that GIP was observed. Products correctly irradiated and labelled as such must show analytical signals proving such a treatment. On the contrary, the product not labelled as irradiated must not show any signal above the background or detection level. However, the latter obviously does not contribute to the enforcement of GIP. It should be noted that by principle a non-treatment cannot be proven analytically regardless of how low the detection limit might be as long as this limit is larger than zero.

7 SPECIAL CONTRIBUTIONS OF ANALYTICAL DETECTION METHODS

As long as regulations are restricted to setting reasonable maximum dose limits taken from practical experience and technical data, identification methods may in the future contribute to enforce adherence to such limits. There is a consensus that any excessive or unnecessary treatment - for any food process - should be avoided. The identification methods could contribute considerably to the estimation of maximum doses applied. Once the correlation between detection effect and the dose applied is well established the resulting data could be used to extrapolate to the maximum dose. For this purpose the characteristics of the resulting frequency distributions determined from dose mapping studies can be compared with the characteristics obtained in analytical detection studies. Obviously, this requires that larger numbers of measurements on a greater series of samples are analysed. Such data are usually rendered from strategically planned, randomized sampling. However, a single estimated dose falling too close to limiting values, or even exceeding such values could be taken as clear evidence of an excessive treatment.

It should be noted that, up to the present, reliable data on shapes and variations of dose distributions in food irradiation under practical conditions have not been collected and published. Operators of irradiation facilities have no interest of their own in such data, the concept of minimum effective and maximum tolerable doses is preferred. If available at all, such data is considered basically proprietary. In a different context, radiation processing of fresh fruit for quarantine purposes, the knowledge of such frequency distributions and their fluctuations would be essential, when it comes to the judgment of the effectiveness of such treatment from reference dose measurements available for actual inspection (cf. IAEA Coordinated Research Programme on "Standardized Methods to Verify Absorbed Dose of Irradiated Fresh and Dried Fruits,

Tree Nuts in Trade" which commenced in 1994). It is to be expected that such dose distributions and their variations are influenced by packaging geometry, size and compositions of the fruit, proportions of the voids *etc*. Dosimetric characterisation of such situations is still under study and also methods of analytical dose estimation, especially the use of DNA methods on the irradiated insects found on the fruits, might emerge in the future.

The case is more difficult when a minimum dose value is to be enforced. If this minimum is well above the detection limit of the analytical method and the response function for the particular signal is accurate and reproducible, then maximum as well as minimum doses may be estimated. The example of the clearance for chicken in the USA and the set limits of 1.5 kGy for the minimum dose and of 3.0 kGy for the maximum dose is a challenge for the reliability of dose estimations through analytical detection methods applicable to chicken meat and bones.

8 DETECTION METHODS VERSUS INSPECTION OF RECORDS

For the control of food irradiation and the enforcement of GIP the inspection of records at the irradiation facility is indispensable. There, at the facility, is the only reliable information available on dose distributions, treatment planning and target dose settings as well as the actual process control measurements. However, this information may often not be available or easily accessible. Typically, this applies to import situations when according to the provisions of Codex Alimentarius the "relevant shipping documents shall give appropriate information to identify the registered facility ..., the date of treatment and lot identification" but obviously not any information about the dose applied. In such cases the considerable effort required to estimate dose values from analytical data and to extrapolate for extreme values can be justified. But also, in a standard situation of sampling from the market, the analytically estimated dose values can be traced back to the log-books of irradiation facilities. From such data collected throughout a prolonged period, a useful link may be established. Such data also, in the hands of a competent authority, renders comprehensive information on the doses applied to food in trade, and hence on 'overall average dose' in the original sense defined by JECFI. It will be seen from the conclusions of this ADMIT conference how much has already been achieved and how analytical detection together with dose estimation methods may contribute in future. There is already one big advantage associated with the analytical methods: they are developed, tested and standardised in national and international cooperation and formally adopted by several competent bodies. In contrast, standardisation of regulations and inspection systems is still rather uncoordinated making exchange of information on the contents of log-books and results of inspections virtually impossible.

9 OUTLOOK

GIP essentially means applying a dose range between specified limits. Such values are usually derived from technical data, for other reasons regulations may impose additional limits. The lawful conduct of an operator of an irradiation facility is always evident from the log-book of the facility. Analytical detection methods for irradiated food and especially those which allow for reliable dose estimates can be used as an additional, independent method for the enforcement of GIP. The availability of such enforcement methods may also bolster consumer confidence in controlling authorities. This is the special value of such methods, but, at the same time, also their limitation.

About 20 years ago, an international meeting (The Identification of Irradiated Foodstuffs, Commission of the European Communities (BCR), Luxembourg, 1974, EUR 5126 EN) arrived at the conclusion that there was a long way to go before reliable identification could be achieved, and any predictions were still impossible; furthermore, no one dared to dream of analytical dose estimation methods. The scientific community, in a big joint effort, has improved analytical methods and instrumentation with great success. Today as a consequence, detection of an irradiation treatment is possible on a food sample for all applications likely to be used commercially, and in many cases surprisingly reliable dose estimations are possible. This conference marking the successful end of the ADMIT programme must also answer the questions about the justification of the enormous funds and means spent for this purpose. Of course, it is of intrinsic value of its own to make available methods to detect fraud and misuses. The real value for society will only emerge after further refinement of the analytical detection methods for irradiated food together with the experience gained and data collected from industrial scale radiation processing of food are combined to verify GIP. So for now this verification is only possible through administrative action, *i.e.* inspection of the records of the irradiation facility. This conference is going to show the potential of analytical detection methods for the verification of Good Irradiation Practice.

Physical Methods
Electron Spin Resonance (ESR) Spectroscopy

EPR SPECTROSCOPY FOR THE DETECTION OF FOODS TREATED WITH IONISING RADIATION

W. Stachowicz,* G. Burlinska,* J. Michalik,* A Dziedzic-Goclawska** and K. Ostrowski**

*Institute of Nuclear Chemistry and Technology, Dorodna 16, 0S195 Warszawa, Poland
**Medical School, Chalubinskiego 5, 02-004 Warszawa, Poland

1 INTRODUCTION

The advantage of electron paramagnetic resonance spectroscopy (EPR or ESR) as a tool for the control of irradiated food lies in its sensitivity and accuracy. Ionising radiation produces, in irradiated materials, paramagnetic species of different kinds, *i.e.* radicals, radical-ions and paramagnetic centres, which can be measured by EPR but most of them are not stable enough to be used for the detection of irradiation. It is because radiation-induced paramagnetic species are thermodynamically less stable then surrounding molecules and take part in fast radiolytic reactions leading to the formation of final diamagnetic products that they are not detectable by the EPR method. Most of organic radicals produced by radiation in the liquid phase are unstable but if the unpaired electron is incorporated into the complex polymeric system as in peptides and polysaccharides and is structurally isolated from the water, its stability is markedly increased. In irradiated solids the excess electrons react with pre-existing traps or vacancies to form ionic species. Stability of paramagnetic centres produced by radiation in solids depends on the structure of the matrix; it is higher in crystalline systems or domains than in amorphous ones.[1] Since 1954[2,3] it is known that ionising radiation produces paramagnetic entities in biological materials, cells and tissues and some are stable enough to be observed by EPR spectroscopy at room temperature.[4,5] A good example related to food is the mineral of bone, tooth, cartilage and blood vessel.[6-8] The specific EPR signal observed after irradiation in mineralized tissues is derived from the crystalline hydroxyapatite fraction and was shown to be stable during several years of storage.[9] The above objectives opened the way to EPR studies on the detection of irradiated food[10] which presently are carried out by researchers from many countries. Most of them co-operate within the BCR and ADMIT programmes co-ordinated by the Joint FAO/IAEA Division of Nuclear Techniques in Food and Agriculture.

The present paper describes and discusses that part of results obtained by this group during the period of ADMIT activity (1989-94) which are original and may be useful to those who will be working in the near future on the development of uniform control systems for the detection of irradiated food. The intention was to focus attention on these facts and data which influence the certainty of the detection in both positive and negative manner.

2 MATERIALS AND METHODS

The following classes of food were examined: pork, beef, poultry (chicken, duck, turkey, goose), fish (carp, roach, trout, shaetfish, cod and some others), egg shell, seeds of fresh fruits and berries (more than 30 species and among them apples, pears, plums, oranges, grapefruits, lemons, black berries, strawberries, wild strawberries, raspberries, blackcurrants, gooseberries, tomatoes), seeds of dried fruits (dates, figs and grapes), an assortment of 25 spices and herbs and among them peppers, paprika, majoram, ginger, cumin, dried vegetables and mushrooms, dehydrated gelatin and macaroni (dried paste) as delivered by a producer to the market. Samples of food were irradiated with ^{60}Co source (initial activity 665 TBq, dose rate 0.7 kGy h^{-1}) as complete specimens (fruits, spices, mushrooms, *etc.*) or cut in pieces (poultry and fish carcasses). After irradiation parts of foods selected for EPR examination (bones, seeds, *etc.*) were dissected, washed in water and/or in methanol-ether mixture to dissolve fats if present and crushed into pieces suitable for EPR measurement. Samples were put into quartz thin-wall EPR tubes (4 mm diameter) then measured with a Bruker ESP 300 spectrometer operating in the X-band. The weight of samples was about 100 mg while sample height was up to 25 mm. The EPR measurements were conducted in the range of magnetic field around g-value equal to 2 and with a sweep width of 20 mT. An important task was to verify the stability of the paramagnetic species as a function of time. For this purpose EPR measurements were repeated with the same samples at different time intervals from one week to 3 months. The EPR signal intensities were measured as peak to peak heights of the most intense EPR lines (first derivative of absorption spectra). In order to ensure the reproducibility of EPR signal intensity measurements, samples were examined at the same instrument settings. In addition, corrections based on the use of internal (Mn^{2+}) and external (irradiated bone powder with constant H_{pp}) standards were adapted. For each sample the mean value from three EPR measurements was calculated and the standard error of the mean (±SE) was estimated.

3 RESULTS AND DISCUSSION

3.1 Bone, Fishbone and Egg

Early EPR studies on bones and teeth[9,11] exposed to the action of ionising radiation documented the appearance of stable and specific EPR signals in these tissues. The characteristic signal is an asymmetric singlet with $g_\perp = 2.0017$, $g_\parallel = 1.9973$ and $\Delta H_{pp} = 0.85$ mT derived from paramagnetic centres in the crystalline lattice of hydroxyapatite,[12] presumably CO_3^{1-}, CO_3^{3-} [13,14] or CO_2^{1-} [15] ion radicals. The first two species could be produced by capture or release of electrons from CO_3^{2-} ions incorporated within the crystal lattice of the bone mineral (hydroxyapatite). Mineralized tissues contains from 3 to 8% by weight of CO_3^{2-} ions. At least part of these ions is located in PO_4^{3-} vacancies in the hydroxyapatite lattice (classified as substitutional impurity). The CO_2^{1-} ion radical, in turn, could be produced through the capture of electrons by CO_2 molecules absorbed on the surface of bone microcrystals or in a more complex surface process. The examination of bone fragments and fishbones dissected from irradiated pieces of poultry, pork, beef and fish shows the appearance of the above described EPR signal in all of them. This is because the mineral part of bone in mammals, birds and fishes has a very similar composition and structure. Similar EPR signals are detected in poultry egg shells supporting the view that paramagnetic entities produced by radiation in mineralised tissues originate from CO_3^{2-} derived ion radicals.[16,29] Specific identification points of this signal in EPR are peaks corresponding to $g_\perp = 2.0030$ and $g_\parallel = 1.9973$, as well as to $g_2 = 2,0007$, if necessary. As additional evidence of this signal identity ΔH_{pp} measurement (0.85 mT) can be done. In the case of low dose irradiation, radiation-induced EPR signals in bone are accompanied by other signals which sometimes are observed in non-irradiated bone as well. This native signal in bone (isotropic singlet with $g_0 = 2008$, $\Delta H_{pp} = 0.57$ mT) absorbs in the same magnetic field range as the radiation-induced one but can be distinguished from the latter by spectral analysis as proposed above.[17]

3.1.1. Poultry and Meat. There are three factors affecting the intensity of EPR signals produced by radiation in bone containing food: (i) origin (animal species), (ii) age of animals, (iii) kind of bone fragments taken for EPR measurement. Fragments of spongy bone dissected from ribs of chicken, calf, ox (bovine) and pig irradiated with 5 kGy were the subject of an experiment done with the use of quantitative standards and sample normalisation. The relationship between normalized EPR signal intensities corresponding to the relevant samples was as follows: 1.0 (chicken): 2.2 (calf): 3.7 (bovine): 4.8 (pig). Distinct difference between samples can be explained in terms of the different content of mineral in spongy bone dissected from ribs of different animals. In a second experiment programmed in a similar way, EPR signals registered in bones dissected from chicken carcasses, veal and beef of different ages were compared. It was shown conclusively that the younger the animal the lower is the normalized intensity of radiation-induced signal observed. The proportion between intensities of EPR signals registered with spongy bone of rib taken from veal (calf 5 weeks old) and beef (ox 12 years old) were as 1.0: 1.7, respectively. Similar results was obtained with chicken. The examination of bone fragments of sternum and femur dissected from one animal body show distinct differences in EPR signal intensity at different doses, as presented in Table 1.

In comparative studies on bones dissected from irradiated poultry and meat it was found that the most intense EPR signals were registered and the best reproducibility of the EPR measurements was obtained when central fragments of thigh bone (femur diaphysis) were used.[18] This is the reason the authors recommend the use of compact bone for the detection of irradiation, whenever it is possible. The intensities of EPR

Table 1 *The Intensities of EPR Signal Registered with Two Kinds of Chicken Bones Irradiated with Doses from 0.5 to 11 kGy.*

Dose (kGy)	EPR Signal Intensity (Arbitrary Units)	
	Sternum	Femur
0.5	0.010	0.015
1.5	0.033	0.050
6.0	0.120	0.190
11.0	0.230	0.350

signals measured with fragments of thigh bone (femur diaphysis) taken from chicken, duck, goose and turkey carcasses irradiated in the frozen state with 3 kGy were not very different; a slight increase of intensity in the case of turkey and goose (about 15%) was within the limit of experimental error. It has been shown previously that relationship between EPR signal intensity and absorbed dose is linear up to 20-25 kGy.[19] It is possible, therefore, to estimate the dose given to chicken using re-irradiation followed by jack-knifing extrapolation.[20] A positive result can be attained when the sources of errors discussed above are eliminated. Our method is based on the standardization of sampling and measuring conditions to ascertain reproducibility of measurements. In a blind test with thigh bone samples dissected from chicken carcasses irradiated with 2, 7 and 12 kGy, the accuracy of the order of 20% in dose estimation was obtained which is satisfactory enough to distinguish the dose ranges cited above. This means that it will be possible to distinguish, by EPR, two processing doses as proposed contemporarily for poultry (1-3 kGy as suitable to prolong the storage time of chilled carcasses and 5-7 kGy to reduce the content of pathogenic microorganisms to the level safe for the consumer), as well as to detect overdosage (>10 kGy) unacceptable according to WHO Codex Alimentarius Standards. As a routine procedure it is recommended that 3 or 4 doses of re-irradiation with 2 kGy are used.[21] The practical value of the EPR dose estimation in poultry carcasses is that it makes it possible to indicate whether radiation treatment was applied in accordance with GIP (Good Irradiation Practice) or not. Detectability tests done with chicken carcasses showed that with samples weighing about 50 mg or more a thigh bone sample irradiated with the dose of 0.1 kGy is detectable with high certainty. If samples are taken from spongy bone (ribs *etc.*) the certainty level of irradiation detection is no worse than 0.3 kGy. Both levels are far below the processing doses recommended for poultry. Aspects of applicability of paramagnetic centres produced by radiation in bone to the detection of irradiation in food have also been studied by other researchers.[22,23]

3.1.2. Fish. In fishbones dissected from frozen fresh fish (carp, roach, trout, shaetfish, cod) irradiated with a recommended processing dose of 1-2 kGy, a similar EPR signal was recorded as the one registered in bones. The relationship between absorbed dose and EPR signal intensity was found to be linear in the range extending from the lowest doses applied (about 0.2 kGy) to those of the order of 15 kGy.[24] The intensities of the EPR signals as measured with bones dissected from different fish are compared in Table 2.

Although, in the above experiment, only the same kind of fishbones were examined, EPR signal intensity from bone of different fishes differed markedly. On the other hand, EPR signal intensity from fishbone of carp is about one third that from chicken compact

Table 2 *Intensities of EPR Signals in Bone Fragments Dissected from Fish Irradiated with 2 kGy*

Species	EPR Signal Intensity (Arbitrary Units)
Carp	0.10 ± 0.02
Roach	0.05 ± 0.02
Trout	0.04 ± 0.01
Sheatfish	0.01 ± 0.005

bone. A relatively stable EPR signal was registered in irradiated fins and scales of carp. In both cases the signal was a singlet with $g_0 = 2.004$ and $\Delta H_{pp} = 0.56$ mT. The signal was found to be about 1.5 times higher in fins than in scales. All EPR signals found in bones, fins and scales of fish were relatively stable but not as stable as in bones of mammals and birds. Quantitative EPR measurements repeated 3 months after irradiation showed a slow decrease of radiation-induced EPR signals in fish resulting in about 25% depression of signal intensity in fish bone and fins compared to a 40% reduction in scales. All 3 signals as recorded with fish bone, fin and scale were still detectable after storage for one year. In deep frozen fish stored at temperatures below -60°C the decrease of radiation-induced EPR signals will be slower.[24]

3.1.3 Poultry Egg Shell. Stable EPR active species have been detected in poultry egg shells.[25] Spectroscopic g-values of the EPR asymmetric singlet recorded in egg shell are close to those of bone, supporting the view that paramagnetic centres produced by radiation in mineralised tissues may originate from the CO_3^{2-} derived ion radicals (see 3.1.1). The normalised intensities of the EPR signals in irradiated egg shell are higher than in bone.[29] It is expected that even small fragments of shells found in egg containing food products could be used for detection of irradiation treatment.

3.2 Fruits, Vegetables, Nuts and Spices

3.2.1. Dried Fruits. The most promising results for dried fruits have been obtained with pressed dates and figs irradiated with doses between 0.6 and 2.0 kGy.[24] In seeds separated from irradiated fruits multicomponent signals were registered. The centre of these signals lies at about $g_c = 2.004$ in fig and at $g_c = 2.005$ in date, while the overall signal widths are comparable in both and equal to about 7 mT. In non-irradiated seeds of fig and date only a weak EPR singlet was found ($g_0 = 2.007$, $\Delta H_{pp} = 0.9$ mT and 2.008, $\Delta H_{pp} = 1.1$ mT, respectively) which did not interfere with irradiation detection. It has been shown that immediately after and a few days after irradiation complex EPR signals in the seeds of dates and figs (to a lesser degree) are predominated by an intense component (probably a singlet) which decays after several days of storage.[24] On the other hand, the multicomponent signal remains only slightly changed and is recorded after a one year storage period. When compared with the EPR spectra of irradiated sugars it is postulated that the multicomponent part of the signals in dates and figs is derived mainly from saccharide radicals[26] embedded within the solid constituents of the seed, thereby, insulating them from the water containing pulp of the fruit. The decay curves of radicals (EPR signal intensities) in seeds of dates and figs are shown in Figure 1. One can distinguish faster decay in the beginning and slow decay of radiation produced species thereafter.

3.2.2 Fresh Fruits and Nuts. It happens quite often that in non-irradiated food native EPR signals are registered which make the detection of irradiation by the EPR technique difficult. It has been found, however, that in fruits and vegetables the intensity of native EPR signals increases after irradiation. The growth of EPR absorption is significant as measured immediately after irradiation but in most cases is not observed any more after several days of storage. More than 30 varieties of fresh fruits and berries have been tested but only in seeds separated from pear, orange, and grapefruit did irradiation

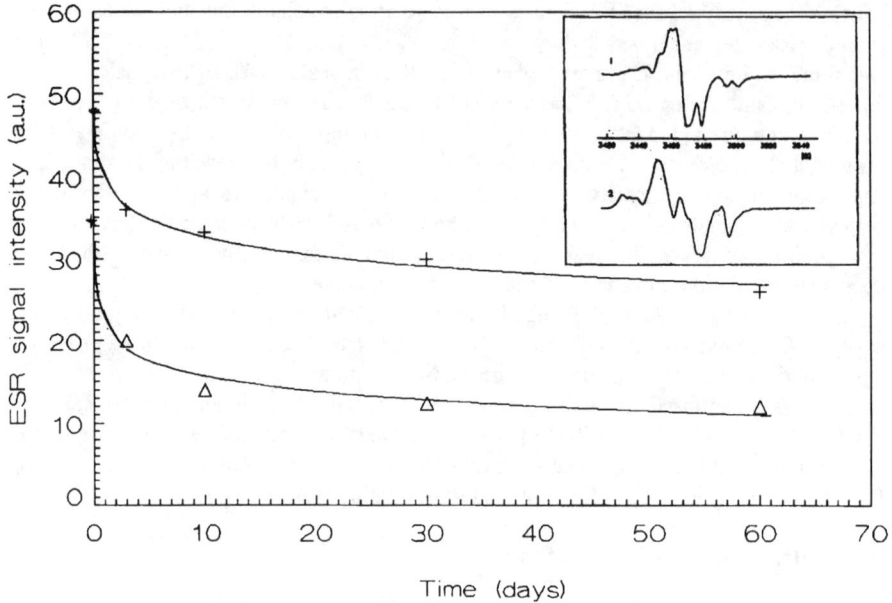

Figure 1 *Upper spectra: Multicomponent EPR signals (first derivative) in seeds of date (1) and fig (2) recorded one month after irradiation with 3 kGy (gamma-rays) Lower graph: Decrease of EPR signals produced by radiation in seeds of dates (+) and figs (Δ) as a function of time*[21]

(3 kGy) increase the EPR signal intensity (single line) by a factor of about 2 in comparison with the non-irradiated samples. The growth of EPR signals has been ascertained 3 months after irradiation.[24] It is suggested that the growth effect of a native signal, described above, can be used as an extra test to prove irradiation in food. It has to be pointed out, however, that radiation-induced EPR singlets are not specific *i.e.* not different from native ones. For that reason the method can be only used to confirm the result obtained by other detection techniques not credible enough in relation to food undergoing examination.

In achenes separated from irradiated strawberries two stable satellite lines belonging probably to a triplet derived from cellulose radical are observed.[27] The lower and higher field components of the triplet lie outside both slopes of a strong EPR singlet assigned to a native EPR signal. The two stable satellite EPR lines of cellulose radicals are very weak and sensitive to microwave power and amplitude modulation variations. They were detected in achenes separated from strawberries harvested in Poland and irradiated with 1.5-3.0 kGy.[21] The spectroscopic character of recorded signal was as follows: $g_0 = 2.004$ and $\Delta H_{pp} = 6.0$ mT for radiation-induced satellite lines and $g_0 = 2.000$ for the native signal. Similar EPR signals were found in pistachio nuts imported to Poland and exposed to the same dose of radiation.

3.2.3. Spices and Herbs. The growth of the native signal and stability of this effect have been examined with the assortment of 25 spices and herbs most popular in Poland. Radiation-induced EPR single lines have been detected in charlock irradiated with a dose

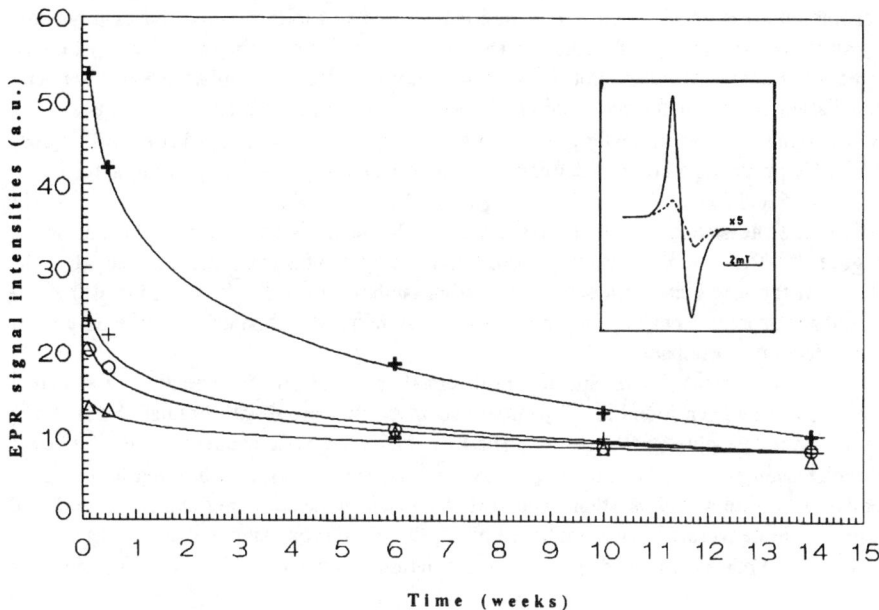

Figure 2 *Upper spectrum: EPR signal (first derivative) of black pepper recorded one day after irradiation with 7 kGy (gamma-rays). Dashed line shows normalised EPR signal in non-irradiated pepper.*
Lower graph: The decay curves of EPR signal intensities in spices: (+) Black pepper, (+) majoram, (o) red paprika, (Δ) ginger. Last EPR measurements were done 14 weeks after irradiation. EPR signal intensities are still higher at this point then in non-irradiated spices.

of 7 kGy after 3 months of storage. The appearance of the EPR singlet is considered in this case as the evidence of radiation treatment since only negligible EPR absorption has been found in non-irradiated samples. This is not the case with other spices and herbs which were examined since all of them have strong native EPR signals when measured before and after irradiation. The growth of a native signal caused by radiation has been ascertained in black pepper, paprika, ginger and cumin following one month of storage. Spectral analysis of EPR signals recorded before and after irradiation showed a slight increase of the EPR line width in irradiated samples by about 10%. With paprika a small shift of g_0 from 2.003 to 2.004 has also been observed.

Kinetic curves showing the decay of radiation-induced components of a single EPR line in spices cited above are given in Figure 2. The growth of a native EPR signal in many spices is reduced by humidity while a radiation-induced drop of signal intensity is sometimes decreased with time. It has been reported that in some irradiated spices a double EPR line assigned to cellulose radicals is observed but its intensity is usually low.[28] The Polish group found satellite lines which could be derived from these species in samples of paprika and ginger irradiated with 7 kGy. However, the certainty of detection as tested with different samples was not satisfactory enough.

3.2.4. *Mushrooms*. In the Polish market three classes of dried mushrooms are available: *Boletus edulis, Boletus badius* and *Agaricus campestris*. *Agaricus campestris* (champignon) is widely used in the food industry as an additive to canned or powdered products as well as to some kinds of cheese. The hygienic quality of dried mushrooms used in industry is often not satisfactory and they need a preservation treatment. Irradiation is one of the most suitable methods. In non-irradiated mushrooms a weak native signal is registered with $g_0 = 2.0046$ and $\Delta H_{pp} = 0.70$ mT. Irradiation with a dose of 7 kGy produces a complex EPR signal[29] in mushrooms in which a strong singlet with $g_0 = 2.005$ and $\Delta H_{pp} = 0.74$ mT predominates (Figure 4 Upper spectrum). After 3 weeks of storage the strong singlet decays and the EPR signal is as shown in the bottom of Figure 3. The overall width of a broad multicomponent signal which is left after the decay of the single line produced by radiation is about 5.4 mT. It is estimated that the signal has at least 5 peaks. The specificity and stability of this signal promote its use for the detection of irradiation.

3.2.5. Processed Food. Specific EPR signals suitable for the detection of radiation treatment have been found in dehydrated gelatin and macaroni both irradiated and stored in packages available on the market. The EPR signal in gelatin irradiated with 7 kGy is a doublet with $g_0 = 2.0032$ and $\Delta H_{pp} = 1.8$ mT. Similar signals are observed in irradiated collagen.[8] The signal is stable and was recorded in gelatin after several months of storage. In dehydrated macaroni (dried paste) there is a broad signal with $g_0 = 2.008$ and $\Delta H_{pp} = 2.1$ mT was recorded after an irradiation treatment of 7 kGy. The signal is

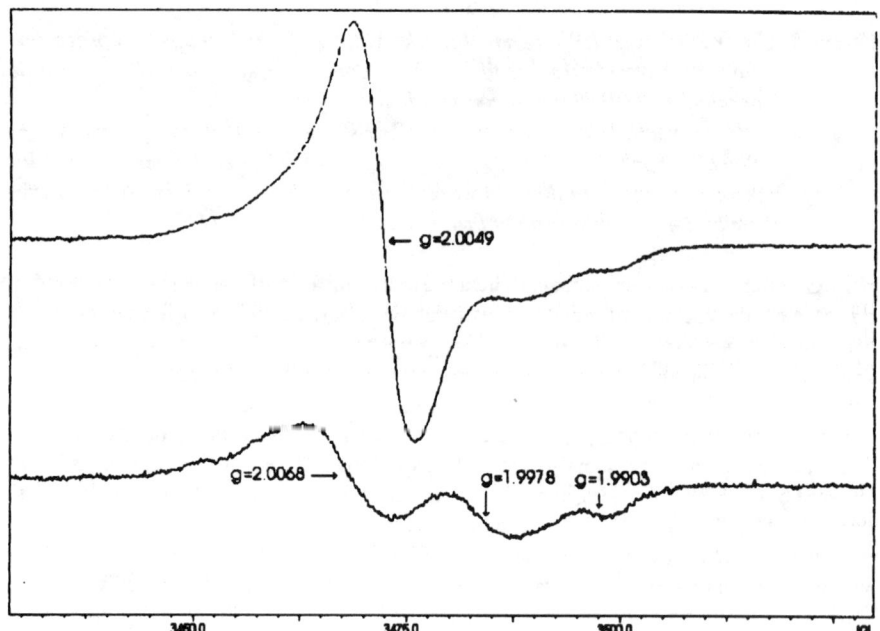

Figure 3 *EPR spectra (first derivative) of dried mushrooms (Agaricus campastris) irradiated with 7 kGy (gamma-rays)*
Upper spectrum: EPR signal recorded one day after irradiation
Lower spectrum: EPR signal recorded 3 weeks after irradiation

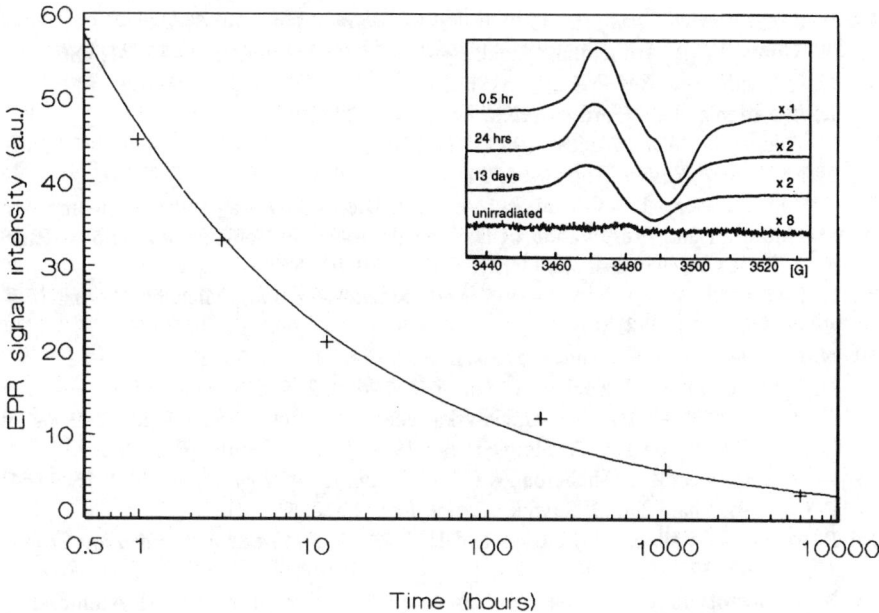

Figure 4 *Upper spectra: EPR spectra (first derivative) in macaroni (dehydrated paste) irradiated with 3.4 kGy (gamma-rays) at different time intervals (0.5 h, 24 h, 13 days after treatment*
Lower EPR spectrum: Non-irradiated macaroni
Lower graph: Decay curve of the EPR signal intensity in macaroni during its storage in commercial packages. Last measurement (last point in the graph) was done after 9 months of storage.[21]

complex but its spectral components are not well resolved. During storage transformation of the signal occurs supporting the view of its complexity. In Figure 4 the decay curve of this signal as a function of time is shown. The stability and specificity of the signal allowed detection of irradiation in macaroni after 9 month of storage.

4 CONCLUSIONS

The results described in this paper prove the applicability of EPR spectroscopy in the detection of various foodstuffs of animal or vegetable origin. The method is based on the detection of asymmetric singlets assigned to radiation-induced paramagnetic centres in the mineral component of bone. The method is specific and sensitive enough to be recommended to perform routine control of irradiation in poultry, meat and fish. The Polish group participated in EPR intercomparison studies on irradiated foodstuffs organized by BCR.[30]

References

1. H. M. Swartz, J. R. Bolton, D. C. Borg, "Biological Application of Electron Spin Resonance," Wiley-Interscience, New York, 1972.

2. B. Commoner, J. Townsed and G. E. Pake, *Nature,* 1954, **174,** 698.
3. W. Gordy, W. B. Ard and H. Shield, *Proc. Nat. Acad. Sci. U.S.,* 1955, **41,** 986.
4. D. E. Smith, *Ann. Rev. Nucl. Sci,* 1962, **12,** 577.
5. R. A. Patten and W. Gordy, *Radiat. Res.,* 1964, **22,** 29.
6. T. Cole, A. H. Silver, *Nature,* 1963, **200,** 700.
7. H. M. Swartz, *Radiat. Res.,* 1965, **24,** 579.
8. W. Stachowicz, K. Ostrowski, A. Dziedzic-Goclawska and A. Komender, "Sterilization and Preservation of Biological Tissues by Ionizing Radiation," Panel Proc. Series IAEA, Vienna, STVPUB/247, 1970, p. 15.
9. K. Ostrowski, A. Dziedzic-Goclawska, W. Stachowicz and J. Michalik, *J. Ann. N. Y. Acad. Sci.,* 1974, **238,** 186.
10. N. J. F. Dodd, F. J. Lee and A. J. Swallow, *Nature,* 1988, **334,** 387.
11. U. T. Slager and M. J. Zucker, *Radiat. Res.,* 1964, **22,** 556.
12. K. Ostrowski, A. Dziedzic-Goclawska and W. Stachowicz, "Free Radicals in Biology," W. Pryor, Ed., Academic Press, 1980, Vol.4, Chapter 10, p. 321.
13. L. G. Glinskaya, M. J. Shcherbakova, J. N. Zanin, *Kristallografiyia,* 1970, **15,** 1164.
14. P. Cevc, M. Shara and G. Ravnik, *Radiat. Res.,* 1972, **51,** 581.
15. P. Moens, F. Callens, P. Matthys and F. J. Maes, *J. Chem. Soc. Faraday Trans.,* 1991, **87,** 3137.
16. M. F. Desrosiers, F. G. Le, P. M. Harewood, E. S. Josephson and M. Montesalvo, *J. Agric. Food Chem.,* 1993, **41,** 1471.
17. W. Stachowicz, J. Michalik, A. Dziedzic-Goclawska and K. Ostrowski, *Nukleonika,* 1974, **19,** 845.
18. W. Stachowicz, J. Michalik, K. Ostrowski, A. Dziedzic-Goclawska and A. Wojtowicz, *J. Radiat. Biol.,*1989, **55,** 879.
19. W. Stachowicz, J. Michalik, A. Dziedzic-Goclawska and K. Ostrowski, *Nukleonika,* 1972, **17,** 7.
20. M. F. Desrosiers, *Nature,* 1990, **345,** 485.
21. W. Stachowicz, G. Burlinska, J. Michalik, K. Ostrowski and A. Dziedzic-Goclawska, "New Developments in Food, Feed and Waste Irradiation," Bundesgesundheitsamtes (BGA), Berlin, SozEp Hefte, *16,* 1993, p. 13.
22. M. H. Stevenson and R. Gray, *J. Sci. Food Agric.,* 1989, **48,** 196.
23. M. F. Desrosiers, *J. Food Sci,* 1991, **56,** 1104.
24. W. Stachowicz, G. Burlinska, J. Michalik, A. Wojtowicz, A. Dziedzic-Goclawska and K. Ostrowski, *J. Sci. Food Agric.,* 1992, **58,** 407.
25. N. Kiyak, *Turk. J. Nucl. Sci.,* 1990, **17,** 83.
26. N. Helle and B. Linke, Bruker Report, ISSN 0724-0185, 1992, 91/92, p. 8.
27. J. Raffi, J.-P. Angel, L. Buscarlet and C. Martin, *J. Chem. Soc. Faraday Trans.,* 1988, **184,** 3359.
28. S. U. Uchiyama, Y. Kawamura and Y. Saito, *J. Food Hyg. Soc. Jpn,* 1990, **31,** 499.
29. W. Stachowicz, G. Burlinska, J. Michalik, A Dziedzic-Goclawska and K. Ostrowski, *Nukleonika,* 1993, **38,** 67.
30. W. Stachowicz, J. Michalik, G. Burlinska, K. Ostrowski, A Dziedzic-Goclawska and A. Wojtowicz, "ESR intercomparison studies on irradiated foodstuffs," Committee of the European Communities (BCR), Brussels, Luxembourg, EUR 13630 EN, 1992.

DETECTION OF IRRADIATION TREATMENT IN CRUSTACEA BY ELECTRON SPIN RESONANCE (ESR) SPECTROSCOPY

E. M. Stewart,[1] M. H. Stevenson,[1,2] and R. Gray[2]

[1]Department of Food Science, The Queen's University of Belfast and
[2]Food Science Division, The Department of Agriculture for Northern Ireland
Newforge Lane
Belfast BT9 5PX
U.K.

1 INTRODUCTION

When the Food (Control of Irradiation) Regulations 1990 came into force in the United Kingdom in January 1991[1] they included provision for the irradiation of Crustacea to an overall average dose of 3 kGy. The treatment of Crustacea with ionising radiation would reduce numbers of potential pathogens and spoilage organisms thus giving a microbiologically safer product with a longer shelf-life at chill temperatures. At present the process is being used in countries such as France and The Netherlands for the decontamination/shelf-life extension of shrimp.[2] Therefore, as for other food products such as poultry, liquid whole egg and fruit, which are also treated with ionising radiation, it is desirable that a suitable test should be available to help in the control of the irradiation process. One such detection method which has been applied to irradiated Crustacea is that of electron spin resonance (ESR) spectroscopy due to the fact that the rigid exoskeleton has a relatively high dry matter so free radicals produced by ionising irradiation can be trapped and are, therefore, sufficiently stable to be detected.

Ideally a detection method should be specific for irradiation, that is, no other process should produce similar changes in the food. Thus, the ESR signals induced in the cuticle of Crustacea should:
(a) be stable or fairly stable during the expected shelf-life of the product,
(b) be clearly distinguishable from the signals of non-irradiated samples, even after a long storage period,
(c) not be induced by other food processes, *e.g.* cooking, and
(d) be proportional to irradiation dose and constant during storage if absorbed dose is to be measured.[3]

In general, it may only be necessary to have a qualitative indication of irradiation, while in some cases an estimate of the actual irradiation dose received by the product may be required. This paper will review work which has been carried out to systematically evaluate the feasibility of using ESR spectroscopy for the detection of irradiated Crustacea, in particular, the Norway lobster (*Nephrops norvegicus*), and to determine if the methodology exhibited the detection test characteristics outlined above.

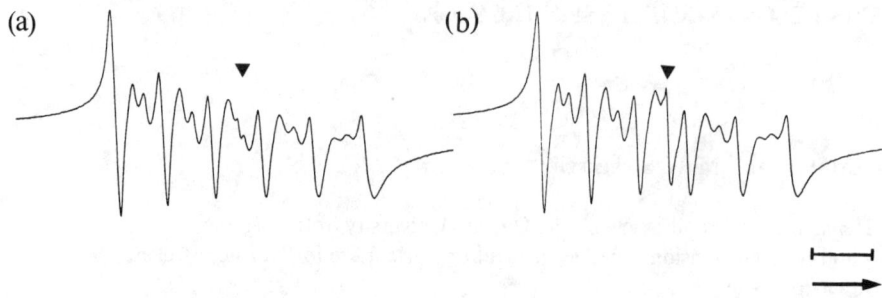

Figure 1 *Wide scan ESR spectra from irradiated (a) and non-irradiated (b) Norway lobster cuticle*
▼ = 349.5 mT; ├─┤ = 10 mT; → = increasing H

(Reprinted from E. M. Stewart, M. H. Stevenson and R. Gray, Detection of Irradiation in Scampi Tails - Effects of Sample Preparation, Irradiation Dose and Storage on ESR Response in the Cuticle, *Int. J. Food Sci. Technol.*, **27**, 125-132, 1992, with kind permission from Blackwell Scientific Publications Ltd., Osney Mead, Oxford, OX2 OEL, UK)

2 IDENTIFICATION OF RADIATION-INDUCED ESR SIGNALS

Initial investigations showed that the ESR signal derived from non-irradiated Norway lobster tail cuticle was characteristic of Mn^{2+} with six equally spaced resonance peaks (Figure 1a). On examination of irradiated cuticle (Figure 1b) it was found that the Mn^{2+} signal was again present but that there was an additional peak in the centre at a field value of 349.5 mT (g = 2.0009). This radiation-induced signal was more obvious when it was isolated from the background Mn^{2+} signal using the software available on the Bruker ESP300 spectrometer. It was found that the most effective separation was achieved using a narrow sweep width (15.0 mT) as opposed to a wide scan (82.0 mT) as fewer Mn^{2+} peaks had to be taken into account in the subtraction process. It is this additional peak which is indicative of irradiation in the Norway lobster and which could possibly be used to qualitatively identify irradiated samples. Storage studies also established that the radiation-induced signal was detectable throughout the expected shelf-life of chilled and frozen whole Norway lobster and could be identified in dried cuticle stored for prolonged periods at ambient temperature.[4]

The shape of the ESR signals (Figure 2 (a-d)) derived from the irradiated and non-irradiated cuticle of pink shrimp (*Pandalus montagui*), tiger prawn (*Penaeus monodon*), king prawn (*Penaeus plebejus*) and Mediterranean crevette (*Palaemon serratus*) were different to those of the Norway lobster. The non-irradiated signal from each of the

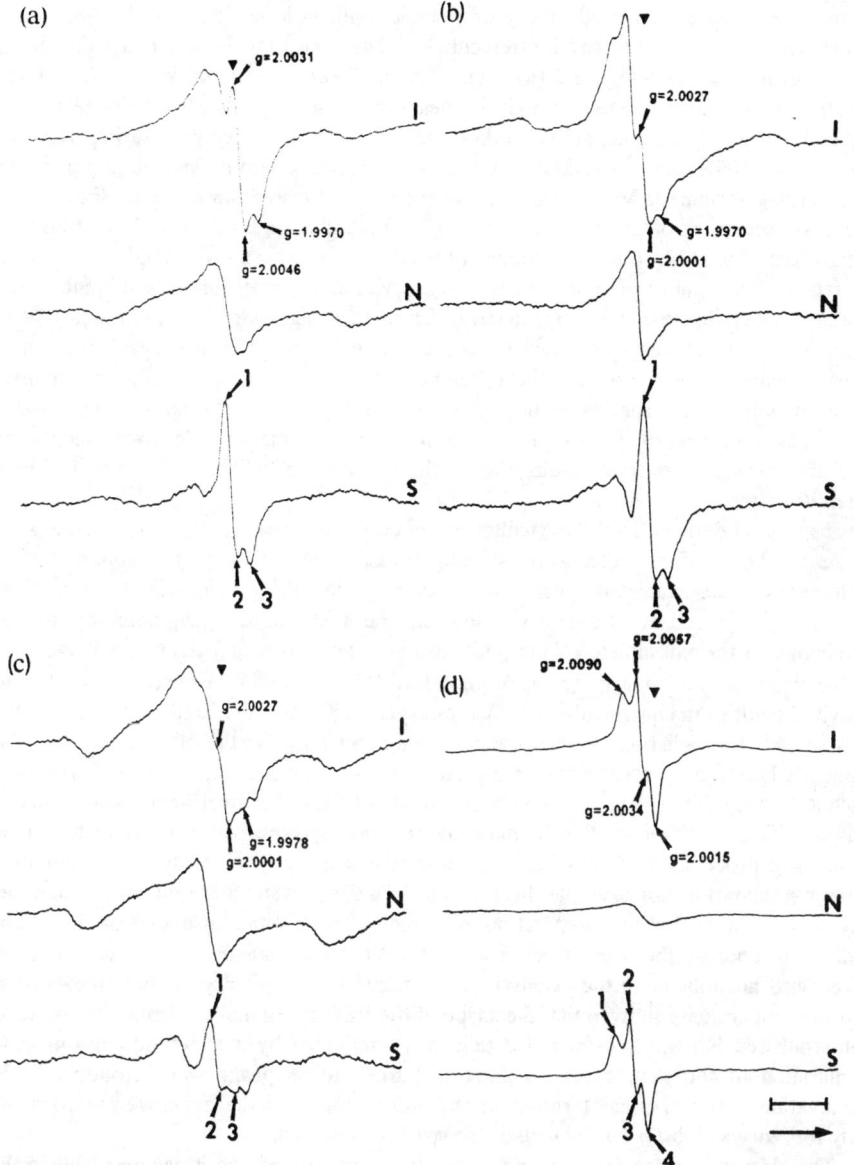

Figure 2 *ESR spectra from pink shrimp (a), tiger prawn (b), king prawn (c) and Mediterranean crevette (d) cuticle*
I = irradiated; N = non-irradiated and S = isolated radiation-induced ESR signal; ↓ = free radical peaks measured; ▼ = 349.5 mT; ⊢⊣ = 2.0 mT; → = increasing H

(Reprinted from E. M. Stewart, M. H. Stevenson and R. Gray, Use of ESR Spectroscopy for the Detection of Irradiated Crustacea, *J. Sci. Food Agric.*, **65**, 191-197, 1994, with kind permission from the Society of Chemical Industry, 14/15 Belgrave Square, London SW1X 8PS, UK)

former four species consisted mainly of a singlet with little evidence of the Mn^{2+} signal which was present in Norway lobster cuticle. The ESR signals obtained from the tail cuticle of pink shrimp (Figure 2 (a)), tiger prawn (Figure 2 (b)) and king prawn (Figure 2 (c)) were similar in shape although the peaks from the latter species were less clearly defined.[5] ESR signals comparable to those derived from irradiated pink shrimp were also obtained by Helle and co-workers[6] from pink shrimp, true shrimp and glass prawns and from brown shrimp by Morehouse and Desrosiers.[7] However, in contrast, the shape of the ESR spectrum obtained by Desrosiers,[8] Morehouse and Ku[9] and Morehouse and Desrosiers[7] from the irradiated cuticle of pink shrimp was more complex due to the presence of a number of free radical peaks. Moreover, three batches of pink shrimp tested by Morehouse and Ku[9] exhibited different ESR signals and it was noted that the spectral features of the Mediterranean crevette actually resembled the signal from one of these batches.[5] The reason for the differences in the results may be due to incorrect identification of the species examined since the term pink shrimp can be used to encompass a number of species. Therefore, it is advisable that in order to reduce the risk of a discrepancy in reported results, the species being examined should be identified by its scientific name.

The signal derived from the Mediterranean crevette (Figure 2 (d)) was of contrasting shape to those of the other species being investigated. Despite the apparent visual differences in the irradiated spectra of the four species, the g values of the peaks were similar. Compared to the Norway lobster, the ESR signals generated by ionising irradiation in the tail cuticle of the pink shrimp, king prawn and tiger prawn were less stable when stored for a three week period at 1°C (Table 1). Consequently, it could prove difficult to unequivocally identify these species if they were irradiated at a low dose (1 kGy) and the radiation-induced signal was not isolated. On the other hand, the signal from Mediterranean crevette was more stable, especially at higher doses of irradiation such as 3 kGy.[5] If the signal induced in the cuticle of Crustacea by irradiation is to be used for identification purposes then it must be radiation-specific and not induced by other processing procedures. It has been demonstrated that many normally used methods of sample preparation, for example, heating and grinding, create free radicals.[10] Therefore, any process used for the preparation of samples, must either produce no interfering radicals to confuse the spectra or must have a recognised standard effect which can be taken into account in further analysis. A comparison of air-drying and freeze-drying followed by grinding showed that the shape of the ESR signal derived from non-irradiated and irradiated Norway lobster tail cuticle was unaffected by the methodology used for preparation of the samples as no additional free radical peaks were produced.[4] This observation was consistent throughout the entire course of the experimental work not only for Norway lobster but with all of the species examined.

The Norway lobster can be sold frozen, chilled or cooked and it has been shown that these different processing treatments do not generate free radicals similar to those produced by irradiation.[11] This provided further evidence of the specificity of the radiation-induced ESR signal. The radiation-induced ESR signals from Norway lobster cuticle irradiated in the frozen, chilled or cooked state, or cooked after completion of irradiation, were essentially similar in shape thereby indicating that ESR spectroscopy has potential to detect not only raw but also cooked Norway lobster. However, it was found that the peak from samples cooked prior to irradiation had shifted to a lower magnetic

Table 1 *Visual Identification Chart for Irradiated Mediterranean Crevette, Pink Shrimp, King Prawn and Tiger Prawn Stored for up to 21 days at 1°C*

(Reprinted from E. M. Stewart, M. H. Stevenson and R. Gray, Use of ESR Spectroscopy for the Detection of Irradiated Crustacea, *J. Sci. Food Agric.*, **65**, 191-197, 1994, with kind permission from the Society of Chemical Industry, 14/15 Belgrave Square, London SW1X 8PS, UK)

Species Examined	Intended Dose (kGy)	Dose Received (kGy)	Storage Period (Days)			
			0	7	14	21
Mediterranean crevette	1	1.2	D	D	D	D
	3	3.4	D	D	D	D
	5	5.8	D	D	D	ND
Pink shrimp	1	1.2	D	ND	ND	ND
	3	3.4	D	D	ND	ND
	5	5.8	D	D	D	ND
King prawn	1	1.2	ND	ND	ND	ND
	3	3.4	D	D	ND	ND
	5	5.8	D	D	D	ND
Tiger prawn	1	1.2	ND	ND	ND	ND
	3	3.4	D	D	ND	ND
	5	5.8	D	D	ND	ND

Visual identifications were made using the irradiated spectrum prior to subtraction of the non-irradiated spectrum. D = detectable and ND = not detectable.

field value in comparison to the signals isolated from cuticle subjected to the other treatments.[11] There also appeared to be additional resonances present but these were not sufficiently intense to interfere with identification of the radiation-induced free radical peak used for detection purposes. The differences observed may be due to a change in the structure of the cuticle caused by cooking, a consequence of which has been an alteration in its ability to trap free radicals. Alternatively, the composition of the cuticle may have been altered in such a way that additional free radicals have been formed.[11]

The initial studies on Norway lobster used only the tail section of the exoskeleton but it may not always be possible to obtain this part of the shell for identification purposes. However, it was found that ESR signals of similar shape to those from the tail section were obtained from the cuticle of irradiated and non-irradiated carapace, claws and walking legs. Thus, irradiated samples of Norway lobster could be identified using the cuticle from a number of sites within the exoskeleton.[12] Similar conclusions were reached by Desrosiers[9] from pink shrimp while Gray and Stevenson[13] showed that the site from which bone was excised in a chicken carcass did not influence the shape of the signal.

The specificity of the radiation-induced signal and its usefulness as a qualitative means of detecting irradiated Norway lobster at levels likely to be used in practice was confirmed by an in-house blind trial where 72 samples of unknown history were examined. Using small fragments or ground samples of the cuticle, all of the samples, either non-irradiated

small fragments or ground samples of the cuticle, all of the samples, either non-irradiated or given doses of 1, 2 or 3 kGy and stored for 0, 7 or 14 days at 1°C, were correctly identified.[14]

Ideally a detection method should be capable of providing a rapid result. A qualitative assessment of whether or not a sample has been irradiated can be carried out within 2 h if air-dried cuticle is used or within 24 h if the samples are freeze-dried. In the case of the Norway lobster, the blind trial showed that the radiation-induced signal can be detected using a sample size of approximately 10 to 20 mg. Another advantage of the ESR technique is that it is non-destructive, therefore, a sample can be analysed more than once.

3 FACTORS AFFECTING ESR SIGNAL INTENSITY

In most cases qualitative identification is all that is required to enforce the labelling regulations. However, if the method is to be used to estimate dose, then it is essential to establish the conditions under which the intensity of the signal would be affected. Therefore, a number of possible variables were systematically evaluated.

3.1 Sample Preparation

Previous work on irradiated bone had shown that the method of sample preparation affected signal intensity[15] and similar effects were obtained with irradiated Norway lobster (Table 2). The radiation-induced free radical peak in freeze-dried samples was significantly more intense than that from air-dried cuticle so the former procedure was used throughout the experimental work.[4] The difference observed may be due to the different dry matter concentrations of the cuticle at the time of measurement as the freeze-dried samples had a dry matter content of 920 g kg^{-1} while the corresponding value for air-dried samples was 860 g kg^{-1}.

3.2 Irradiation Dose

It will only be feasible to use ESR spectroscopy to quantify the irradiation dose received by samples of Norway lobster if the intensity of the radiation-induced signal is dose dependent. Initial studies on Norway lobster indicated that ESR signal intensity was significantly affected by irradiation dose. The signal derived from tail shell given an irradiation dose of 5 kGy was approximately four times greater than that derived from samples which had received a 1 kGy dose.[4] Further evidence of a dose dependent response was obtained from larger trials carried out using whole tails of Norway lobster given doses of 1 to 5 kGy where the response to dose was significantly linear (Figure 3).[14] Throughout the entire work on Norway lobster, ESR signals of equivalent intensity were obtained from samples of tail cuticle treated in a similar manner, thereby, verifying the reproducibility of the methodology.

As for the Norway lobster, a linear response to irradiation dose was observed for Mediterranean crevette and pink shrimp over the dose range 1 to 5 kGy but in the case of the king prawn and tiger prawn the response to dose was quadratic.[5] A linear response

Table 2 *Effect of Drying Method on ESR Signal Strength (Arbitrary Units) in Norway Lobster Cuticle given 5 kGy*

Drying Method	Signal Strength (Arbitrary Units)	SEM/Sig
Air-dried	6620	375/***
Freeze-dried	9870	

Signals are corrected for sample size and dry matter
Signal strengths are the mean of 6 measurements
SEM = Standard Error of the Mean
Sig. = Significance of effect where *** = P<0.001

(Reprinted from E. M. Stewart, M. H. Stevenson and R. Gray, Detection of Irradiation in Scampi Tails - Effects of Sample Preparation, Irradiation Dose and Storage on ESR Response in the Cuticle, *Int. J. Food Sci. Technol.*, **27**, 125-132, 1992, with kind permission from Blackwell Scientific Publications Ltd., Osney Mead, Oxford, OX2 OEL, UK)

within the dose range 1 to 3 kGy has also been reported for pink shrimp by Desrosiers[8] who quantified the signal using integration of the area under the absorption curve. There was also an apparent difference in signal intensity between the four species, with the Mediterranean crevette exhibiting the most intense signal at each dose while the spectrum from king prawn was the weakest. Morehouse and Desrosiers[7] suggested that the differences between species could be directly related to the thickness of the exoskeleton. They found that the signal for the thick, rigid prawn shell was more intense than that for the thin, flexible shrimp shells. This, however, was not the case for the species of prawn and shrimp examined in this work. The ESR signals from varieties with the thinnest shells, that is the pink shrimp and Mediterranean crevette, were the most

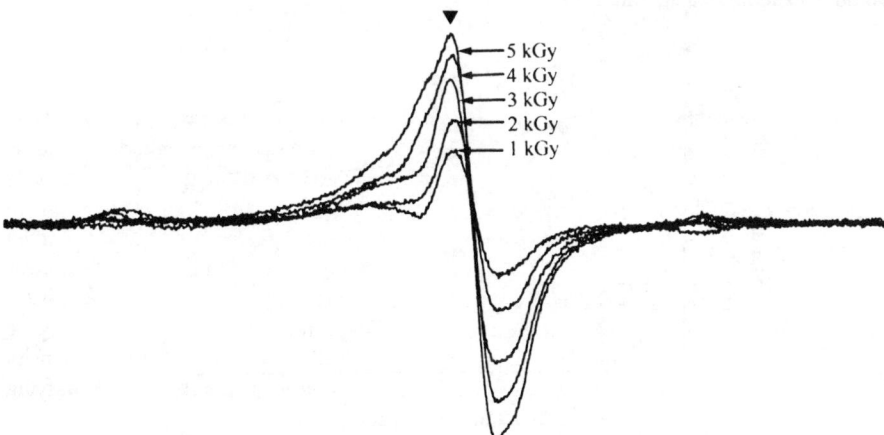

Figure 3 *Effect of irradiation dose on the intensity of the radiation-induced ESR signal from Norway lobster cuticle*
▼ = 349.5 mT; ⊢⊣ = 1.0 mT; → = increasing H

intense while those from the tiger prawn and king prawn, which possessed the thickest exoskeletons, exhibited the weakest signals.[5]

3.3 Storage Time and Temperature

The intensity of the radiation-induced ESR signal from Norway lobster cuticle was found to be relatively stable (Figure 4) over a four week period by which time samples stored at a chill temperature had deteriorated to an unacceptable quality. A small decay in signal strength was exhibited by samples stored at -20°C in comparison to those kept at +5°C but this was not unexpected since the opportunity for free radical movement will be minimised in frozen material.[4] The signal induced by a 1 kGy dose exhibited a greater stability compared to that measured in cuticle given higher doses and this was apparent in all the exoskeleton components examined. At the dose of 3 kGy all the exoskeleton components showed a decrease in signal strength during storage, the extent of decay being dependent on the part of the exoskeleton examined. These latter observations could possibly be related to the initial intensities of the respective signals since the signal from claw cuticle, as well as being the strongest, also exhibited the greatest decay during storage.[12]

The signal induced in the cuticle of pink shrimp, tiger prawn, king prawn and Mediterranean crevette by a dose of 1 kGy was less stable during storage at 1°C than that from samples given the higher doses.[5] These results are contrary to the observations for Norway lobster stored at the same temperature. The stability of the signal derived from the cuticle of pink shrimp was also different to that reported by Desrosiers.[8] He found that the radiation-induced peak from pink shrimp cuticle given a dose of 1 kGy and kept at -20°C showed no decay even after 43 days while the signal from cuticle given higher doses was reduced in strength. More recently, Morehouse and Desrosiers[7] showed that the ESR signal from pink shrimp and brown shrimp shell given a dose of 4 kGy and kept at -20°C was stable for several months. It is therefore apparent that frozen storage would enhance signal stability.

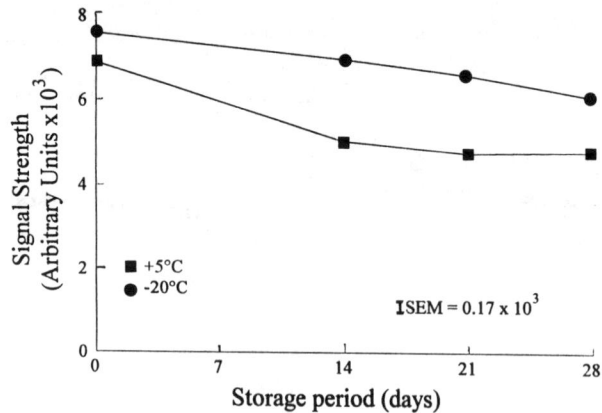

Figure 4 *Effect of storage at -20°C and +5°C on the intensity of the radiation-induced ESR signal from Norway lobster cuticle*
(SEM = Standard Error of the Mean)

In practice, samples received for testing in a laboratory may be prepared but not analysed immediately. In the case of Norway lobster, the signal from cuticle samples stored in a dried, ground form decreased slowly during storage at ambient temperature.[4] The greatest decline occurred during the first 7 days and thereafter signal intensity showed a gradual reduction up to 84 days.[4] Morehouse and Desrosiers[7] also showed that the ESR signal from pink shrimp decreased slowly when stored in the dried, ground state and had an expected half-life of approximately two weeks.

3.4 Exoskeleton Components

Suggestions by Desrosiers[8] that the part of the shell examined influenced ESR signal intensity were confirmed by the findings of the present work.[12] The intensity of the radiation-induced ESR signal derived from the cuticle of tail, carapace, walking legs and claws of Norway lobster were different, with that obtained from the claws being the strongest while the signal in the walking legs was the weakest.[12] As far as the Norway lobster is concerned, the claws are the strongest part of the exoskeleton while the walking legs are the most fragile and perhaps differences in structure/composition may affect the component's ability to trap radicals.

3.5 Processing methods

The way in which the Norway lobster is processed significantly affects the intensity of the radiation-induced ESR signal. Although freeze-drying did not cause any significant decrease in signal strength, cooking tails following completion of irradiation caused a reduction in intensity while cooking prior to irradiation substantially increased signal strength (Figure 5).[11] Similar effects of cooking after irradiation have been reported previously. Goodman and co-workers[16] showed that boiling scampi following irradiation resulted in the signal intensity falling to approximately 25% of its former level. It was also observed that boiling samples before irradiation caused considerable destabilisation of the radiation-induced radical which consequently led to a rapid decrease in signal strength within a day of irradiation. In the present work, the effect of storage on the intense signal obtained after irradiating cooked samples was not investigated.

A number of possible explanations for the behaviour of the ESR signal in samples cooked prior to irradiation have been put forward for chicken bone where effects similar to those observed for Norway lobster have been reported.[13,17] It has been proposed that cooking reduces the water content of the bone so that during subsequent irradiation there is less opportunity for free radicals to react with water. Consequently, more radicals are trapped and the outcome is a more intense signal. A change in the structure or crystallinity of the bone has been suggested as an alternative explanation.[13] A reduction in water content and possible changes in cuticle structure when samples are cooked before irradiation may also be feasible explanations for the results obtained with Norway lobster.

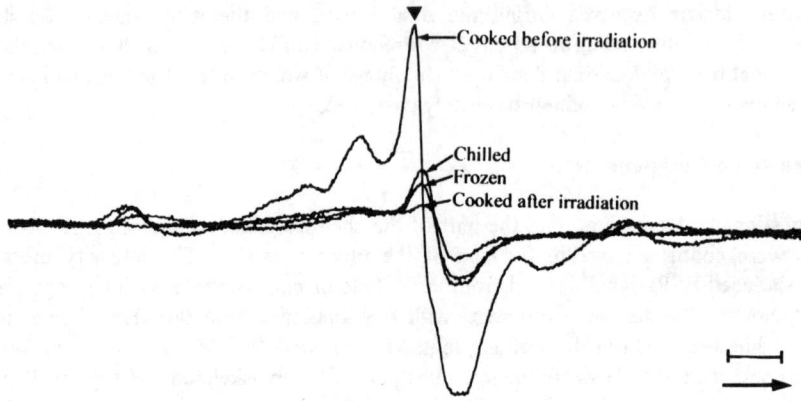

Figure 5 *Effect of different processing methods on the radiation-induced ESR signal in Norway lobster cuticle given 3 kGy*
▼ = 349.5 mT; |—| = 1.0 mT; → = increasing H

(Reprinted from E. M. Stewart, M. H. Stevenson, R. Gray and C. H. McMurray, *Radiat. Phys. Chem.*, **42**, 367-370, 1993, with kind permission from Elsevier Science Ltd., The Boulevard, Langford Lane, Kidlington OX5 1GB, UK)

3.6 Sex and Size

The Norway lobster grows by a process of moulting where the old exoskeleton is cast off to be replaced by the new cuticle which has been developing underneath the old. Associated with this effect are various changes in the cuticle structure and biochemistry. The frequency of the moulting varies according to the sex and size of the Norway lobster and, in general, the larger and older the animal is, the less frequently it moults. Mature females moult only once a year during spring while mature males tend to moult more frequently from spring to autumn. It was, therefore, presumed that these factors may lead to differences in the structure of the exoskeleton thereby affecting the intensity of the radiation-induced ESR signal. However, the latter proved to be unfounded as neither the size or sex of the Norway lobster was found to affect the intensity of the radiation-induced ESR signal from either tail or claw cuticle.[14]

4 CONCLUSIONS

In conclusion, the investigations undertaken demonstrated the feasibility of using ESR spectroscopy as a detection method for irradiated Norway lobster having an attached cuticle. The technique exhibited many of the characteristics of a reliable detection method[18] and the studies confirmed that the ESR signal induced in Norway lobster cuticle by gamma irradiation is:

(a) specific to ionising irradiation and not produced or destroyed by drying, grinding, chilling, freezing or cooking,
(b) detectable in the cuticle of different components of the exoskeleton,
(c) detectable in a sample size of 10 mg at doses liable to be used in commercial practice,
(d) sufficiently stable to facilitate detection throughout the expected shelf-life of the Norway lobster,
(e) capable of providing an estimate of the actual dose of irradiation received by the product,
(f) reproducible giving comparable intensities when samples received doses of the same magnitude under similar conditions, and
(g) capable of rapid and repeated measurement.

The detection of radiation-induced signals in the cuticle of pink shrimp, tiger prawn, king prawn and Mediterranean crevette confirmed that the ESR methodology could also be applied to other species. However, the instability of the induced peaks, especially at low doses of irradiation, may limit the period over which successful detection can be accomplished.

Further research is needed to evaluate the shape of the radiation-induced signal in a wider range of Crustacea and to study the factors which could influence signal intensity and hence quantification of absorbed dose in these species. The performance of the ESR technique should also be assessed in collaborative blind trials involving participants from a number of laboratories.

Acknowledgements

The authors wish to acknowledge the Ministry of Agriculture, Fisheries and Food (MAFF) for partial funding of this work, and also Mr William D Graham and Ms Emma McCarville for technical assistance.

References

1. Anon, The Food (Control of Irradiation) Regulations 1990, No. 2490, 1990. Her Majesty's Stationery Office, London.
2. Anon, Supplement to Food Irradiation Newsletter, 1995, 19, IAEA, Vienna, Austria.
3. L. Saint-Lèbe and J. J. Raffi, "Health Impact, Identification and Dosimetry of Irradiated Foods," Report of a WHO Working Group, K. W. Bögl, D. F. Regulla and M. J. Suess, Eds., Neuherberg/Munich, WHO, Copenhagan, Denmark, 1988, p. 139.
4. E. M. Stewart, M. H. Stevenson and R. Gray, *Int. J. Food Sci. Technol.*, 1992, **27**, 125.
5. E. M. Stewart, M. H. Stevenson and R. Gray, *J. Sci. Food Agric.*, 1994, **65**, 191.
6. N. Helle, B. Linke, O. Popoola, K. W. Bögl and G. A. Schreiber, "New Developments in Food, Feed and Waste Irradiation," Proceedings of the 3rd German Meeting on Food Irradiation and the Working Group "Radiation Technology" of the 23rd Meeting of the European Society for New Methods in Agricultural Research, G. A. Schreiber, N. Helle and K. W. Bögl, Eds., Bundesgesundheitsamtes (BGA), Berlin, SozEp-Heft 16, 1993, p. 99.
7. K. Morehouse and M. F. Desrosiers, *Appl. Radiat. Isot.*, 1993, **44**, 429.
8. M. F. Desrosiers, *J. Agric. Food Chem.*, 1989, **37**, 96.

9. K. M. Morehouse and Y. Ku, *J. Agric. Food Chem.*, 1992, **40**, 1963.
10. H. Chandra and M. C. R. Symons, *Nature*, 1987, **328**, 833.
11. E. M. Stewart, M. H. Stevenson, R. Gray and C. H. McMurray, *Radiat. Phys. Chem.*, 1993, **42**, 367.
12. E. M. Stewart, M. H. Stevenson and R. Gray, *Appl. Radiat. Isot.*, 1993, **4**, 433.
13. R. Gray and M. H. Stevenson, *Int. J. Sci. Food Technol.*, 1989, **24**, 447.
14. E. M. Stewart, "The Use of ESR Spectroscopy for the Identification of Irradiated Crustacea with Particular Reference to Nephrops norvegicus (Norway lobster)," PhD Thesis, The Queen's University of Belfast, Northern Ireland, 1993.
15. M. H. Stevenson and R. Gray, *J. Sci. Food Agric.*, 1989, **48**, 261.
16. B. A. Goodman, D. B. McPhail and D. M. L. Duthie, *J. Sci. Food Agric.*, 1989, **47**, 101.
17. N. J. F. Dodd, J. Haishun, J. S. Lea and Swallow, *Int. J. Food Sci. Technol.*, 1992, **27**, 371.
18. H. Delincée, *Radiat. Phys. Chem.*, 1993, **42**, 351.

TIME COURSE STUDY OF THE EPR SPECTRA OF SEEDS OF SOFT FRUIT IRRADIATED IN WET AND DRY STATES

S. M. Glidewell, N. Deighton, A. E. Morrice and B. A. Goodman

Scottish Crop Research Institute
Invergowrie
Dundee DD2 SDA
U.K.

1 INTRODUCTION

Irradiation by ionising radiation leads to the formation of free radicals which are directly detectable by electron paramagnetic resonance (EPR) spectroscopy. The vast majority of these react very rapidly in aqueous systems such as predominate in food. The harder parts of food, such as bone, cuticle, seeds *etc.* however, provide matrices in which such radiation-induced radicals can be trapped and stabilised. These trapped radicals often

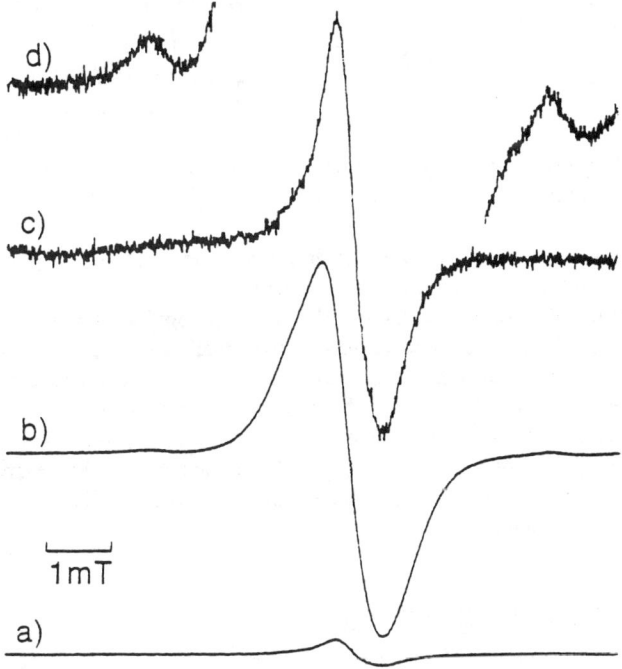

Figure 1 *EPR spectrum of a) non-irradiated strawberry seeds; b) irradiated (3 kGy) strawberry seeds; c) (a) x 16; d) (b) x 16.*

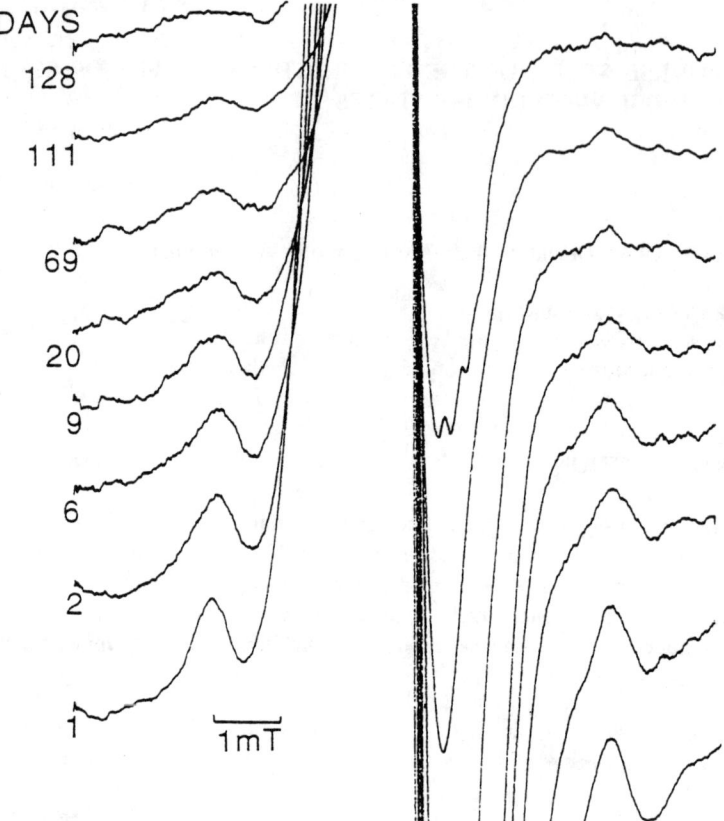

Figure 2 *EPR spectrum of irradiated strawberry seeds at increasing intervals after irradiation*

have sufficiently long lifetimes to allow their direct observation over a period which can sometimes extend to several months after irradiation.

The detection of irradiated seeds by EPR spectroscopy[1] is based on the radiation-induced signal from cellulose which comprises two small peaks - 6.0 mT apart, on either side of a large central feature at g - 2.0027.[2] This central feature increases markedly in intensity upon irradiation but, unlike the two smaller side features, is not radiation-specific[3,4] (Figure 1). These relatively small cellulose-derived signals decrease in intensity with time over a period of some weeks (Figure 2) and hence provide a useful method for the detection of irradiation (if not dose estimation) over the shelf-life of soft fruits such as strawberry and raspberry.

2 MATERIALS AND METHODS

Fresh and frozen strawberries (*cv* Melody), fresh blackcurrants (various cultivars), strawberry seed (*cv* Begota), raspberry seed (*cv* Glen Lyon), blackcurrant seed (various cultivars) and winter barley grains (*cv* Magie) were irradiated to a dose of 3 kGy at the

Scottish Universities Research and Reactor Centre (SURRC), East Kilbride. Seeds were removed from the whole fruits and dried on filter paper overnight at 50°C. Husks were removed manually from a portion of the barley grains. Seeds and husks were placed in quartz EPR tubes (diameter 4.2 mm) and capped and stored at 20°C. The samples were kept in these tubes for the duration of the experiment so that the quantity of seed examined remained constant for that period.

Initially, EPR spectra were recorded daily, and then at increasing intervals over a total period of 18 weeks. The microwave power was the relatively low value of 0.25 mW as this has been shown to be the optimum power for detection of radiation-induced peaks of interest.[5] Other instrumental parameters were: 10 mT sweep width; 4.0 G modulation amplitude; 100 kHz modulation frequency; the gain was adjusted to maximise the largest peak on scale. Peak intensities were obtained from double integrations of the entire recorded spectra for whole barley and dry-irradiated blackcurrant seeds and from peak to peak heights or manual measurements of peak *extrema* from a baseline estimated by eye, of spectra filtered by a moving average filter of width 33 data points (corresponding to 0.32 mT) for the other samples. The relative intensity for each particular peak is presented as a multiple of its intensity in the first recorded spectrum of each sample.

3 RESULTS AND DISCUSSION

In order to observe EPR spectra, there must be a low water content in the microwave cavity; therefore, seeds must be removed from their fruits and dried before EPR spectra are recorded. In initial experiments, seeds were removed and dried before irradiation but subsequent measurements involved irradiation of intact fruits followed by the removal and drying of the seeds before monitoring the decay of the free radical content by EPR spectroscopy. Figure 3 shows that the side peaks decay at similar rates in both cases, but if the central peak is monitored, as shown in Figure 1, a big difference is observed between the seeds which were irradiated dry (*i.e.* moisture content <15%) and those which were irradiated *in situ, i.e.* wet.

The EPR signals of those seeds which were irradiated *in situ* increased over a period of approximately one month after which they declined, reaching their initial intensity after about 3 months for blackcurrant. Strawberry seeds still gave an EPR signal greater than the initial value after more than 4 months. In contrast, the intensities of the central peaks of seeds which were irradiated dry decreased progressively from day 1. In addition to the increase in intensity of the central feature, there are also qualitative differences between wet and dry irradiated seeds for blackcurrant but not for strawberry (Figure 5).

The side peaks in barley and blackcurrant cannot be observed due to the presence of the large, broad signal thought to originate from starch, as comparison with the EPR spectra of irradiated wheat and rice starches shows (Figure 6).

The starch signal in blackcurrant decays much more rapidly in the presence of the rest of the fruit than it does in isolated dried seeds. This more rapid decay of the starch-derived radical(s) in a wet environment is such that it is not visible *per se* in seeds with a relatively low starch content such as strawberry or raspberry. Strawberry seeds irradiated dry do have a broader central feature than those irradiated *in situ* (Figure 7) hinting at the same phenomenon as is observed in blackcurrant.

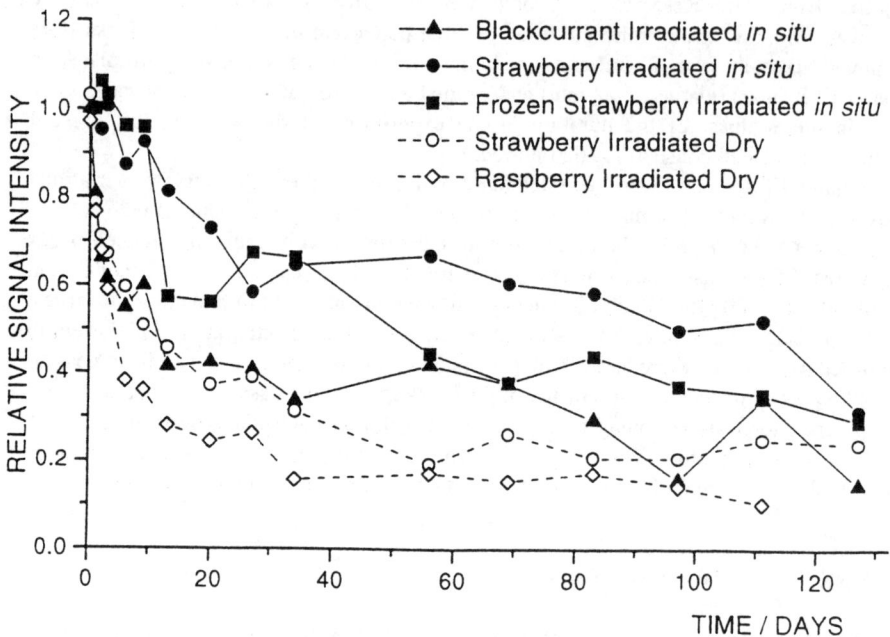

Figure 3 *Time course of intensity of side peaks of EPR signal from irradiated seeds*

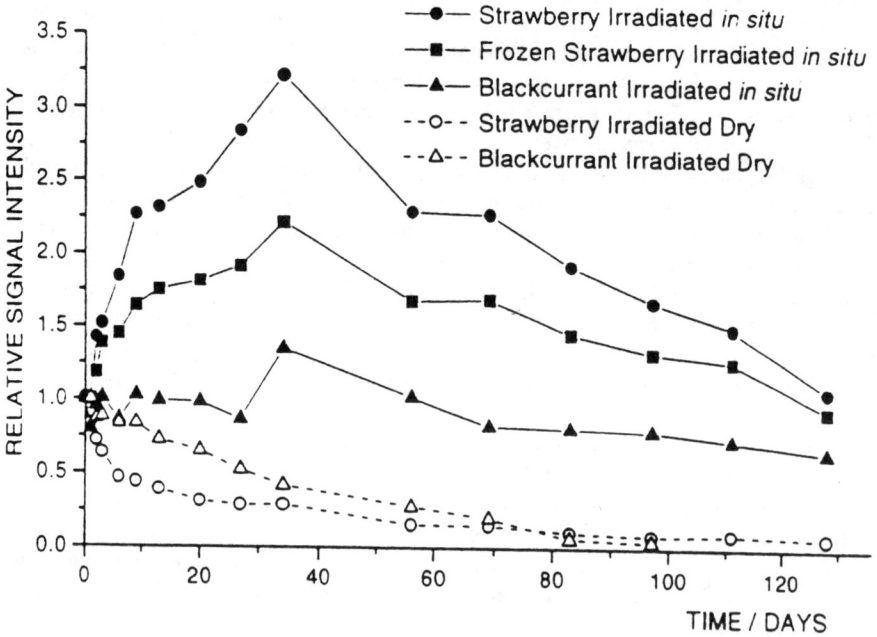

Figure 4 *Time course of intensity of central peak of EPR signal from irradiated seeds*

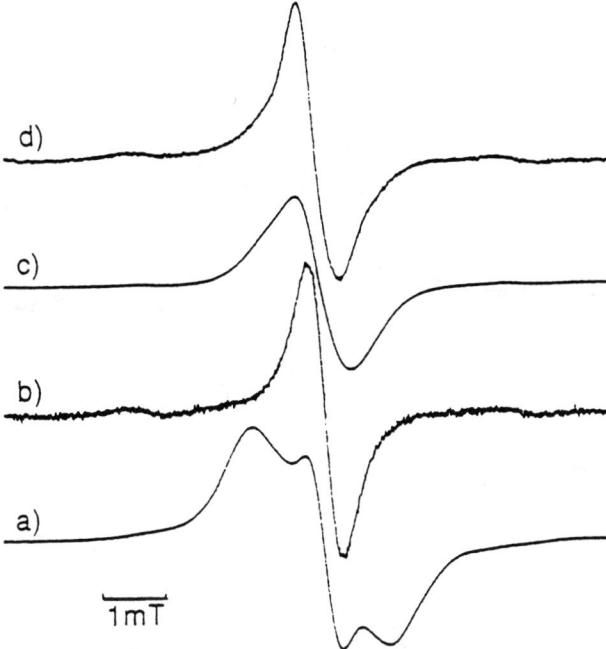

Figure 5 *EPR spectra of a) blackcurrant seeds irradiated dry; b) blackcurrant seeds irradiated in situ; c) strawberry seeds irradiated dry; d) strawberry seeds irradiated in situ*

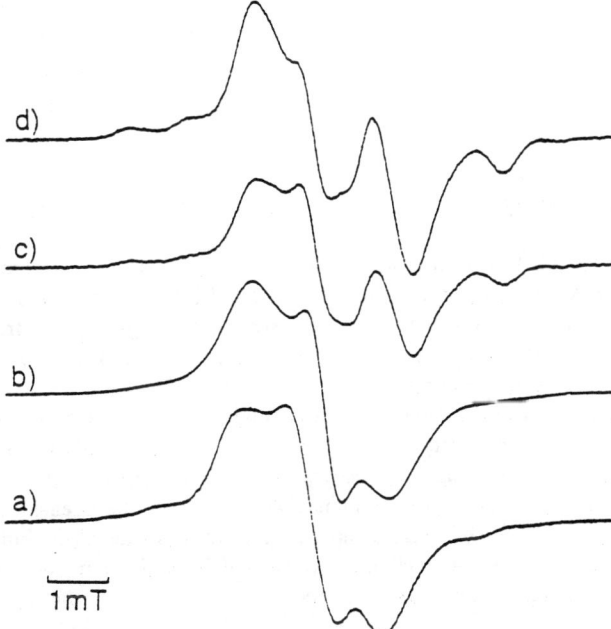

Figure 6 *Comparison of EPR spectra of irradiated a) barley and b) blackcurrant seeds with those of irradiated c) wheat starch and d) rice starch*

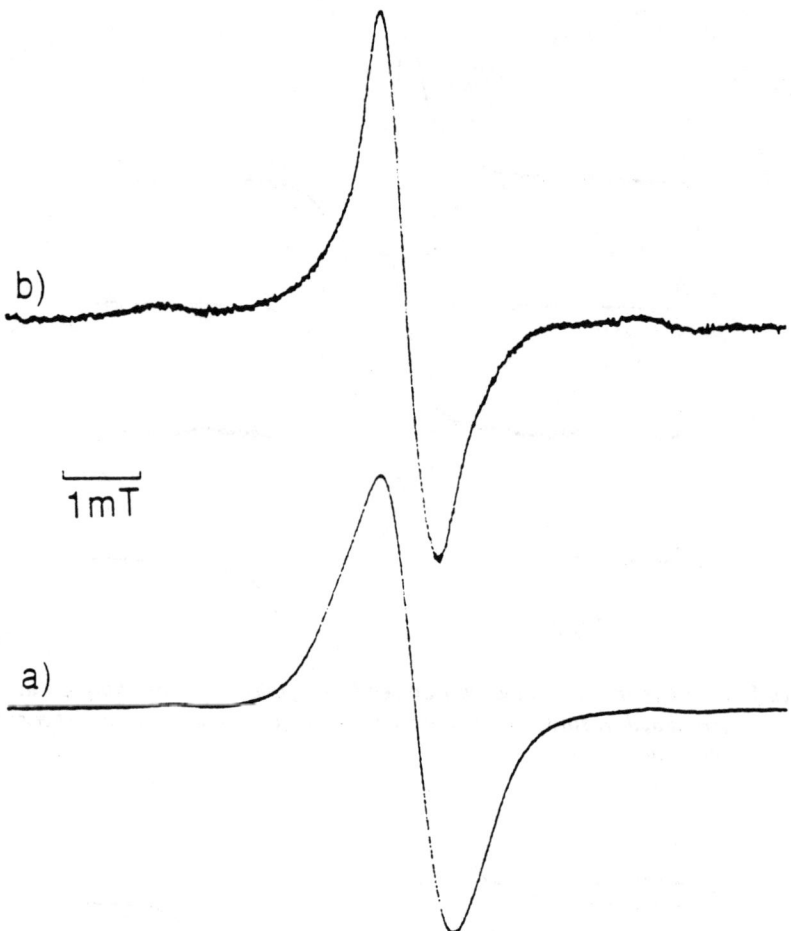

Figure 7 *Comparison of EPR spectra of strawberry seeds irradiated a) dry and b) in situ*

In order to be able to measure the side peaks in irradiated barley, it was necessary to separate the husks from the grains and record the EPR spectra of the former alone. There is little difference between the decay rates of the EPR signals from irradiated barley whole grain and husk, or their difference (Figure 8), which is attributed to starch by analogy with the results presented above.

The presence of varying levels of starch could explain some of the qualitative differences observed in the various spectra. It is, however, of no assistance in explaining the rise in the observed EPR signal in seeds which were irradiated *in situ i.e.* irradiated in the manner in which they would be commercially. A rise in the intensity of the central feature has also been observed in the stalks and skins of irradiated citrus fruits.[6]

The free radical content of cellulose which had been dried under vacuum before irradiation was reduced by 70% for cellulose I and 90% for cellulose II on

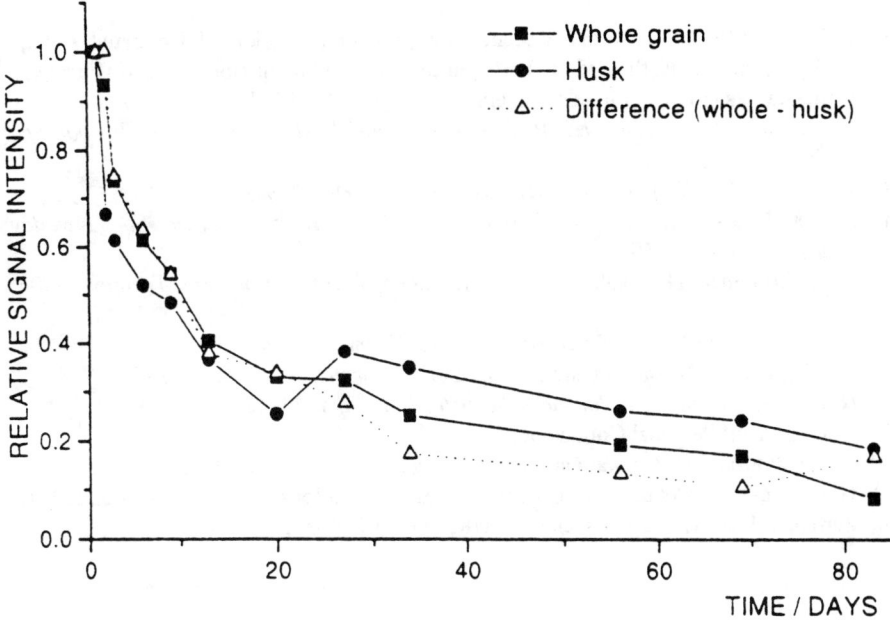

Figure 8 *Time course of EPR signal from components of irradiated barley*

re-introduction of water vapour.[7] The EPR spectrum of irradiated cellulose has been shown to have a minimum intensity at 5% water content and higher values on either side of this.[8] The water content of air-dried seeds used in the present work is of the order of 15%,[9] and would not have varied appreciably over the timescale of the measurements. Thus, the water content of the sample would not have been expected to have a significant effect on the cellulose-derived EPR signal, as was indeed observed.

The presence of water in seeds means that in addition to directly formed cellulose radicals,[10,11] there will be considerable quantities of the extremely reactive hydroxyl radicals which will react almost indiscriminately with the surrounding matrix to form other radical products. Such reactions would be expected to be of greater importance for seeds irradiated *in situ* because of their higher water content. It appears that the products of these reactions continue to change over the weeks following irradiation, giving rise to the initial increase in radical load before decaying in a manner similar to the cellulose radicals. Hence, there are reactions involving the formation as well as the removal of free radical species occurring well after irradiation, throughout the most likely period of consumption.

Acknowledgment

This work was funded by the Scottish Office, Agricultural and Fisheries Department.

References

1. J.-J. Belliardo and J. Raffi, "Recent Advances on Detection of Irradiated Food," M. Leonardi, J. Raffi and J.-J Belliardo, Eds., Commission of the European Communities (BCR), Brussels, Luxembourg, EUR 14315 EN, 1993, p. 203.
2. N. Deighton, S. M. Glidewell, B. A. Goodman and I. M. Morrison, *Int. J. Food Sci. Technol.*, 1993, **38**, 45.
3. J. Raffi and J.-P. Agnel, *Ann. Fals. Exp. Chem.*, 1989, **82**, 979.
4. J. J. Raffi, J.-P. Agnel, L. A. Buscarlet and C. C. Martin, *J. Chem. Soc. Faraday Trans.*, 1988, **84**, 3359.
5. B. A. Goodman, N. Deighton and S. M. Glidewell, *Int. J. Food Sci. Technol.*, 1994, **29**, 23.
6. B. J. Tabner and V. A. Tabner, *Radiat. Phys. Chem.* 1994, **41**, 545.
7. O. Hinojosa, Y. Nakamura and J. C. Arthur, *J. Polymer Sci.: Part C*, 1972, **37**, 27.
8. N. Hon, *J. Polymer Sci. Polymer Chemistry Edition*, 1975, **13**, 955.
9. J. R. Hillman, Personal Communication.
10. Y. Nakamura, S Y. Ogiwara and G. O. Phillips, *Polymer Photochem.*, 1985, **6**, 135.
11. G. O. Phillips, "Cellulose Chemistry and Its Applications," T. P. Newell and S. H. Zeronian, Eds., Howard, Chichester, UK, 1985, Chapter 12, p. 290.

THE USE OF ESR SPECTROSCOPY FOR THE DETECTION OF IRRADIATED MECHANICALLY RECOVERED MEAT (MRM) IN TERTIARY FOOD PRODUCTS

M. H. Stevenson,[1,2] E. Marchioni,[3] R. Gray,[1] E. M. Stewart,[2] M. Bergaentzle[3] and F. Kuntz[4]

[1]Food Science Division, The Department of Agriculture for Northern Ireland and
[2]The Queen's University of Belfast, Newforge Lane, Belfast BT9 5PX, U.K.
[3]Département des Sciences de l'Aliment, Université Louis Pasteur, Strasbourg, France
[4]Aérial, Schiltigheim 67300, France

1 INTRODUCTION

The use of electron spin resonance (ESR) spectroscopy as a detection method for irradiated food has been well documented in recent years, the research dealing mainly with primary products such as chicken, shellfish and fruit.[1-4] An example of how the technique can be applied to processed food products was demonstrated when ESR spectroscopy was used to differentiate between samples of commercially irradiated and non-irradiated mechanically recovered turkey meat (MRM).[5] The latter refers to meat which has been mechanically removed under high pressure from irregularly shaped bones subjected to hand-boning operations. The paste-like product contains small fragments of bone which can be extracted[5,6] and used for the purposes of ESR spectroscopy provided a sufficient quantity is recovered.

On a commercial scale MRM can be incorporated into other food, such as beef burgers, to produce a tertiary product. The aim of this experimental work, carried out simultaneously in Belfast and Strasbourg, was to compare two methods for extracting bone fragments from such tertiary products containing MRM at varying inclusion rates and to determine if ESR spectroscopy could be used to qualitatively detect the presence of small amounts of irradiated MRM. In addition, a number of coded foods containing either non-irradiated or irradiated MRM were examined in order to establish the feasibility of applying the method to samples, the processing history of which was unknown.

2 MATERIALS AND METHODS

2.1 Preparation of Beef Burgers

The burgers were prepared in the Belfast laboratory using the following procedure. Two 20 kg blocks of MRM (containing approximately 0.3 g bone 100 g^{-1} meat), one commercially irradiated (5.2 kGy) and one non-irradiated, were obtained in the frozen state from a French food company. The irradiation treatment had been carried out using a linear accelerator (MeV Industrie, Paris, France) with an average energy of 7.5 ± 0.5

MeV and intensity of 1.2 mA. The absorbed dose received by the MRM was measured using Far West dosimeters calibrated against the French National Reference (alanine).

Approximately 250 g of the non-irradiated and 750 g of the irradiated MRM were left to thaw overnight at a chill temperature of 2°C. Lean beef (20 kg) was obtained from a local retailer in Belfast, passed through a mincer and stored at 2°C until required. Burgers, each weighing approximately 100 g, were prepared using the minced beef with MRM being incorporated at inclusion rates of either 1, 3, 7 or 10 g 100 g^{-1} beef.

At an inclusion rate of 1 g 100 g^{-1}, a total of 20 burgers were prepared using non-irradiated MRM while 60 were prepared using the irradiated MRM. Using an inclusion rate of 3 g 100 g^{-1}, 12 burgers were made using non-irradiated MRM and 36 with the irradiated meat. At both the 7 and 10 g 100 g^{-1} inclusion rates, a total of 8 burgers were prepared with the non-irradiated MRM and 24 using the irradiated MRM. Following preparation, the burgers were frozen at -20°C. Half of the samples for each inclusion rate were retained in Belfast while the remainder were packed in polystyrene boxes containing dry ice and forwarded to the laboratory in Strasbourg where they arrived within 48 h. Each laboratory received 48 burgers for each of two extraction procedures. The samples were stored in the frozen state until required for analysis.

2.2 Preparation of Pre-cooked Meals

Approximately 1 kg of both irradiated and non-irradiated MRM (obtained from the same source as the MRM used in the previous section) was left to thaw overnight at a chill temperature of 4°C. Turkey meat was purchased from a local retailer in Strasbourg and forcemeat balls, each weighing approximately 100 g, were prepared by a catering college in Strasbourg using the following method. Turkey meat weighing 1 kg (plus the appropriate weight of MRM, that is, either 1, 3, 7 or 10 g 100 g^{-1}) was minced, after which cream (500 g), beaten egg whites (80 g), salt (20 g) and pepper (8 g) were added. The forcemeat balls were baked for 15 min at 160°C after which they were cooled and frozen at -20°C. For the purposes of this experimental work, a total of 38 samples were prepared using MRM at an inclusion rate of 1 g 100 g^{-1}, with half being made with irradiated MRM and the remainder with non-irradiated MRM. In a similar manner, 36 samples were prepared using an inclusion rate of 3 and 7 g 100 g^{-1} while 22 samples were made with 10 g 100 g^{-1} of MRM. Following preparation the samples were frozen at -20°C until analysed by the Strasbourg laboratory.

2.3 Preparation of Blind Samples (Pre-cooked Meals)

Twenty-four samples were prepared at each inclusion rate of 1, 3, 7 or 10 g 100 g^{-1} using both irradiated and non-irradiated MRM after which each forcemeat ball was individually packed in a coded polyethylene bag. Both the Belfast and Strasbourg laboratory received a total of 24 samples for each extraction method. Each laboratory was asked to identify the samples containing the irradiated MRM.

2.4 Extraction Methods

The two extraction methods employed to obtain bone fragments from the burgers, pre-cooked meals and blind samples were as follows.

2.4.1 Alcoholic KOH Extraction. Each 100 g sample was minced and placed in a 500 ml flat bottomed flask along with 200 ml of alcoholic KOH and digested at approximately 85°C for 90 min under reflux in order to avoid excessive dehydration.[5] Following digestion the contents of the flask were filtered under vacuum through a sintered glass crucible (porosity 1, pore diameter 100-160 µm) (Belfast) or through a nylon sieve cloth (100 µm porosity) (Strasbourg), which was then washed with hot distilled water followed by acetone prior to drying for 1 h at 100°C (Belfast) or 50°C (Strasbourg). Following drying, the crucible or nylon sieve cloth plus contents were allowed to cool after which the weight of bone extracted from the food was measured.

2.4.2 Alcalase Extraction. The second extraction method used food quality 'Alcalase 2,4 LFG' (Novo Nordisk, Denmark), an enzyme produced by *Bacillus licheniformis*. The main constituent is an endoproteinase (Subtilisine Carlsberg) which has been especially developed for the hydrolysis of proteins.[7] A phosphate buffer (pH 8.0) containing 1% sodium dodecyl sulphate (SDS) in 0.2 M solution of Na_2HPO_4 and NaH_2PO_4 was prepared. Each sample was minced and placed in a 1.0 litre beaker along with 400 ml of the phosphate SDS buffer and 100 ml of the Alcalase enzyme. The contents of the beaker were heated quickly to 60°C in a microwave oven. The beaker was then placed on a magnetic stirrer hotplate in order to mix the contents and to maintain the temperature at 60°C. Care was taken that the temperature did not exceed 80°C as this would have denatured the enzyme. During the 40 min required for hydrolysis of the meat, the pH was maintained at 8.0, if necessary, by the addition of small amounts of 10 M NaOH. Following hydrolysis the contents of the beaker were filtered under vacuum and treated as described previously.

2.5 ESR Spectroscopy

The extracted bone fragments were introduced into a 5 mm (internal diameter) ESR tube which was placed in the magnetic cavity of a Bruker ESP300 spectrometer (Belfast) or a Bruker ECS106 spectrometer (Strasbourg) ensuring that the sample was in the area of the active length.[1] The Belfast laboratory altered a number of the parameters for examination of the blind samples. A receiver gain of 5×10^4 was used along with a modulation amplitude of 0.4 mT and a conversion time of 2.56 msec. The resolution of field axis was 4096 points and each sample was scanned 16 times.

Following derivation of the ESR spectra the peak height values of the signals obtained from the irradiated samples of beef burgers and pre-cooked meals were measured and the values corrected for sample weight.

2.6 Statistical Analysis

Results of the experimental work were subjected to analysis of variance.

Table 1 *Operating Conditions for the ESR Spectrometers*

	Belfast	Strasbourg
Centre field	349.4 mT	342.0 mT
Sweep width	10.0 mT	20.0 mT
Receiver gain	5×10^5	1×10^5
Modulation frequency	100 kHz	100 kHz
Modulation amplitude	0.26 mT	0.2 mT
Microwave frequency	9.79 GHz	9.79 GHz
Microwave power	15.8 mW	40 mW
Conversion time	40.96 msec	81.92 msec
Time constant	40.96 msec	655.36 msec
Sweep time	41.94 sec	83.89 sec
Resolution	1024 points	1024 points
Number of scans/determination	5	5

3 RESULTS AND DISCUSSION

3.1 Beef Burgers

The average weights of bone obtained from the beef burgers containing MRM at varying inclusion rates are presented in Figure 1 and Table 2.

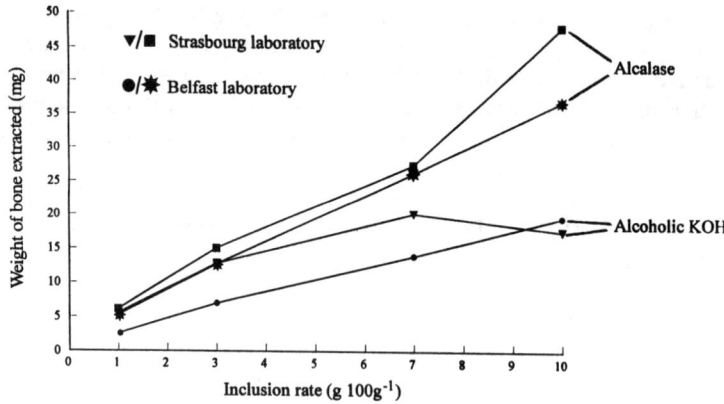

Figure 1 *Recovery of bone fragments from beef burgers using two extraction methods*
  1 g 100 g^{-1} inclusion rate (n=20) SEM=0.71
  3 g 100 g^{-1} inclusion rate (n=12) SEM=1.93
  7 g 100 g^{-1} inclusion rate (n=10) SEM=2.46
  10 g 100 g^{-1} inclusion rate (n= 8) SEM=4.98
  SEM = Standard Error of the Mean

As expected the highest weight of bone recovered was from burgers containing MRM at an inclusion rate of 10 g 100 g^{-1} while the least bone was obtained from samples prepared using 1 g 100 g^{-1}. Statistical analysis showed that the weight of bone extracted increased with increasing inclusion rate, the response being significantly linear (P<0.001). Both extraction procedures proved to be successful as bone fragments were recovered from all burgers examined by both the Belfast and Strasbourg laboratories. However, it was found that by using the Alcalase enzyme procedure, a significantly greater (P<0.001) quantity of bone was extracted than that obtained using the alcoholic KOH methodology. The lower yield of bone fragments from the alcoholic KOH procedure could possibly be explained by a partial dissolution of the bone matrix during the chemical procedure.

There was also a wide variation associated with both extraction methods in the amount of bone obtained from the burgers at each inclusion rate (Table 2). For example, at the 1 g 100 g^{-1} inclusion rate, the minimum quantity of bone extracted by the Belfast laboratory, using the Alcalase method, was 1.4 mg while 13.1 mg was the maximum amount obtained. In the case of the Strasbourg laboratory, at the same inclusion rate, 1.6 mg was the least amount of bone recovered while the highest weight was 13.7 mg. The variations in bone weight could, however, be simply due to the distribution of bone fragments in the MRM resulting in some beef burgers containing less bone fragments than others. In burgers prepared using a low inclusion rate of irradiated MRM this could present a problem as it may not be possible to extract sufficient bone for analysis by ESR spectroscopy.

It is also evident from Table 2 that, on average, there was a significant difference (P<0.001) in the weight of bone extracted using both methods by each laboratory, with the Strasbourg laboratory exhibiting higher yields than the Belfast laboratory. This could be accounted for, at least in part, by differences in the porosity of the filtering devices used, the crucible in Belfast having a greater porosity (100-160 µm) than the nylon sieve

Table 2 *Weight of Bone Extracted from Beef Burgers Containing MRM at Varying Inclusion Rates*

Inclusion Rate (g 100 g^{-1})	Weight of Bone Extracted (mg)				n*	SEM
	Belfast Alcoholic KOH	Strasbourg Alcoholic KOH	Belfast Alcalase	Strasbourg Alcalase		
1	2.56	5.55	5.22	6.10	20	0.71
Max/Min values	(3.9-1.7)	(20.3-0.4)	(13.1-1.4)	(13.7-1.6)		
3	6.99	12.78	12.70	15.03	12	1.93
Max/Min values	(12.9-4.1)	(43.2-4.1)	(19.7-5.9)	(23.9-6.3)		
7	13.90	20.10	26.12	27.40	10	2.46
Max/Min values	(23.0-9.80)	(29.8-10.2)	(40.4-18.9)	(40.5-11.0)		
10	19.50	17.50	36.80	47.90	8	4.98
Max/Min values	(27.0-13.7)	(41.7-3.2)	(58.3-28.0)	(85.0-23.0)		

Significance of effect where: *** = P<0.001; ** = P<0.01
Inclusion rate   ***
Extraction method   ***   SEM = Standard Error of the Mean
Laboratory   **   n* = replication for each experiment

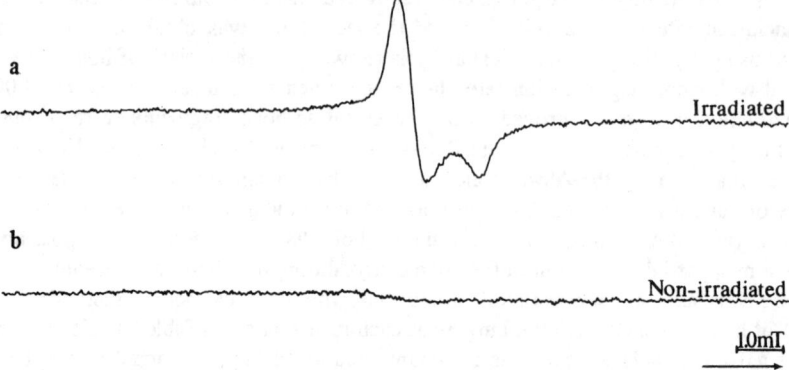

Figure 2 *ESR spectra derived from bone fragments extracted from beef burgers*

cloth (100 μm) used in Strasbourg. The filtering device used by the latter laboratory would be expected to retain a greater quantity of bone.

The ESR signal derived from the bone fragments extracted from the beef burgers containing irradiated MRM was characteristic of irradiated bone (Figure 2a) while the bone from the control samples (Figure 2b) exhibited no such signal. Using the Wilcoxan Rank-Sum Test[8] it was found that there was no significant difference in the signal strength of bone fragments recovered using either of the two extraction procedures. The ESR signal connected to the mass of bone remained quite constant irrespective of the inclusion rate of MRM. The values presented in Table 3 are the average intensities of the specific ESR signal derived from the bone extracted from all the samples. Examination of the coefficients of variation (% CV) showed that, in general, the signal intensity values given by the Belfast laboratory were less variable than those obtained by the laboratory in Strasbourg (Table 3). However, it was observed that not all of the bone samples extracted by the alcoholic KOH procedure from the burgers containing irradiated MRM exhibited the characteristic ESR signal. At an inclusion rate of 1 g 100 g^{-1} only one of the bone samples examined by the Strasbourg laboratory gave the radiation-induced ESR signal while the Belfast laboratory detected 8 out of 15 samples at the same inclusion rate. The Belfast laboratory also encountered problems at the higher inclusion rates of 3 and 7 g 100 g^{-1} as one sample at each of these rates did not give an identifiable signal. On the other hand the Strasbourg laboratory found that all of the bone fragments extracted by the Alcalase method, exhibited the specific ESR signal. This was also the case for the Belfast laboratory except for the bone samples extracted from the burgers containing irradiated MRM at the lowest inclusion rate of 1 g 100 g^{-1} where the specific ESR signal was not detected in 5 of the 15 samples examined.

3.2 Pre-cooked Meals

The experiment carried out using the pre-cooked meals was similar to that of the beef burgers but it involved only the Strasbourg laboratory. The findings of this work were

Table 3 *Effect of Extraction Procedure on the Strength of the Characteristic ESR Signal*

Laboratory	Extraction Method	Signal Strength (Arbitrary Units)	Standard Deviation	% CV
Belfast	Alcoholic KOH	6921.4	3437.2	49.66
	Alcalase	5941.5	1834.1	30.87
Strasbourg	Alcoholic KOH	1548.2	804.0	51.93
	Alcalase	1832.3	1064.9	58.12

similar to those for the burgers in that the Alcalase enzyme extraction was the most efficient (P<0.001) method for the recovery of bone fragments and the weight of bone extracted increased linearly (P<0.001) with increasing inclusion rate. However, in contrast to the burgers it was found that the extraction procedures not only recovered bone fragments but other non-protein components. When subjected to analysis by ESR, the extracted samples produced a strong singlet which masked the characteristic bone signal.

Examination of the samples under a binocular microscope showed that the extracted material contained small bone fragments, considerable quantities of pepper and fragments of minerals, probably adhering to the pepper. As a consequence, the bone fragments were separated by hand from the other materials prior to derivation of the ESR spectra, but in doing this the amount of bone recovered by both methods was very low. It therefore proved difficult to derive a definitive ESR signal from the samples when the inclusion rate of MRM was below 3 g 100 g^{-1} for the Alcalase procedure and less than 7 g 100 g^{-1} for bone obtained using the alcoholic KOH method. These results were confirmed by analysis of the blind samples

3.3 Blind Samples (Pre-cooked Meals)

As for the burgers and pre-cooked meals, it was found that the Alcalase extraction procedure was the most successful for recovering bone fragments from the coded blind samples and as for the pre-cooked meals, the bone fragments had to be manually removed from the other extracted material. Following completion of the blind trial, it was revealed that the coded samples had contained MRM at inclusion rates of 1, 3, 7 or 10 g 100 g^{-1} and that half of the samples at each inclusion rate contained irradiated MRM (5.2 kGy) while the remainder were prepared using non-irradiated MRM.

Using the alcoholic KOH methodology, the Belfast laboratory correctly identified 18 of the 24 coded samples while the Strasbourg laboratory made 16 correct identifications (Table 4). On the other hand, the former laboratory identified 2 non-irradiated samples as being irradiated whereas the Strasbourg laboratory correctly detected all reference non-irradiated samples.

A significant improvement in the success rate of identification was achieved when bone fragments were extracted using the Alcalase enzyme procedure (Table 5). The minimal rate of inclusion which gave a 100 % correct rate of identification was 3 g MRM 100 g^{-1}. At the 1 g 100 g^{-1} inclusion rate, only 1 of 3 coded samples containing irradiated

Table 4 *Results of Blind Test using Bone Samples Extracted by Alcoholic KOH*

Laboratory	Inclusion Rate (g 100 g^{-1})	Treatment	Rate of Success
Strasbourg	1	3 Irradiated	1/3
		3 Reference	no false positives
	3	3 Irradiated	0/3
		3 Reference	no false positives
	7	3 Irradiated	2/3
		3 Reference	no false positives
	10	3 Irradiated	1/3
		3 Reference	no false positives
Belfast	1	3 Irradiated	2/3
		3 Reference	no false positives
	3	3 Irradiated	2/3
		3 Reference	no false positives
	7	3 Irradiated	1/3
		3 Reference	two false positives
	10	3 Irradiated	3/3
		3 Reference	no false positives

Table 5 *Results of Blind Test using Bone Samples Extracted by the Alcalase Enzyme*

Laboratory	Inclusion Rate (g 100 g^{-1})	Treatment	Rate of Success
Strasbourg	1	3 Irradiated	2/3
		3 Reference	no false positives
	3	3 Irradiated	3/3
		3 Reference	no false positives
	7	3 Irradiated	3/3
		3 Reference	no false positives
	10	3 Irradiated	3/3
		3 Reference	no false positives
Belfast	1	3 Irradiated	1/3
		3 Reference	no false positives
	3	3 Irradiated	3/3
		3 Reference	no false positives
	7	3 Irradiated	3/3
		3 Reference	no false positives
	10	3 Irradiated	2/3
		3 Reference	no false positives

MRM was correctly identified by the Belfast laboratory while 2 out of 3 samples were correctly detected by the Strasbourg laboratory. At this low inclusion rate, the weight of bone recovered was obviously very low in some cases and this may have accounted for the inability to detect irradiated samples. One of the samples containing irradiated MRM at the highest MRM inclusion rate of 10 g 100 g^{-1} was incorrectly identified as non-irradiated by the Belfast laboratory. The spectrum of the bone fragments extracted from this sample were re-examined but the characteristic signal from irradiated bone was not evident. One explanation could be a possible error in the preparation or labelling of the sample or simply a variation in the distribution of bone fragments in the MRM from one meal to another.

4 CONCLUSIONS

Although both the alcoholic KOH and the Alcalase digestion procedures were both effective in extracting bone fragments from food products containing MRM at inclusion rates below 10 g 100 g^{-1} the Alcalase method was significantly more efficient especially for samples containing MRM at inclusion rates as low as 1 g 100 g^{-1}. However, a greater sensitivity in both methods is observed when samples of MRM itself are examined as opposed to when it is incorporated into a food product. If samples of unknown history are being examined there appears to be a higher probability of making a correct identification using bone fragments extracted using the Alcalase procedure and in fact, all samples containing as little as 3 g irradiated MRM 100 g^{-1} food were correctly identified. Although it is feasible that a food processor could include levels of inclusion as low as 1 g MRM 100 g^{-1} food, it seems more likely that the inclusion levels would be higher. Nevertheless, multiple sampling of products with very low inclusion levels would enhance the efficiency of detecting irradiated MRM. This work has therefore confirmed the potential of the ESR technique for the detection not only of primary food products but also food products containing levels of MRM well below 10 g 100 g^{-1}.

Acknowledgements

The authors wish to thank Professor J. Michel Truchelut and the students of the master of the catering college of Illkirch, France who prepared the pre-cooked meals for this study and also Ms Paula Kirk and Mr Gary Houston for technical assistance.

References

1. M. H. Stevenson and R. Gray, *J. Sci. Food Agric.*, 1989, **48**, 261.
2. E. M. Stewart, M. H. Stevenson and R. Gray, *J. Sci. Food Agric.*, 1991, **55**, 653.
3. J. J. Raffi, M. H. Stevenson, M. Kent, J. M. Thiery and J. J. Belliardo, *Int. J. Food Sci. Technol.*, 1992, **27**, 111.
4. K. M. Morehouse and M. F. Desrosiers, *Appl. Radiat. Isot.*, 1993, **44**, 429.
5. R. Gray and M. H. Stevenson, *Radiat. Phys. Chem.*, 1989, **34**, 899.
6. P. G. H. Bijker, P. A. Koolmees and J. G. van Logtestijn, *Meat Sci.*, 1983, **9**, 257.
7. J. A. Nissen, *J. Agric. Food Chem.*, 1979, **27**, 1256.
8. G. W. Snedecor, W. G. Cochran, "Statistical Methods," The Iowa State University Press, Ames, Iowa, 6th Edition, 1967, Chapter 5, p. 128.

ESR DOSIMETRY OF IRRADIATED CHICKEN LEGS AND CHICKEN EGGS

S. Onori and M. Pantaloni

Istituto Superiore di Sanità, Physics Laboratory
Viale Regina Elena 299
00161 Rome
Italy

1 INTRODUCTION

Ionising radiation induces stable free radicals in chicken bones and in the shell of chicken eggs which can be detected, by the electron spin resonance (ESR) technique, well beyond the shelf-life of the food and can be used for dosimetry.[1-3] The method usually adopted to evaluate "a posteriori" the dose given during the ionising radiation treatment of food, is the dose additive method. To assess the dose, the ESR signal amplitude of the irradiated food (bone or egg shell in the present case) is measured and then the dose-effect relationship is obtained by re-irradiating the sample with some additive doses (usually of 1 kGy). The dose-effect curve is back-extrapolated and the initial given dose determined. Despite the wide consensus on the method, it is the opinion of the authors that the value of the dose additive method is questionable. As a matter of fact, (1) it implies the need for a suitable ionising radiation source for the re-irradiation steps to be installed in every ESR evaluation laboratory, and (2) the uncertainty in dose estimate is, in some cases, so high that it makes the method useless.[4] It is clear that the first point should be analysed in a risk-benefit context, while the second should be investigated in more detail.

At the Istituto Superiore di Sanità (ISS), Rome, Italy, a research programme was approved two years ago aimed to, (1) study new methodological approaches for ESR dose assessment, and (2) analyse the factors which may influence the ESR readout of irradiated chicken bones and chicken egg shells.

In relation to the first aim, the goal was to find an ESR dose assessment method which was less time consuming and more reliable than the dose additive method. The idea was to test if the use of a calibration curve could solve the problem. The method is based on the evaluation of the dose-effect relationship by the analysis of a limited number of samples of the food under study and on the assumption that the behaviour observed is representative of that specific type of food. Of course, the more homogeneous the ESR readout for different samples of a food irradiated at the same dose is, then the lower the uncertainty in dose estimate will be. The tests performed up until now have given evidence on the feasibility of the method for chicken bones[5] and chicken egg shells.[3] In this paper the main results will be discussed and the two ESR dose evaluation procedures, the dose additive and the calibration curve methods, compared in some dose assessment exercises.

With regard to the second aim, the goal was to reduce the uncertainty in dose estimate through the analysis of the possible factors which may influence the ESR readout. Some goals have already been achieved, such as the use of a standardized procedure for sample preparation and readout[5] or the use of the proper dose-effect relationship for dose assessment.[6] Other possible influencing factors such as dose rate, temperature during irradiation, and storage and fading could play an important role in increasing uncertainty and should be investigated in more detail.

2 MATERIALS AND METHODS

All the results of the present paper refer to chicken legs and chicken eggs bought randomly in different local markets in Rome. Food was assumed fresh and previously not exposed to freezing or irradiation.

ESR measurements were performed on chicken bone fragments and powdered egg shells.

The bones, obtained from chicken drumsticks, had the marrow removed, were rinsed with water and then dried overnight under vacuum. To increase the readout reproducibility, (1) bone samples were taken only from the central part of the femur and cut into fragments of 10 x 4 mm² size, and (2) the samples were accurately positioned in the microwave cavity by a PTFE sample holder.[5]

The powdered chicken egg shell samples were prepared by first removing the shell from the eggs, cleaning it in water and drying with filter paper. Then the egg shell was powdered in a mortar. For ESR readout, the samples were put in quartz tubes which in turn were inserted in a quartz double-wall sample holder.

Irradiation were performed with a ^{60}Co gamma source (Gammacell, Nordion Inc., Canada) at a dose rate of about 0.2 Gy s^{-1}. The uncertainty in the nominal dose values was about 3%, while reproducibility was better than 99%. Irradiation of whole eggs was also performed. In the latter case the configuration of the isodose curves in the irradiation chamber of the Gammacell source did not allow a uniform dose distribution on the whole sample. Dose variation, as high as 5%, in the sample volume could be present.

ESR measurements were carried out with Varian E112 (Varian, Palo Alto, CA, USA) and Bruker ESP300 (Bruker, Karlsruhe, Germany) ESR spectrometers working in X band (9.3 GHz) and equipped with standard TE102 microwave cavity. Measurements were performed at room temperature.

3 RESULTS AND DISCUSSION

The results obtained during the last two years work at ISS will be reviewed and discussed in two separate sections. Section A, is devoted to discussion of the suitability of the calibration curve method for dose assessment in irradiated chicken bones and chicken egg shells. Section B is devoted to the analysis of some of the possible factors which may contribute to the uncertainty in dose estimate using the ESR technique.

3.1 Dose Evaluation Methods

The possibility of using a calibration curve for dose assessment is related to the influence that factors such as sex, age, feeding and others have on the yield of the free radicals induced by radiation. Gray and co-workers[7] showed a 20% variation, at maximum, with chicken age while to the authors' knowledge, no data is available for the other parameters or for chicken eggs.

In the absence of an extensive study on the subject, and as a first approach to the problem, it was decided to have an indication of the biological variability using chickens and eggs bought in the local markets. The samples to be tested were simultaneously irradiated to the same dose and, to avoid fading problems, the ESR readouts were performed on the day of irradiation. Figure 1 shows the variation in the ESR response for different samples irradiated at 5 kGy.

Figure 1 shows that (1) the variability in the ESR readout was lower for eggs than for chicken legs, and that (2) in the worst case a deviation from the mean value of about 30% was observed in bones while, typically, the coefficient of variation (CV) was about 10% for chicken legs and 3% for chicken eggs. It follows that the biological variability (for the moment identified as the variability found in the food market) seems to be of no great practical relevance. As a matter of fact, it can be "a priori" hypothesized that, at the worst, the biological variability adds a stochastic component of about 30% to the overall uncertainty in dose assessment. And this figure could be acceptable, if compared to the accuracy usually achieved with the dose additive method.[4,8]

From a general point of view, the dose assessment by a calibration curve can be done using the expression:

$$D = F(h) \times \Pi K_{ro} \times \Pi K_{env} \times K_{beam} \quad (1)$$

where D is the dose to be evaluated, h the ESR signal amplitude, F the calibration function of the food for a reference radiation quality (usually ^{60}Co), ΠK_{ro} the product of all the correction factors for readout, ΠK_{env} the product of all the correction factors

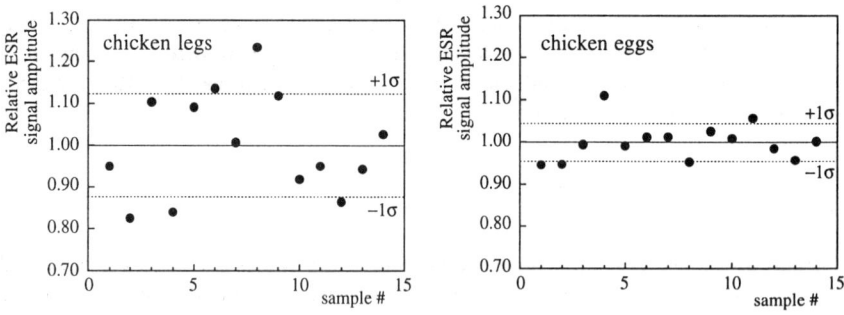

Figure 1 *Distribution of the relative ESR signal amplitude for samples irradiated at 5 kGy*

for environmental effects during irradiation and storage, and K_{beam} the correction factor for beam quality or type. In practice many correction factors can be assumed to be one, while others such as fading, dose rate, temperature, beam quality or type could have a significant influence on the ESR readout and then on dose assessment.

Figure 2 shows the dose-effect curve for chicken leg bones and chicken egg shells. The dose-effect relationship is linear up to 10 kGy for chicken eggs while exponential for chicken leg bones.

3.2 Influencing Factors

No matter which method is chosen for dose evaluation, many systematic and random sources of uncertainty may influence the ESR dose assessment. Amongst these, experimental reproducibility, dose-effect relationship and fading are considered and discussed here.

3.2.1 Experimental Reproducibility. ESR readout of chicken bone fragments is usually characterized by a low reproducibility. This is mainly due to the fact that often irregular bone fragments are used whose crystallinity makes their position and orientation in the microwave cavity important. The use of a standardized readout procedure is then necessary to increase reproducibility. As a matter of fact, in the case of bone fragments, the use of properly sized and shaped bone samples, accurately positioned in the microwave cavity by a home-made sample holder[5] lowered the fluctuations of repeated ESR readouts of the same sample from the initial 20% to about 1.5%.

For both powdered chicken egg shells and bone fragments the fluctuations from sample to sample are of great concern if the calibration curve method is employed. To increase homogeneity and then accuracy, the amplitude of the ESR signal, normalized to the mass of the corresponding sample, has to be evaluated.

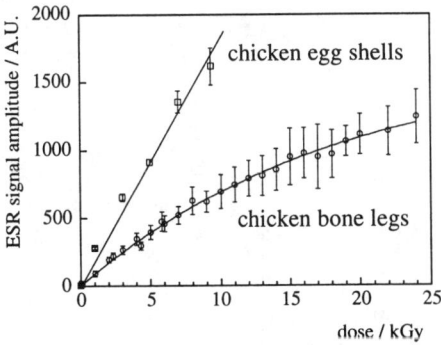

Figure 2 *ESR signal amplitude as a function of absorbed dose (^{60}Co irradiation) for chicken leg bones and chicken egg shells. Each point is the mean value of all samples. Error bars represent one standard deviation. Solid lines are the best fits of the data. The sensitivity of egg shells is about 50 times that of chicken leg bones.*

3.2.2 Dose-Effect Behaviour. It is usually accepted that the dose-effect relationship is linear up to 10 kGy for both chicken bones and egg shells. Consequently, in the dose additive method the initial dose given is determined by a linear back extrapolation. Many authors[3,5,9] have shown that the linear behaviour is correct for egg shells while for chicken bones the dose-effect relationship is exponential. It follows that for bones the dose is systematically overestimated using a linear back extrapolation. In an extensive test performed at ISS[6] on irradiated chicken legs, it was shown that the linear approximation overestimated the dose of about 35% at 1 kGy; that figure increased with dose being about 70% at 10 kGy. Figure 3 summarizes the overall result of the test.

Two main points should be underlined. The first is the above mentioned overestimation; the second one is the width of the distribution, which reflects the large fluctuations found in dose estimate among different samples irradiated at the same dose.

The same chicken bone samples used for the test of Figure 3a were used to check the exponential dose-effect relationship. The results are reported in Figure 3b. In this case the overestimation decreased significantly, while the width of the distribution remained practically unchanged. It follows that the use of the proper exponential dose-effect relationship greatly improves the reliability in dose estimate, lowering the systematic component of uncertainty. Nevertheless, the overall uncertainty is still high because of the large distribution width, which is a measure of the random component of error.

Finally, the feasibility of the calibration curve method for dose assessment was tested. The results are reported in Figure 4.

The distribution is nearly symmetric around zero and sharper than those obtained with the dose additive method. However, large fluctuations are still present whose cause must be understood to increase the reliability of dose assessment by ESR technique. Similar tests were also performed on chicken eggs and similar results were obtained. A positive difference with respect to chicken bones was that the width of the distributions of the evaluated doses were sharp; typically a CV of about 10% was found.

3.2.3 Fading. The stability of the free radicals induced by ionising radiation in the food is a pre-condition for a retrospective dose assessment by ESR technique. Otherwise

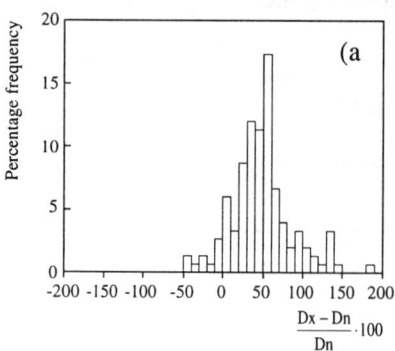
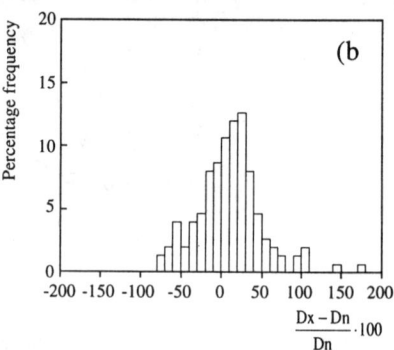

Figure 3 *Observed frequency of the percentage deviations of the evaluated dose, Dx, from the nominal one, Dn, in the 1-10 kGy range. Dx was evaluated using the dose additive method. Figure 3a uses a linear extrapolation, while in Figure 3b an exponential extrapolation was used.*

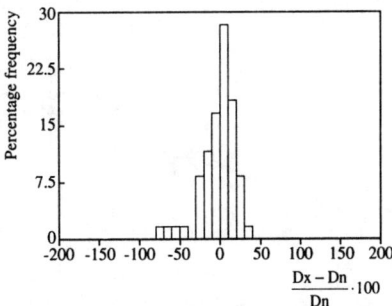

Figure 4 *Observed frequency of the percentage deviations of the evaluated dose, Dx, from the nominal one, Dn, in the (1-10) kGy range. Dx was evaluated using an exponential calibration curve.*

a correction factor must be determined and applied. Usually the fading problem is not taken into account in chicken bone dosimetry, while the results of Gray and Stevenson[10] together with Desrosiers and Le[11] strongly indicate the need to go more deeply into the subject. As a matter of fact a loss of about 30% in the ESR signal amplitude of irradiated bone was observed in about 3 weeks. Also, some data on stability of free radicals in irradiated chicken eggs were recently obtained.[2,3] In approximately 10 days the signal faded by about 25% reaching a plateau thereafter. In other words, on the basis of the available results, it may be stated that fading cannot be ignored in ESR retrospective dose reconstruction of chicken bones and eggs.

From a practical point of view two main problems must be faced, (1) correction for "long-term" decay, and (2) for "short-term" decay. Referring only to egg shells, for which more data is available, long-term decay means the quasi constant level that ESR signal amplitude reaches after about 10 days. Usually the operator does not know the date of irradiation so he cannot apply any correction factor for fading. A possible solution could be to perform the ESR readout only when the signal has reached a good stability (after about 10 days). The use of the long-term correction factor is all that is needed when the calibration curve method is chosen. On the contrary, if the dose additive method is applied, the ESR readout and re-irradiation should be performed all in the some day. If this procedure cannot be followed (and it is often the case), the dose assessment takes about two or three days. In that period a short-term decay (5-10%) takes place that should be taken into account for a more accurate dosimetry. Indeed, Figure 5 shows what could be the situation in the hypothesis that the additive doses were given one per day. An overestimation of the given dose should be systematically present.

To be more confident about the influence that long- and short-term decay could have on dose assessment, some dose evaluations were performed on irradiated powdered egg shells and on shells of irradiated whole eggs. The dose additive method was applied soon after the irradiation and after 15 days. The additive doses were given one per day. The results shown in Table 1 refer to a 3 kGy initially delivered dose; 5 samples obtained from different eggs were used for each test.

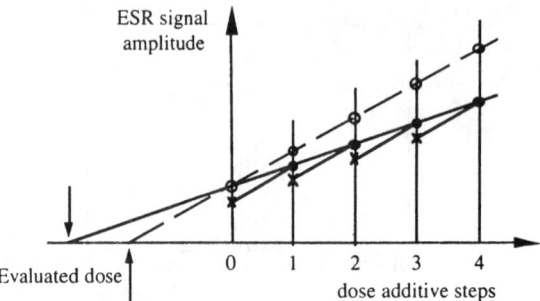

Figure 5 *Dashed line represents the dose-effect behaviour as determined in the dose additive procedure when no fading is present. Solid line was obtained in the hypothesis that in the time elapsed between two subsequent irradiations a constant short-tern decay takes places. The two situations lead to different dose estimates*

Table 1 clearly shows that both long- and short-term decay have a marked influence on dose assessment but that an accurate dose evaluation can be done using the proper correction factors.

The procedure adopted for sample preparation did not make use of Soxhlet treatment for organic component extraction, so it is not clear if the fading behaviour observed has to be related to the hydroxyapatite or to the organic material. Work is in progress to understand the role of the organic component in the fading behaviour looking also for the best sample preparation procedure.[12]

3.2.4 Other Influencing Factors. Some other influencing factors could contribute to the overall uncertainty in ESR dosimetry of chicken bones and eggs. Among these, dose rate and temperature during irradiation and storage should be studied and eventually

Table 1 *Influence of Fading on ESR Dose Assessment.*

Sample Type	0 Days		15 Days	
	Evaluated Dose kGy (a)	Evaluated Dose kGy (b)	Evaluated Dose kGy (a)	Evaluated Dose kGy (c)
Whole egg	4.0	3.5 CV= 11%	2.3	3.1 CV= 9%
Powdered egg shell	3.8	3.3 CV = 5%	2.1	2.7 CV = 10%

a) no correction factors applied
b) correction factor for short-term decay applied
c) correction factors for long- and short-term decay applied

taken into account. While the available data[10] seem to exclude the presence of any significant effect of dose rate on ESR signal amplitude in the (5-600) kGy h^{-1} range, to the authors' knowledge no data exist on temperature effects. And this last point could be of importance when the radiation process is performed at low temperature.

4 CONCLUSIONS

On the basis of the available results the following conclusions can be drawn:
(1) The ESR technique can be used for retrospective dose assessment in chicken bones and egg shells. Suitable approaches are the dose additive method as well as the calibration curve method. Nevertheless, the calibration curve method should be tested in a more systematic way to find out the factors which may influence the ESR signal. Some interlaboratory trials should also be organized.
(2) The overall uncertainty in dose estimation can be greatly reduced if the proper dose-effect relationship is used and if some care is taken to increase the ESR readout reproducibility. Evidence has been reached that the dose-effect relationship is exponential for the chicken leg bones while the available data for chicken eggs (up to 10 kGy) suggests a linear behaviour.
(3) To get a more accurate dose estimate some correction factors should be applied. Long-term decay leads to an underestimation of the dose while short-term decay overestimates the dose to an extent that has to be taken into account. The role of dose rate and temperature should be investigated.

References

1 J. S. Lea, N. J. F. Dodd and A. J. Swallow, *Int. J. Food Sci. Technol.*, 1988, **23**, 625.
2. M. F. Desrosiers, F. G. Le, P. M. Harewood, E. S. Josephson and M. Montesalvo, *J. Agric. Food Chem.*, 1993, **41**, 1471.
3. S. Onori and M. Pantaloni, *Int. J. Food Sci. Technol.*, 1995, **29**, 641.
4. J. Raffi, M. H. Stevenson, M. Kent, J. M. Thiery and J.-J. Belliardo, *Int. J. Food Sci. Technol.*, 1992, **27**, 111.
5. F. Bordi, P. Fattibene, S. Onori and M. Pantaloni, *Appl. Radiat. Isot.*, 1994, **44**, 443.
6. F. Bordi, P. Fattibene, S. Onori and M. Pantaloni, *Radiat. Phys. Chem.*, 1994, **43**, 487.
7. R. Gray, M. H. Stevenson and D. Kilpatrick, *Radiat. Phys. Chem.*, 1990, **35**, 284.
8. M. F. Desrosiers, W. L. McLaughlin, L. A. Sheahen, N. J. F. Dodd, J. S. Lea, The late J. C. Evans, C. C. Rowlands, J. J. Raffi and J.-P. L. Agnel, *Int. J. Food Sci. Technol.*, 1990, **25**, 682.
9. M. F. Desrosiers, *Appl. Radiat. Isot.*, 1992, **42**, 617.
10. R. Gray and M. H. Stevenson, *Int. J. Food Sci. Technol.*, 1991, **26**, 669.
11. M. F. Desrosiers and F. G. Le, *Appl. Radiat. Isot.*, 1994, **44**, 439.
12. A. Wieser, E. Haskell, G. Kenner and F. Bruenger, *Appl. Radiat. Isot.*, 1994, **45**, 525.

ESR DETECTION OF FREE RADICALS IN GAMMA IRRADIATED SPICES AND OTHER FOODSTUFFS

J. R. Pilbrow, G. J. Troup, D. R. Hutton (Physics Department) and
C. R. Hunter (Anatomy Department)

Monash University
Clayton, Victoria
Australia 3168

1 INTRODUCTION

Irradiation of various food products, including vegetables, fruits, meats, seafoods, herbs, spices and seeds by appropriate doses of γ-rays[1] has for many years been suggested as a means of killing or sterilizing bacteria, viruses and pests and, therefore, as a means of preserving the foods.

The position of food irradiation has been under review in Australia generally,[2] through consumer organisations[3] and by a Federal Government (House of Representatives) inquiry. From these reviews and inquiries, recommendations for irradiation, packaging, and labelling *etc.*, are emerging, with, for example, an NH and MRC recommended maximum dose of 10 kGy for Australia, with 6 kGy being a minimum dose for grains and spices.[1]

Some concern has been expressed about the production of radiolytic products, which are unnatural, and usually unidentified chemicals produced by the reaction of free radicals and other ionised and activated atoms and molecules, created earlier by γ-irradiation. Some reports suggest that some of these may be toxic or carcinogenic.[4]

In early studies, electron spin resonance (ESR) spectroscopy was used to detect stable free radicals in bone and cuticle[5] and it was demonstrated that γ-irradiation breaks down proteins and DNA.[6] Earlier studies[7] suggested that induced free radical signals in spices rapidly decayed to negligible levels after three weeks, especially if some moisture was present.

Although the members of the Monash group do not carry out research formally in the area of food technology, participation in the ADMIT program was appropriate given the availability of suitable ESR and ^{137}Cs irradiation facilities and interest both politically and amongst consumer groups regarding food irradiation. Amongst our many interests in applications of ESR is alanine dosimetry, and we are convinced that foodstuffs that are irradiated should contain an alanine pellet attached to the package or embedded in the packaging material. This matter is not covered by the ADMIT program. There is a separate IAEA program for dosimetry. It is our view that this matter should not be overlooked.

2 EXPERIMENTAL

Small quantities of the foods studied were irradiated with an Isomedix Gammacell 1000 ^{137}Cs source capable of delivering a dose of 0.25 kGy h^{-1}.

Samples (usually 0.2 ml) were transferred to standard 3 mm quartz ESR tubes, which were flushed with N_2 gas and sealed.

ESR measurements were carried out with Varian V4500 and E12 spectrometers and more recently with a Bruker ESP380E CW/FT spectrometer. Most measurements were conducted with X band TE_{102} cavities operating at room temperature.

Experimental conditions were carefully set to ensure repeatability and comparability between runs. In particular, the microwave power was kept low enough to avoid saturation, the modulation narrow enough to avoid line broadening and the time constant and sweep speed adjusted to give good sensitivity without line distortion.

Samples whose dielectric properties significantly changed the cavity Q were accompanied by another standard sample, containing one or more transition metal ions, fixed in position in the cavity for the duration of the experiments. The most useful of these samples was a copper sulphate single crystal, and small samples of ruby (Cr^{3+} in alumina) or emerald (Cr^{3+} in beryl) whose orientation was such that the magnetic field was perpendicular to the crystal axis, thus producing ESR lines well away from the g = 2 free radical lines produced in the foods by irradiation.

For dry samples, which did not perturb the cavity unduly, the system was calibrated with either a standard Varian weak pitch sample, or with alanine powder which had been previously irradiated to a suitable level.

A protocol of ESR operating conditions at X-band, using alanine as a dosimeter, was developed. DL-α-Alanine powder was tested with irradiation doses ranging from 50 Gy to 100 kGy using X-rays of peak energy 8.05 keV, ^{137}Cs γ-rays of energy 0.662 MeV and ^{60}Co γ rays of energy approximately 1.2 MeV. On the basis of a series of experiments, it was shown that in order to avoid power saturation and modulation broadening of the irradiated alanine spectra at X-band, the maximum microwave power should be 1 mW, and the maximum modulation amplitude 1 G peak to peak (p-p). A relationship relating the ESR p-p signal height (S), measured as the number of small squares on the ESR paper, to the absorbed dose (D) in Grays, has been determined to be $S = 9290[1 - \exp(-8.99 \times 10^{-6}D)]$ for irradiation with ^{137}Cs γ-rays.[8] It was also shown that, for higher accuracy, the use of alanine pellets with a suitable binder should be used instead of just alanine powder alone.

For use as an ESR calibration standard and for comparison with the food signals standard 0.2 ml samples of DL alanine powder irradiated with doses of 10 and 100 Gy were found most appropriate. A 1 Gy sample was also used to regularly check the sensitivity limits of the spectrometer. Figure 1 illustrates the 1 Gy signal and the 10 Gy signal which is comparable with natural levels of free radicals in foods.

It is our contention that measurements of radiation dose, by means of the protocol worked out at the two previous ADMIT Workshops, should always be supplemented by parallel examination of alanine samples.

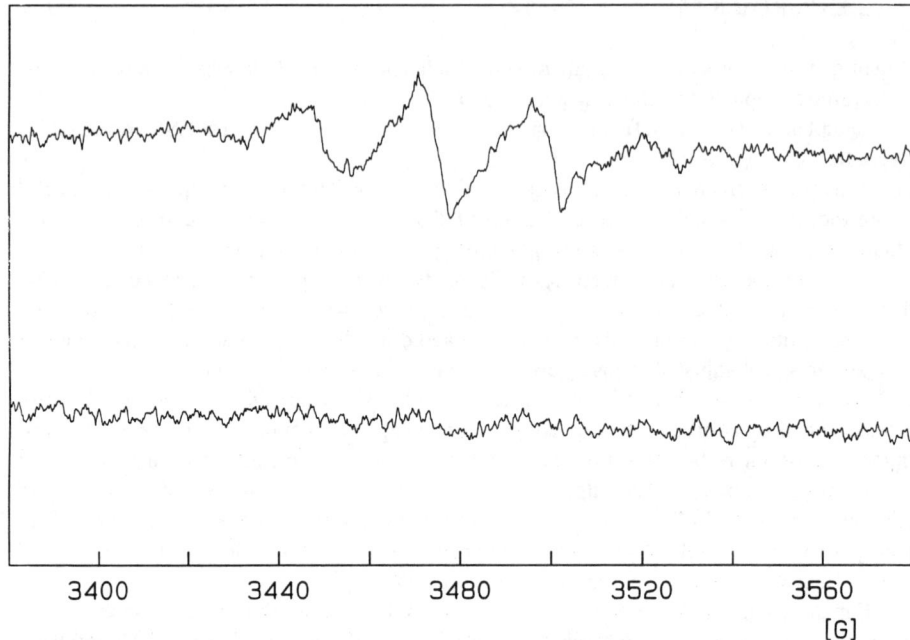

Figure 1 *ESR Signal from 1 Gy and 10 Gy irradiated alanine (5 g, 0.5 mW, 1 s)*

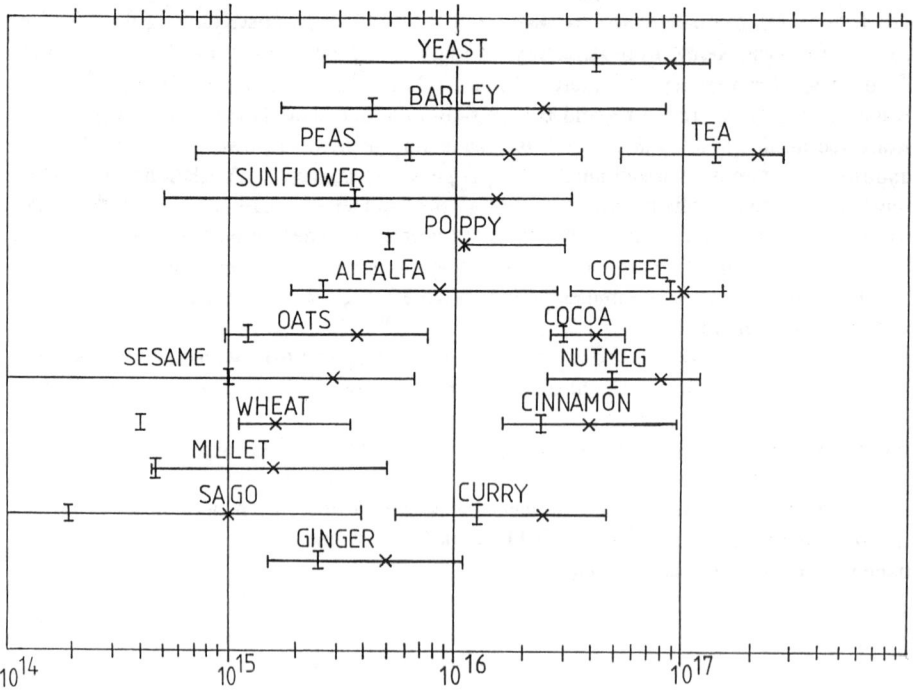

Figure 2 *Free radical numbers in 0.15 ml food (4 day(x) 30(I) are indicated)*

3 BENCH TOP INSTRUMENT FOR DOSIMETRY/RADIATION DIAGNOSIS

Since the experiments conducted at this, and at other laboratories have convincingly demonstrated that ESR can be used as a diagnostic tool in testing the radiation history of some foods, especially those involving bone, or preserved in a very dry state, a low cost diagnostic spectrometer is desirable. Three of the authors (D R Hutton, J R Pilbrow and G J Troup) have contributed to an earlier development of a miniature microwave bridge suitable as the centre-piece of compact spectrometer[9] operating at 2.5 GHz though this has not been taken up commercially at this point in time. Commercial dosimeter spectrometers are now being marketed by Bruker (Germany) and MicoNow (USA).

4 EARLY EXPERIMENTS - FOODS IN GENERAL

In early experiments a range of seeds and spices was given a dose of 6 kGy and the decay of the free radicals produced was monitored regularly, thereafter, for 4 months.[10,11]

Figure 2 shows the range of foods studied and indicates that most of these foods have initially low but detectable levels of free radicals, presumably produced during photosynthesis, drying, *etc.* of the plants. This was confirmed for a number of herbs (tansy, oregano, parsley, rosemary) which were measured at liquid nitrogen temperatures, both freshly picked, and after air drying in the dark.

Figure 2 also shows that all foods suffered increases in free radical concentration on irradiation, with the sensitivity to irradiation varying from low (~3, *e.g.* wheat, poppy seeds) to quite high (~50, *e.g.* yeast, barley, split peas, sunflower seeds).

Levels of the radicals 4 and 40 days after irradiation are also shown in Figure 2. For most samples, radical concentrations had decayed to about 10% of initial post-irradiation levels in about 30 days, with decay curves as illustrated in Figure 3. This shows that initial decay rates are very high with a 50% loss occurring in about 4 days, but that the apparent half-life steadily increases as time goes on. This suggests the presence of several types of free radicals with different half-lives. Some samples, most notably yeast, peas, sago and sunflower seeds, showed long-lived persistent radicals (Figure 4) which remain at levels enhanced 10 fold over natural levels after one month and enhanced 5 fold after 3 months. Both coffee and tea had levels enhanced 3 fold one month after irradiation.

These early ESR measurements showed that most dry seeds and spices, when irradiated by γ-rays, have free radicals induced in them. Initially, levels of radicals are enhanced about 10 fold over natural levels, then decay initially rapidly but increasingly slowly as time goes on. Most samples gave signals which were still elevated 2 fold after one month, but had less than a 10% elevation after 4 months. In some samples significantly elevated levels remained after many months, and the more extensive spice study was begun.

Preliminary studies on seeds and or skins from fresh fruits, *e.g.* avocado, strawberries and mandarins, produced poor correlations between γ-irradiation dose and ESR signal especially after 24 h and presumably because of moisture content. Poultry egg shells gave persistent ESR signals proportional to the dose of γ-rays up to the 7 kGy required to destroy Salmonella bacteria, a major contamination problem in the poultry industry and to human health. Irradiated and control whole eggs poached in boiling water revealed that

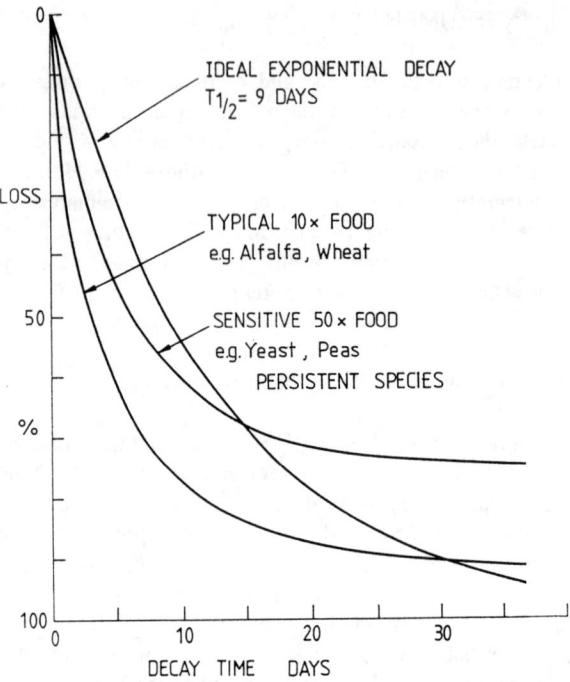

Figure 3 *Typical decay curves for free radicals generated in γ-irradiated food*

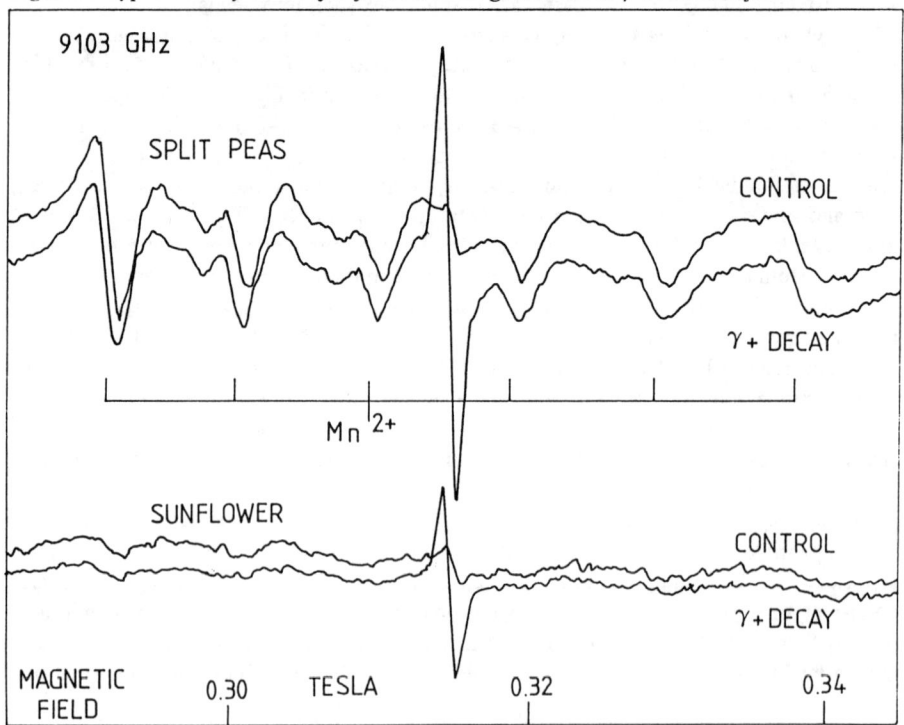

Figure 4 *Seed ESR signals, both non-irradiated controls and 138 days post irradiation (Mn^{2+} lines provide a suitable reference level)*

the odour, appearance and texture of the irradiated eggs was unacceptable and further experiments were terminated. Ground poultry bones subjected to ^{137}Cs γ-rays gave ESR signals proportional to dose at low doses (less than 2 kGy) but non linear above this dose.[12]

Preliminary experiments were also conducted on γ-irradiated dry food pellets which may be fed to farmed animals such as poultry, pigs and rabbits. In Australia such foods are commercially irradiated to 30 kGy using a ^{60}Co source to sterilise them and minimise pathological bacterial contamination. ESR results of this work indicate that changes to long-lived free radical and transition element signals are indeed useful indicators of exposure to γ-irradiation for these food pellets.

5 LONG TERM STUDY OF FREE RADICALS IN SPICES

Spices are widely used in the food industry and in domestic cooking, and since they often suffer high microbiological contamination, they are attractive candidates for irradiation.[13] Since many have a low moisture content, there seemed to be a reasonable chance of finding prolonged elevated levels of free radicals, at least, allowing a check for the occurrence of recent irradiation, if not of dose levels.

Twenty-five different spices were unpacked, bottled and supplied by the Australian Consumers Association (ACA). Samples from each of these were then γ-irradiated to a dose of 20 kGy. The ESR spectra of all samples were measured at 2, 8, 22, 42, 72 and 403 days and then yearly, post-irradiation, using a carefully standardised technique. In addition to standard spectra recorded in the g = 2 region, wide low-field scans were also run to check for transition metal ions.

The early part of the experiment was repeated a year later using a second batch of samples supplied by ACA. These were resealed under N_2 gas, irradiated with 16 kGy of ^{137}Cs γ-rays, and the ESR spectra regularly monitored for 3 months, then yearly, post-irradiation. For all samples, non-irradiated controls were also monitored over these periods for comparison with the irradiated samples.

The results are illustrated in Figure 5 (Batch 1) and Figure 6 (Batch 2). For Batch 1 samples, in only one (15, garlic) was there a significant (60%) fall in the control free radical level over the year. In two samples the control signal level rose (8, satay, up 100%; 20, coriander, up 400%). After one year, 5 samples still had signals significantly higher than their corresponding controls (5, fenugreek, 2x; 8, satay, 3x; 11, fennel, 4x; 15, garlic, 8x; 20, coriander, 4x).

Generally the decay between 8 and 80 days caused a drop by a factor of between 2 and 3, and between 80 and 403 days by a factor of between 1.3 and 2.0. The measurements at days 22 and 42 showed that for many samples, the half-life of the induced free radicals is about 20 d over the first few months, and that one month after irradiation, most samples have signal levels falling by approximately 2% per day.

The ESR for some samples demonstrate persistent stable changes induced by the γ-irradiation in the state of transition metal ions. In particular, the oxidation or reduction of copper and iron shown through the creation or annihilation of Cu^{2+} and Fe^{3+} was particularly striking, as illustrated in Figure 7.

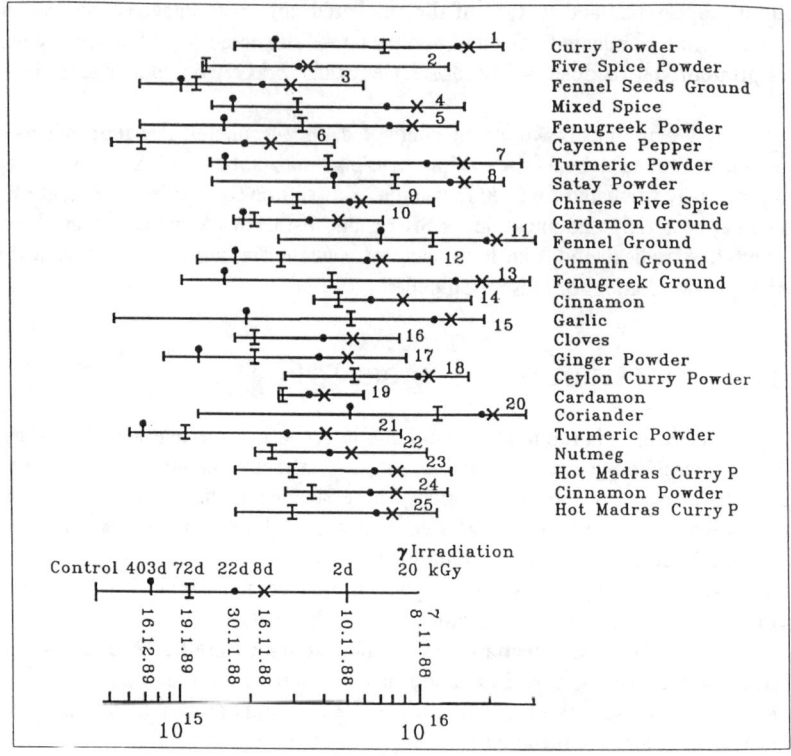

Figure 5 *Free radical spins as measured by ESR (Batch 1)*

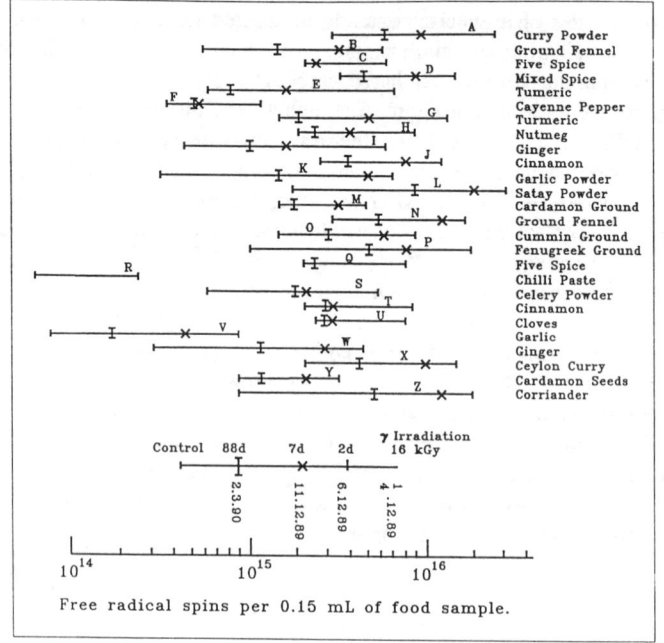

Figure 6 *Free radical spins as measured by ESR (Batch 2)*

Physical Methods: Electron Spin Resonance (ESR) Spectroscopy

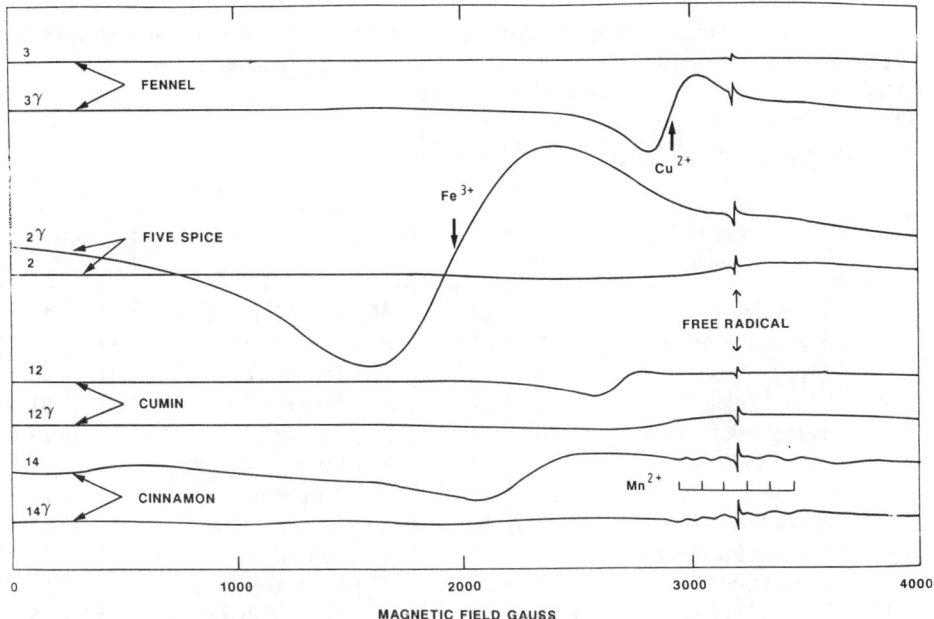

Figure 7 *Wide scan ESR signals from control and γ-irradiated spices after 1 year*

The spectra for Batch 2 confirmed the results for Batch 1. All samples, with the one exception of cayenne pepper which showed a doublet, gave a broad single ESR radical line which decayed significantly with time.

After 2 days, radical enhancement varies considerably from the sensitive coriander (up 30x) to insensitive cayenne pepper and cardamom (up only 3x). Often, the same spice from different sources showed differing sensitivity and absolute radical levels after the same γ-irradiation treatment. However, high levels persisted in similar spices from both batches, 3 months after irradiation. Once again, permanent transition metal ion signal changes were produced by the irradiation. It was also noted that the typical ESR spectrum of Mn^{2+} was unchanged by γ-irradiation of the sample if manganese was present in the spice. This spectrum was also consistent over the period of a year thus and served as a useful standard, (*e.g.* sample 14, illustrated in Figure 7).

Continued monitoring over 5 years has confirmed that in none of these samples do significant levels of γ-induced free radicals persist, nor are satellite lines (± 30 G from $g = 2$) detectable, which differentiate between γ-irradiated and non-irradiated spices. Table 1 summarises the most recent data taken 5 years after irradiation. It demonstrates that levels of radicals in these natural foods differ over a range of 2 orders of magnitude from low (*e.g.* garlic, ginger, cayenne pepper) to high (cinnamon, nutmeg, fennel, curry).

Table 1 *Peak to Peak Free Radical Signals Recorded for 52 ACA Spices, May, 1994, as Compared to Varian Weak pitch.*

Experimental Conditions   (Bruker ESP380E Spectrometer, Standard rectangular cavity)
Receiver Gain 4×10^4,   Field   3480/200 g,   Frequency   9.74 Ghz
Modulation Amp. 5 G,   Sweep Time   3 min,   Power   0.5 mW
Time constant   1.3 sec
(Varian weak pitch   2.2 p-p)   (Alanine 10 Gy 2.0 p-p)

	Batch 1 Spices	p-p Signal C	γ		Batch 2 Spices	p-p Signal C	γ
1	Curry Powder	15	18	A	Curry Powder	41	58
2	Five Spice Powder	32	25	B	Ground Fennel	11	15
3	Fennel Seeds Ground	17	10	C	Five Spice	41	39
4	Mixed Spice	21	24	D	Mixed Spice	55	61
5	Fenugreek Powder	10	12	E	Turmeric	13	13
6	Cayenne Pepper	9	8	F	Cayenne Pepper	7	9
7	Turmeric Powder	21	19	G	Turmeric	20	19
8	Satay Powder	48	46	H	Nutmeg	56	45
9	Chinese Five Spice	49	41	I	Ginger	7	12
10	Cardamom Ground	24	25	J	Cinnamon	50	53
11	Fennel Ground	47	55	K	Garlic Powder	5	5
12	Cumin Ground	38	25	L	Satay Powder	30	72
13	Fenugreek Ground	18	18	M	Cardamom Ground	19	24
14	Cinnamon	55	40	N	Ground Fennel	45	42
15	Garlic	4	5	O	Cummin Ground	26	34
16	Cloves	34	36	P	Fenugreek Ground	18	25
17	Ginger Powder	14	13	Q	Five Spice	38	39
18	Ceylon Curry Powder	54	44	R	Chilli Paste	6	5
19	Cardamom	27	24	S	Celery Powder	16	20
20	Coriander	36	28	T	Cinnamon	37	48
21	Turmeric Powder	15	14	U	Cloves	41	42
22	Nutmeg	55	40	V	Garlic	0.5	0.6
23	Hot Madras Curry P.	27	25	W	Ginger	5	6
24	Cinnamon Powder	49	48	X	Ceylon Curry	42	46
25	Hot Madras Curry P.	29	20	Y	Cardamom Seeds	10	12
				Z	Coriander	24	35

Generally there is a close correspondence between the ESR signals from both control and γ-irradiated samples of the same spice, and for different samples of that spice in the different batches. Some samples do show persistent elevated levels (up to 100%) but these are not repeated in all samples of this spice from both batches. On occasion (*e.g.* 3 and 12, fennel and cumin of Batch 1) the free radical levels in the irradiated sample was lower than the control (perhaps indicating scavenging of natural radicals by radiation-induced radicals) but once again not all samples of this spice show this behaviour.

The long term study demonstrates the permanence of the persistent stable changes induced by the γ-irradiation in the valance state of some of the transition metal ions and as shown in Figure 7. However Mn^{2+} ions usually remain unaltered by irradiation. A study has been made of a number of different samples of two spices, ginger and cinnamon. In no spice so far studied are the transition metal signals, or their changes, consistent from differently sourced samples of the same spice. Hence, it is deduced that such signals are provenance (probably soil) dependent, and not spice dependent and originate in the inorganic material in the spice.

In conclusion, spice free radical levels as detected by ESR vary over a large range in the natural materials. Sensitivity of spices to γ-irradiation and radical decay rates are also very variable. Free radical levels remain elevated for many samples for a few months, for a few samples for one year, but for no samples after 4-5 years. For some samples transition metal ion signals permanently indicate irradiation. Thus for spices, irradiation may be detected by ESR by the following phenomena:
(1) Rapidly falling signals, detected by repeat measurements spaced about 20 days apart,
(2) ±30 G satellite lines (up to one year),
(3) elevated signals (up to one year, control needed), and
(4) transitions ion signal changes (control needed); if such changes can be caused by the follow up radiation dose then the sample has probably not previously been irradiated.

6 CO-OPERATIVE PRAWN TRIAL WITH THE DSIR, NEW ZEALAND

In 1990 a co-operative trial program was agreed with Dr P. B. Roberts, New Zealand for the interchange of prawn samples between the two countries.

The Australian samples of 6 king prawns (cooked, or red) and 6 school prawns (raw, or green) were purchased at a local fish retailer in a frozen condition. Four of each were γ-irradiated at melting ice temperature to approximate dose of 0.5 and 1 kGy. Controls were also kept at melting ice temperature. The prawns were shelled at room temperature (carapace and 6 pleonites) and dried overnight (for approximately 15 hours) under vacuum at room temperature. The shells from 6 pairs of prawns were ground to form a "powder" in a glass mortar and pestle at room temperature. No difference in odour, colour, textural appearance (or taste of the cooked king prawns) could be detected between the control and 0.5 kGy and 1 kGy dose samples. The 6 samples prepared and labelled Red 1-3 and Green 1-3, *i.e.* control and those gamma-irradiated to 0.5 kGy and 1 kGy (not necessarily in that order), were sent to New Zealand in plastic vials. ESR measurements were all made at room temperature.

Likewise samples prepared in New Zealand were received in Australia. Testing by both laboratories gave 100% correct identification of irradiated samples and the results are summarised in the following table.

These samples have recently been re-measured and it can be reported that ESR can still detect the irradiated samples 2 years later, *e.g.* see Figure 8.

The signals are weak (cf. weak pitch), but the irradiation produced line, 5 G lower in field, is clearly visible. Similar results are obtained from the Australian prawn samples

Table 2 *Results from Co-operative New Zealand(NZ)/Australian(Aust.) Trial*

Australian Shrimp			New Zealand Shrimp		
Sample	Dose	Test (NZ)	Sample	Dose	Test (Aust.)
R1	1	I (st)	A	3	I (st)
R2	0.5	I (wk)	B	0	N
R3	0	N	C	0	N
G1	0	N	D	3	I
G2	1	I (st)	E	2	I
G3	0.5	I (wk)			

I = irradiated; N = non-irradiated; st = strong; wk = weak

with the characteristic line being better defined in the green prawn samples (Figure 9), than in the red cooked prawns.

7 ADMIT CO-OPERATIVE TRIALS

Five trial sets of samples were received for interlaboratory comparison to determine whether or not the samples had been irradiated.[14] The results of the Monash group were consistently successful in all evaluations.

A mid-1994 re-evaluation of samples previously received from 1991-1993 was carried out and clearly demonstrated the long term validity of bone test results (Figure 10). Samples of pistachio nut shells 18 months post-irradiation could be clearly differentiated on the basis of the ±30 G satellite lines as shown in Figure 11 and also less reliably on the height of the g=2.004 line as shown in Figure 12. Paprika samples of the same age could, with difficulty, only be distinguished by these two features, after 18 months. Experiments with samples 9 months old showed that black pepper signals could not be used for reliable diagnosis after this time, but that egg shell signals allowed a reliable dose estimate after this time (Figure 13).

Physical Methods: Electron Spin Resonance (ESR) Spectroscopy

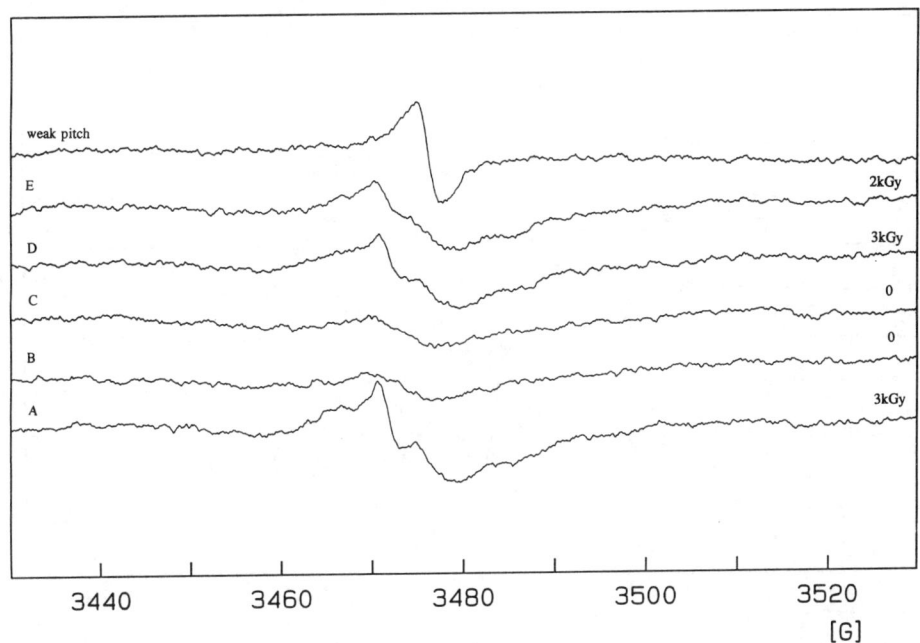

Figure 8 *ESR spectra from New Zealand prawn samples 2 years post-irradiation*

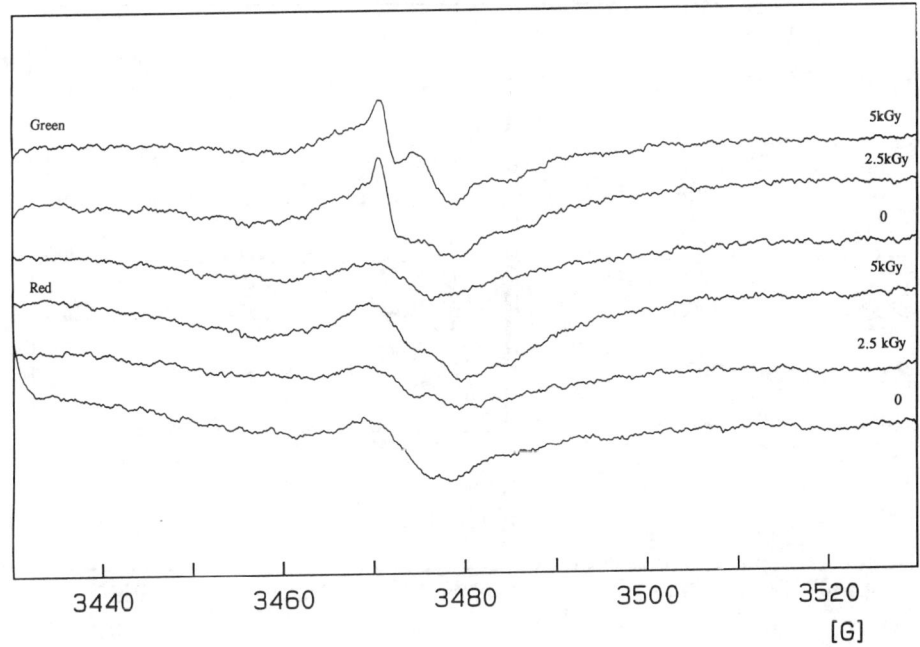

Figure 9 *ESR spectra from Australian prawn samples 2 years post-irradiation*

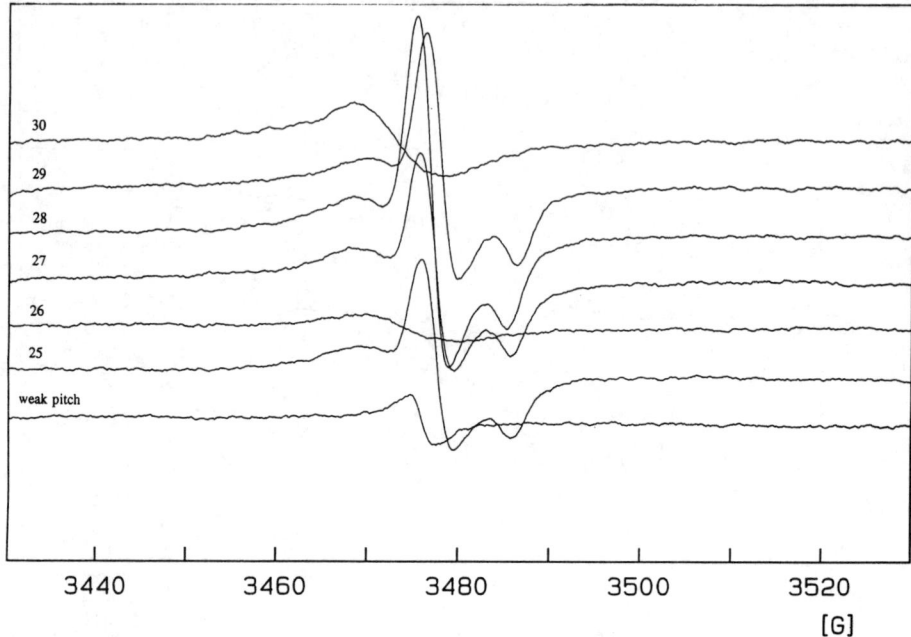

Figure 10 *ESR spectra of pork bones 3 years post-irradiation*

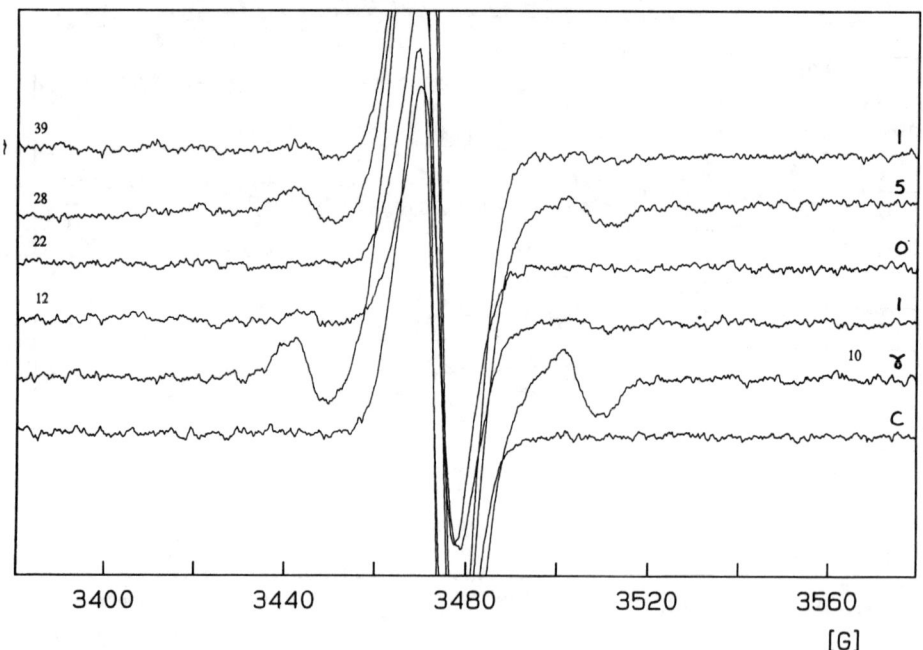

Figure 11 *High gain ESR spectra of pistachio nut shells 1 year post-irradiation*

Physical Methods: Electron Spin Resonance (ESR) Spectroscopy

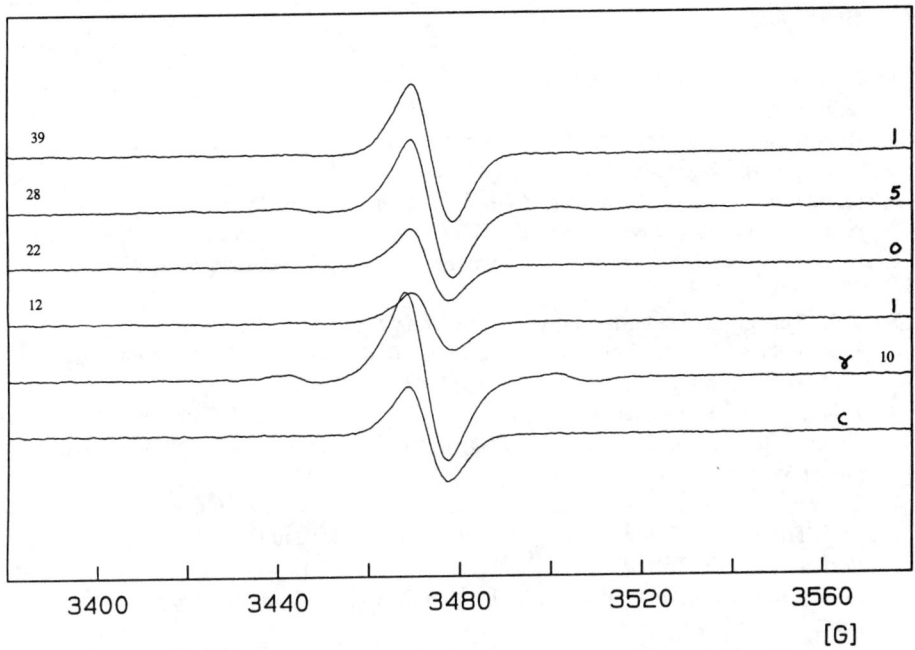

Figure 12 *Low gain ESR spectra of pistachio nut shells 1 year post-irradiation.*

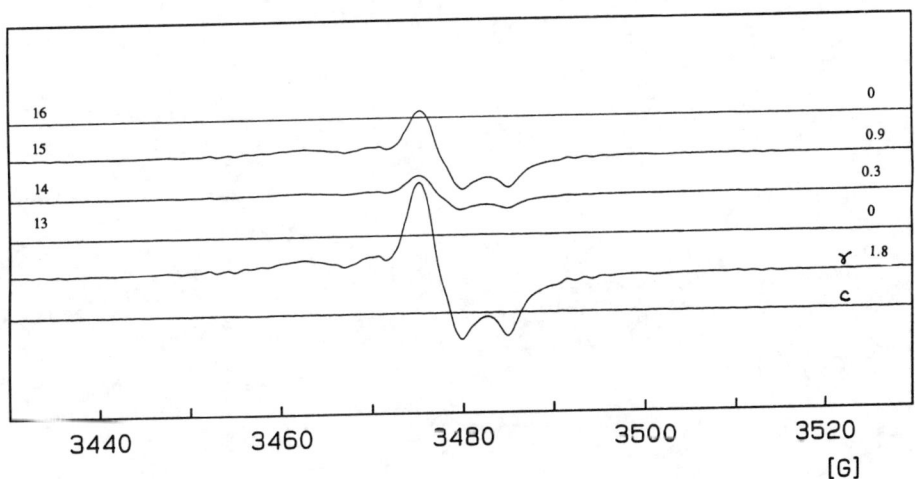

Figure 13 *Low gain ESR spectra of egg shells 9 months post-irradiation.*

References

1. P. A. Wills, *Nuc. Spect.*, 1986, **2**, 5.
2. G. Fisher, *CSIRO Food Res. Quart.*, 1985, **45**, 55.
3. Australian Consumers Association, Food irradiation (ACA Inquiry), *Choice*, 1987, **28**, 18. See also *Nucl. Spec.*, 1987, **3**, 609.
4. C. Bhaskaram and G. Sadasivan, *Aust. J. Clin. Nutr.*, 1975, **28**, 130.
5. N. J. F. Dodd, A. J. Swallow and F. J. Ley, *Radiat. Phys. Chem.*, 1985, **29**, 451.
6. G. D. D. Jones, J. S. Lea, M. C. R. Symons and F. A. Taiwo, *Nature*, 1987, **330**, 772.
7. T. B. Tjaberg, B. Underdal and G. Lunde, *J. Appl. Bact.*, 1972, **35**, 473.
8. J. R. Pilbrow, G. J. Troup, D. R. Hutton, G. Rosengarten, Y. C. Zhong and C. R. Hunter, *Appl. Radiat. Isot.*, 1993, **44**, 413.
9. J. R. Pilbrow, L. Gruner, D. R. Hutton, N. R. McLaren, A. Z. Tirkel and G. J. Troup, "Electron Spin Resonance Spectrometer," US Patent 4803624, Australian Patent 599149.
10. C. R. Hunter, D. R. Hutton and G. J. Troup, *Search*, 1988, **19**, 198.
11. G. J. Troup, J. R. Pilbrow, D. R. Hutton, C. R. Hunter and G. L. Wilson, *Appl. Radiat. Isot.*, 1989, **40**, 96.
12. M. F. Desrosiers, G. L. Wilson, C. R. Hunter and D. R. Hutton, *Appl. Radiat. Isot.*, 1991, **42**, 7.
13. J. Farkas, "Preservation of Food by Ionising Radiation," Josephson, E. S., Peterson, M. S. Eds., CRC Press, Volume 3, p. 110.
14. M. F. Desrosiers, *J. Agric. Food Chem.*, 1990, **37**, 96.

IDENTIFICATION OF γ-IRRADIATED SPICES BY ELECTRON SPIN RESONANCE (ESR) SPECTROMETRY

S. Uchiyama,* M. Murayama,[+] Y. Kawamura[+] and Y. Saito[+]

* Food and Drug Safety Centre, Hatano Research Institute, 729-5, Ochiai, Hatano-city, Kanagawa, 257, Japan
[+] National Institute of Health Sciences, 1-18-1, Kamiyoga, Setagayaku, Tokyo, 158, Tokyo, 158, Japan

1 INTRODUCTION

Irradiation of various foods is increasingly being used internationally, for example, as a substitute for the use of post harvest pesticides.[1] The ESR method seems to be one of the most useful and promising identification methods for irradiated spices as well as the thermoluminescence (TL) method.[2-8] The authors joined the ADMIT (Analytical Detection Methods of Irradiation Treatment of Food) research programme of the FAO/IAEA on the development of the ESR method and have participated in international blind trials from 1990 to 1994.[9] This paper deals with a final report on the ESR method for detection of irradiated spices on the basis of a signal unique to γ-irradiation.[7,8]

2 EXPERIMENTAL

2.1 Spices and Irradiation

Spices used in the experiment were mainly red pepper, white pepper, black pepper, cinnamon and allspice which were obtained through the courtesy of a Japanese spice company and the company guaranteed that they had not been irradiated. Most of the samples were 60-meshed powder and other spices were ground and filtered at 60 mesh.

Spices (5 g) in a glass tube sealed with a glass stopper were irradiated in the range of 3 to 50 kGy at an absorbed dose rate of 10 kGy h^{-1} with a γ-irradiator (^{60}Co source). The irradiated spices in the sealed glass tubes were stored at room temperature in the dark.

2.2 Measurement of ESR and Computer-Simulation

Powdered spices were subjected to ESR measurement with a JEOL ESR spectrometer, Model TES-FE2XG under the following conditions: field, 3369 ± 50 G; sweep time, 2 min; mode, 6.3; modulation amplitude, 100-1000; power, 1 mW; temperature, 23°C. Relative radical intensities of spices were corrected with MnO and represented as the distance from the top of the signal to the bottom. The g-value was calculated by use of MnO as a g-value standard. Numbers of radical spins were calculated by use of 2,2,6,6,-tetramethyl-piperidine-N-oxyl (TEMPOL) in benzene (10^{-6} M) as a spin standard.

The simulation analysis of ESR spectra was performed with a JEOL ESR-data treatment system, Model ESPRIT-330.

2.3 Heating, Photo-Exposure and Storage at High or Low Humidity

To examine the effects of heating, non-irradiated spice (200 mg) in an ESR measurement tube was heated at 100°C for 1 h. With regard to photo-exposure, non-irradiated spice (ca., 1 g) was spread 3 mm in depth on a Petri dish covered with polyvinylidene chloride film (thickness:10~12 μm) and kept for 1 month in a place exposed to sunlight.

Storage at relative humidities of 32.0% and 79.8% was performed by allowing spice in a small beaker to stand at 35°C for 7 days in a chamber with 15 ml of a saturated magnesium chloride solution and a saturated ammonium sulfate solution, respectively.

3 RESULTS AND CONSIDERATIONS[7,8]

3.1 ESR Spectra of Irradiated Spices

Red pepper, white pepper, black pepper, cinnamon and allspice irradiated up to a dose of 50 kGy showed a principal signal (signal I) with a g-value of 2.0040-2.0050, and γ-irradiation of red pepper, cinnamon and allspice gave a minor signal (signal II) 30 G to the low field side of signal I on the ESR spectrum (Figures 1 and 2). On the other hand, white pepper irradiated even at 50 kGy showed no signal II (Figure 1B). Although black pepper produced a small minor signal by irradiation, occasionally it was difficult to clearly detect it.

Signal II could also be detected in irradiated paprika and chilli pepper, which are the same capsicum genus as red pepper.

3.2 Dose Response Curve of Signal Intensity

The intensities of signal I of red pepper, cinnamon and allspice increased convexly with an increase in dose (Figure 3A), but the intensities of signal II increased linearly and the lines almost passed through the origin of the coordinate as shown in Figure 3B. Signal II might exhibit a quantitative relationship between dose and signal intensity.

3.3 Stability of Radicals

The radical stability in irradiated spices was examined by monitoring signal intensities during storage for 1 year at room temperature (average 25°C). Figure 4 presents the decay curves of both signal I and signal II in irradiated cinnamon. The intensities of both signals declined rapidly immediately after irradiation up until 2 months, with the decrease

Physical Methods: Electron Spin Resonance (ESR) Spectroscopy

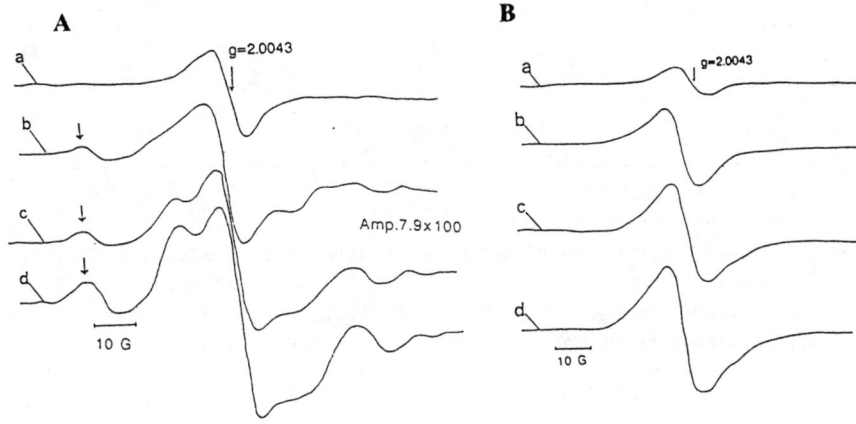

Figure 1 ESR spectra of irradiated red pepper and white pepper[7]
A: red pepper; B: white pepper; a: non-irradiated; b: irradiated at 10 kGy;
c: irradiated at 30 kGy; d: irradiated at 50 kGy
The arrows show the minor signal

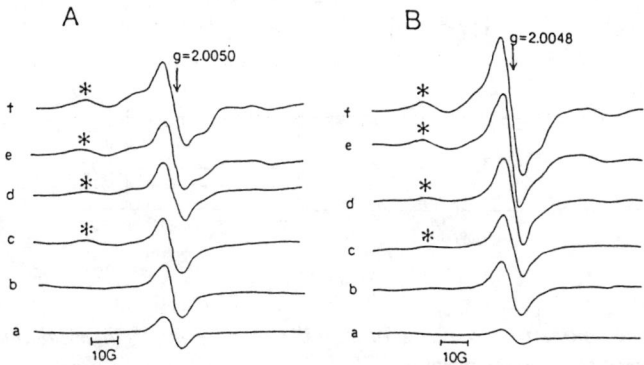

Figure 2 ESR spectra of irradiated allspice and cinnamon[8]
A: allspice; B: cinnamon; a: non-irradiated; b: irradiated at 3 kGy;
c: irradiated at 5 kGy; d: irradiated at 10 kGy; e: irradiated at 30 kGy;
f: irradiated at 50 kGy.
* indicates signal II

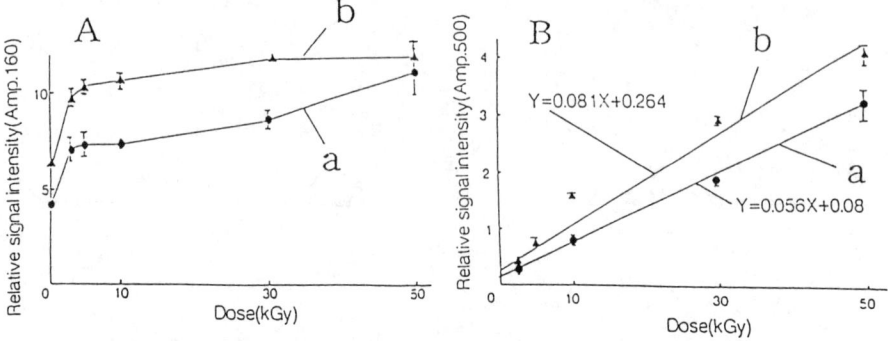

Figure 3 *Dose-response curves of signal I and signal II in irradiated allspice and cinnamon*[8]
A: signal I; B: signal II; a: allspice; b: cinnamon
Values are means ± SD for 4 trials

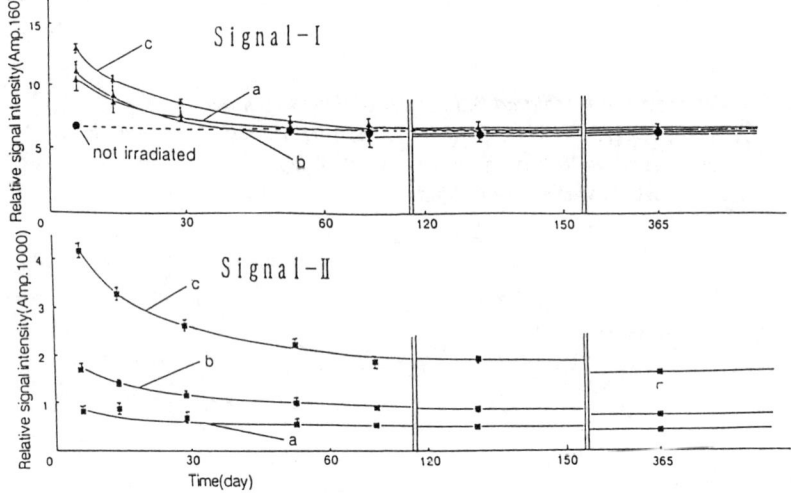

Figure 4 *Time-course of intensities of signal I and signal II in irradiated cinnamon stored at room temperature*[8]
a: irradiated at 5 kGy; b: irradiated at 10 kGy; c: irradiated at 30 kGy
Values are means ± SD for 4 trials

being slower thereafter. Signal I of cinnamon irradiated up to 30 kGy was similar to the intensity of the non-irradiated sample following 6 months storage but signal II of cinnamon irradiated at a dose level of 5 kGy or greater could be readily detected.

3.4 Radical Production by Heating and Photo-Exposure

Radical production in heated spices was examined in comparison with intact samples as shown in Figure 5. The signal intensities of spices heated at 100°C for 1 h were increased by 2.7 times for allspice, 1.8 times for cinnamon, 2.0 times for red pepper but signal II was not found in any of the heated samples.

The influence of photo-exposure on radical production was examined under sunlight for 1 month. In 2 weeks, signal I had increased 19.8 times for allspice, 3.7 to 19.5 times for cinnamon but showed no change for red pepper. Signal II in all of these samples was not detected under sunlight as is shown in Figure 6.

3.5 Influence of Humidity on Radicals Produced by γ-Irradiation

In order to examine the influence of humidity on radicals in irradiated spice, the irradiated spices were kept at 3 different relative humidities of 32.0%, 79.0% and the average room humidity (45%) for 7 days. The weight of sample stored under the conditions became constant after 7 days.

The intensity of signal I in allspice irradiated at 10 kGy and 30 kGy increased by 2-fold under the low humidity of 32.0% as compared with that of the spice under room humidity as shown in Figure 8. Under the high humidity of 79.8%, the intensity at 10 kGy decreased by 12% and that of 30 kGy by 20%. The intensity of signal II seemed to be constant even under different humidities. The influence of humidity on both signals in irradiated cinnamon was similar to that for allspice. These results suggest that signal I is affected by a high humidity but that humidity does not have an effect on signal II.

3.6 Computer-simulation of ESR Signals

Radical identification of irradiated red pepper was performed by a computer simulation method. Several trials of simulation showed that the spectrum of irradiated red pepper was identical to that of a mixture of 3 types of radicals from methyl, phenoxyl and peroxyl groups. Table 1 summarises the factors used for the computer simulation. Radical numbers 1, 2 and 3 in the table correspond to the radicals from methyl, phenoxyl and peroxyl groups, respectively. Each signal was expressed as a curve. In Figure 8, the simulated signal of the methyl radical is shown by spectrum A. The combined signal of methyl and phenoxyl radicals is shown by spectrum B and finally the combined signal of methyl, phenoxyl and peroxyl radicals is shown by spectrum C which is very similar to that of red pepper irradiated at 30 kGy (spectrum D). It can be concluded that the minor signal (signal II) is attributable to the methyl radical and the principal signal (signal I) is mainly attributable to a combination of phenoxyl and peroxyl radicals.

Signal II in irradiated allspice and cinnamon was not affected by humidity but signal I was significantly affected. The intensity of signal I also increased more by extraction with methanol than that of signal II. These results suggest that the radical corresponding to signal II seems to be located in a different position of the spice matrix than that of signal I.

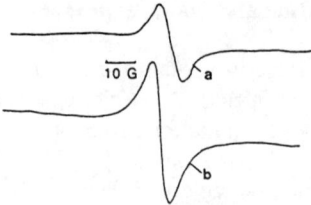

Figure 5 *ESR spectra of red pepper heated at 100°C for 1 h*[7]
a: intact; b; heated

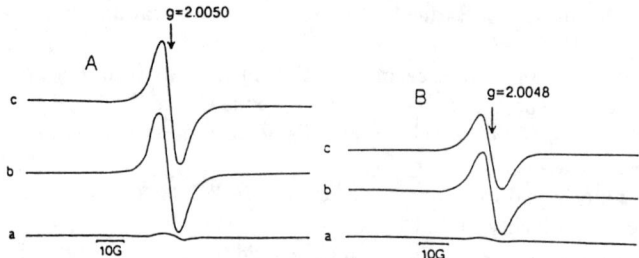

Figure 6 *Change in ESR spectra of allspice and cinnamon after exposure to sunlight*[8]
A: allspice; B: cinnamon; a: not exposed; b: exposed for 2 weeks; c: exposed for 1 month

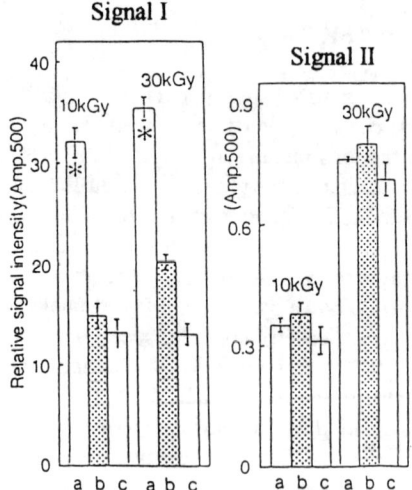

Figure 7 *Influence of humidity on intensities of signal I and signal II in irradiated allspice*[8]
a: 32.0%; b: room humidity (average 45%); c: 79.8%. humidity
Values are means ± SD for 4 trials.
*: significantly different (p<0.001) from the value at room humidity

Table 1 *Factors for Computer-simulation of Radical Assignment in Irradiated Red Pepper*[7]

		Radical Number		
		1	2	3
Line width [mT]		0.7000	0.9200	1.5000
Intensity		30	75	90
Lorentz (%)		100	100	100
Gauss (%)		0	0	0
Centre field [mT]		337.00	336.90	336.74
g-Value		2.0034	2.0040	2.0050
Spectral range		333.49	336.90	336.74
		340.51	336.90	336.74
Nuclear	Spin	0.5	0.5	0.5
	Number	3	1	1
	Hfcc [mT]	2.340	0.000	0.000
Frequency [MHz] 9450.000, Data length 2000, 331.90 < Sweep width < 341.90				

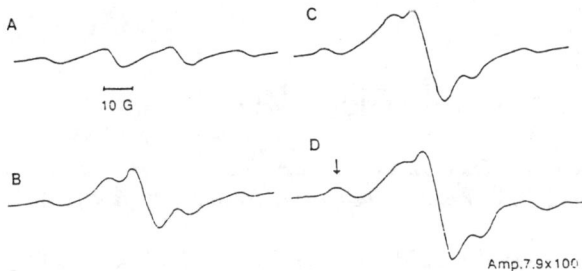

Figure 8 *Identification of radicals in irradiated red pepper by a computer-simulation method*[7]
A: simulated spectrum of methyl radical; B: simulated spectrum of combined methyl and phenoxyl radicals; C: simulated spectrum of combined methyl phenoxyl and peroxyl radicals; D: spectrum of red pepper irradiated at 30 kGy The arrow in spectrum D shows the minor signal (signal II)

3.7 Production of Minor Signals in Other Spices

Other spices found to produce the minor signal were celery leaf, clove, turmeric and coriander and the signal intensities also increased with increasing dose. However, further detailed studies of radical stability are needed.

4 CONCLUSION

The ESR method was effective in the identification of irradiated spices, especially by observation of a minor signal (signal II) unique to γ-irradiation. Signal II is 30 G to the lower field of a principal signal (signal I, g-value 2.0048-2.0058) which appears in almost all kinds of irradiated spices. The intensity of signal II increased linearly with increasing dose. Red pepper, cinnamon and allspice irradiated at doses greater than 5 kGy were detected for 1 year on the basis of the production of signal II. The signal was not found in irradiated white pepper. Although black pepper produced a small minor signal by irradiation, it could be difficult to clearly confirm in an ESR spectrum. Using a computer-simulation method it was speculated that signal II could be attributed to a methyl radical.

Signal II was not produced by heating at 100°C for 1 h or by exposure to sunlight for 1 month. The intensity of the signal was not affected by storage for 1 week under different humidities of 32.0%, 45.0% and 79.8%. Other spices which produced signal II on irradiation were celery leaf, clove, turmeric and coriander.

Acknowledgments

The authors wish to thank the Science and Technology Agency (Japan) for financial support and the Japan Atomic Energy Research Institute for cooperation with γ-irradiation. The figures and a table quoted from references 7 and 8 are acknowledged by the Journal of Food Hygienic Society of Japan.

References

1. World Health Organization (WHO), "Food Irradiation, A Technique for Preserving and Improving the Safety of Food," WHO, Geneva, Switzerland, 1988, p. 33.
2. N. J. F. Dodd, A. J. Swallow and F. J. Ley, *Radiat. Phys. Chem.*, 1985, **26**, 451.
3. A. Wieser and D. F. Regulla, *Inslitut für Strahlen Hygiene* (ISH-Heft), 1988, **125**, 155.
4. M. F. Desrosiers and W. L. McLaughlin, *Radiat. Phys. Chem.*, 1990, **35**, 321.
5. J. J. Raffi and J.-P. Agnel, *Radiat. Phys. Chem.*, 1989, **34**, 891.
6. R. Gray and M. H. Stevenson, *Radiat. Phys. Chem.*, 1989, **34**, 899.
7. S. Uchiyama, Y. Kawamura and Y. Saito, *J. Food Hyg. Soc. Japan,* 1990, **31**, 499.
8. S. Uchiyama, A. Sugiki, Y. Kawamura, M. Murayama and Y. Saito, *J. Food Hyg. Soc. Japan,* 1993, **34**, 128.
9. Joint FAO/IAEA Division of Nuclear Techniques in Food and Agriculture: A Report of the First Research Co-ordination Meeting, Warsaw (Jachranka), Poland, 25-29 June, 1990; Second Research Co-ordination Meeting, Budapest, Hungary, 15-19 June (1992); Third Research Co-ordination Meeting, Belfast, United Kingdom, 20-24 June (1994), Joint FAO/IAEA Division of Nuclear Techniques in Food and Agriculture.

ESR IDENTIFICATION OF IRRADIATED FOODSTUFFS: LARQUA RESEARCH

J. Raffi

Laboratoire de Recherche sur la Qualite des Aliments (LARQUA)
Faculte de Saint-Jerome, Avenue Escadrille Normandie Niémen
F- 13397 Marseille cedex 20
France

1 RESEARCH STUDIES OF LARQUA DURING THE ADMIT PROGRAMME

As electron spin resonance (ESR) spectroscopy is the leading method for identification of irradiated foodstuffs,[1] meat[2,3] and fish bones, fruit[4] and relative products (of vegetable origin),[5,6] sea-food[7] *etc.* were studied (Table 1).

In order to prepare a large co-trial on ESR identification of irradiated foodstuffs, experiments were carried out at LARQUA, especially on fruits and vegetables. The radicals induced in the fruit pulp are not stable because the water content of fruit is generally high, but ESR can be used with dried fruit or dry components such as achenes, pips or stones.[4-6] Different responses are observed, depending on the fruit. In "sugar type" fruits (papaya, dried grapes), an ESR multicomponent signal is radio-induced, but the non-irradiated fruit presents no ESR signal or a single line, while in "cellulose type" fruits (pistachio nut, berries), a triplet is induced. But a six line signal due to Mn^{2+} and a central single line may also be present both in irradiated and non-irradiated samples.

In the case of aromatic herbs, the proposed official CEN protocol for irradiated food containing cellulose was used.

1.1 The ESR Protocol used for Aromatic Herbs

The herbs are directly put in an ESR tube and the spectra recorded with the following main parameters: frequency, 9.5 GHz; power, 0.4 mW; magnetic field, 348 ± 10 mT; modulation amplitude, 0.4 mT. One typical spectrum is given in Figure 1.

1.2 ESR Study

Just after irradiation, there was no problem in finding the two satellite lines "typical" of an irradiated food containing cellulose. However, mainly due to the water content of those herbs, the signal intensity decreases with storage time. Consequently, ESR can not be used to show proof of irradiation; however, with the aim of establishing a screening method, the influence of storage time was studied in order to compare thermoluminescence (TL)[8] and ESR measurements.

Table 1 *LARQUA ESR Studies on Irradiated Foodstuffs*

Foodstuff	Type of Esr Signal Reference Irradiated (Except Reference)		Life Time of the ESR Signal	Reference Number

frog legs
  RT
  frozen
	0 / M		0.5 kGy:
			    > 6 months
			    > 2 years
				9, 2

lamb	-	-		
rabbit	-	-		
chicken	-	-		
pork				
sardine	M	HP	1 kGy > 40 days (RT)	
brown trout	-	-	-	
eel	-	-	-	
pike-trech	-	-	-	
onion	B	C	1.5 kGy: ~ 60 days 8 kGy: ~ 150 days	10, 11 12, 13
apricot	B	m, C	1.5 kGy: < 17 days 8 kGy: ~ 90 days	
dried grapes	B	m	1.5 kGy: > 441 days 8 kGy: > 4 years	
papaya	0	m	1.5 kGy: > 441 days 8 kGy: > 4 years	
pistachio nuts	B	C	1.5 kGy: ~ 388 days	
dried prunes			8 kGy: > 4 years	
dried banana	B	C	8 kGy: > 2 years	
coconuts	B	mm	8 kGy: > 2 years	
figs	B	m, C	8 kGy: > 2 years	
berries: fresh (a)	B	C	8 kGy: > 2 years	
frozen	B + Mn	C	1 kGy: ~ 25 days	14, 15
dates	B + Mn	C	1 kGy: > 2 years	
	B + Mn	C, m	1 kGy: > 2 years	4
Norway lobster	Mn	B		7
scallop shell	0/M	mm	> 2 years at RT	16
mussels				
clam	0/M	mm	- - -	
	0/M	mm	- - -	
egg shell	0	mm	- - -	
parsley	B ± Mn	C	5 kGy: ~ 7 weeks	8
thyme			- > 3 months	
savory			- 8/9 weeks	
sage			- 6/7 weeks	
rosemary			- 10/12 weeks	
origanum			- 5/6 weeks	
fennel			- > 5 months	

Legend of Table 1

(a) strawberries, raspberries, red currants and bilberries
RT: room temperature   0: no signal
M: "marrow" signal   HP: "hydroxyapatite" signal   Mn: manganese signal
B : single line signal   C: "cellulose" signal (triplet)   m: "multicomponent" signal, type sugar (glucose, *etc.*)
     mm: unknown multi component signal

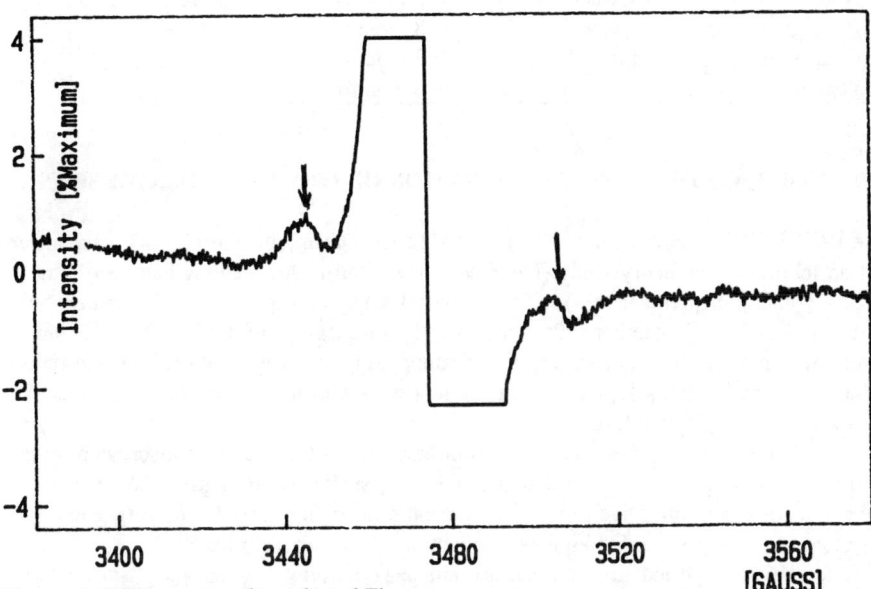

Figure 1 *ESR spectra of irradiated Thyme*
Arrows show the characteristic lines which are not present in the non-irradiated sample

1.3 Comparison of TL and ESR Results with the Aim of the Establishing a Screening Method for Identification of Irradiated Aromatic Plants

If the maximal time of identification of irradiated products using TL (on the whole plant or on minerals) is compared to that of ESR (Table 2), the following conclusions can be drawn:
- TL done on minerals is always the best method, *i.e.* the only one leading to proof of irradiation; in that case it is important to remember that the "official" method needs to re-irradiate the sample and to compare the TL signals before and after re-irradiation;
- depending of the plant, TL on whole plant or ESR may be the best method for screening.

Table 2 *Comparison of Results Obtained by ESR and TL*

Herb	TL Analysis		ESR Analysis
	On the Whole Plant	On Minerals	
Parsley	> 8 months	> 8 months	7 weeks
Thyme	> 8 months	> 8 months	> 3 months
Savory	3 weeks	> 8 months	8/9 weeks
Sage	6/7 weeks	> 8 months	6/7 weeks
Rosemary	1/2 weeks	> 8 months	10/12 weeks
Origanum	> 5 months	> 8 months	5/6 weeks
Fennel	> 5 months	> 8 months	> 5 months

2 PARTICIPATION IN AND ORGANISATION OF THE INTERCOMPARISONS

The BCR ESR intercomparison[10,11] involved 22 European laboratories and was carried out as follows: preliminary study (November '89 - April '90), mailing and recording of results (June - September '90), first statistical analysis (October - November '90), preliminary report (December '90 - January '91), discussion and technical evaluation of results with participants (Cadarache, 19 February '91), final statistical analysis and report (March - July 91). Six samples of each product were sent to each of the 21 laboratories involved in the qualitative test.

Results on meat and fish bones erre good and the protocol now in discussion by the European Committee for Standardization (CEN) is similar to the initial one. However, the results of the quantitative test carried out on poultry bones were not very good and, therefore, it is not preferable to publish a quantitative protocol at present.

The results for dried grapes (raisins) and papaya were conclusive. The minimum detectable doses were low, 0.12 kGy for grapes and 0.35 for papaya, which guarantees a high success rate for irradiation detection. The multi-line spectra allows a high level of correct identifications with a very easy protocol but other fruits need to be studied before these results can be extrapolated and this is one reason why CEN has preferred to delay the publication of this protocol.

The results for pistachio nuts were more complicated as some laboratories had a high success rate while others achieved minimal success. The changes in spectral shape,[4,15] that is the two satellite lines, on irradiation are small and very dependent on instrument variables. Thus new experiments have been carried out with the participation of two other laboratories. The complementary report[12] showed that the relative protocol requires training of technicians but, here too, the protocol now in discussion by CEN is similar to the initial one used for this intercomparison.

Acknowledgments

We are indebted to the French Research Ministry (Grant N° 88.G.1014), to the Community Bureau of Reference (BCR, Brussels) (Agreements 5348/1/5/340/ 90/4-

BCR F[10] and 5415/1/5/340/90/11-BCR-F[10]) and to IAEA (Agreement N°5154/CF) for financial support and helpful discussions during the meetings they organised.

References

1. J. Raffi and J-P. Agnel, in "ESR Applications in Organic and Bioorganic Materials," B. Catoire, Ed., Springer-Verlag, Berlin, 1992, p. 135.
2. M. Desrosiers, W. McLaughlin, L. Shean, N. Dodd, J. Lea, J. Evans, C. Rowlands, J. Raffi and J-P. Agnel, *Int. J. Food Sci. Technol.*, 1990, **25**, 682.
3. A. M. Rossi, G. Poupeau, O. Chaix, J. Raffi and J-P. Agnel, "ESR Applications in Organic and Bioorganic Materials," B. Catoire, Ed., Springer-Verlag, Berlin, 1992, p. 151.
4. J. Raffi, J-P. Agnel and S. H. Ahmed, *Food. Tec.*, 1991, **3/4**, 26.
5. J. F. Bayonove, J. Raffi and J-P. Agnel, *Adv. Space Res.*, 1994, **14**, 1053.
6. J. Raffi and S. Benzaria, *J. Radiat. Steril.*, 1994, **September**, 281.
7. J. Raffi and J-P. Agnel, *Sci. des Alim.*, 1990, **10**, 387.
9. J. Raffi, J. Evans, J-P. Agnel, C. Rowlands and G. Lesgards, *Appl. Radiat. Isot.*, 1989, **40**, 1215.
10. J. Raffi, "ESR Intercomparison Studies on Irradiated Foodstuffs," Commission of the European Communities (BCR), Brussels, Luxembourg, EUR 13630 EN, 1992.
11. J. Raffi, M. H. Stevenson, M. Kent, J. Thiery and J-J. Belliardo, *Int. J. Food Sci. Technol.*, 1992, **27**, 111.
12. J. Raffi, M. H. Stevenson, J-P. Agnel, R. Gray and G. Burlinska, Progress Report to BCR on Contract 5348/1/5/340/90/4/BCR-F(10), 1992.
13. J. Raffi, J-J. Belliardo, J-P. Agnel and P. Vincent, *Appl. Radiat. Isot.*, 1993, **44**, 407.
14. J. Raffi, J-P. Agnel, L. Buscarlet and C. Martin, *J. Chem. Soc., Faraday Trans. I*, 1988, **84**, 3359.
15. J. Raffi and J-P. Agnel, *Radiat. Phys. Chem.*, 1989, **34**, 891.
16. J. Raffi and J-P. Agnel, *Ann. Fals. Exp. Chim., Paris*, 1989, **82**, 279.

INTERLABORATORY TESTS TO IDENTIFY IRRADIATION TREATMENT OF VARIOUS FOODS VIA GAS CHROMATOGRAPHIC DETECTION OF HYDROCARBONS, ESR SPECTROSCOPY AND TL ANALYSIS

G. A. Schreiber, N. Helle, G. Schulzki, B. Linke, A. Spiegelberg, M. Mager and K. W. Bögl

BgVV - Federal Institute for Health Protection of Consumers and Veterinary Medicine
FGr 21/ FG 212
Postfach 33 00 13
D-14191 Berlin
Germany

1 INTRODUCTION

The gas chromatographic (GC) analysis of radiation-induced volatile hydrocarbons (HC) and 2-alkylcyclobutanones, the ESR spectroscopic detection of radiation-specific radicals and the thermoluminescence (TL) analysis of silicate minerals are the most important methods for identification of irradiated foods.[1-4] After successful performance in interlaboratory studies on meat products, fish, spices, herbs and shells of nuts, all or some of these methods have been approved by national authorities in Germany and the United Kingdom.[5-14] Recently, draft European Standards have been elaborated for approval by member states of the European Committee for Standardization (CEN).

Several research laboratories have shown that these methods can be applied to various foods not yet tested in collaborative studies.[15-26] However, for an effective application in food control it is necessary to prove their suitability in interlaboratory studies. Therefore, in 1993/94, various interlaboratory tests were organised by the BgVV. In an ESR spectroscopic test, shrimps and paprika powder were examined. Shrimps were also the subject of examination in a TL test. Finally, GC detection of radiation-induced hydrocarbons in the fat fraction of foods was used in another test to identify irradiated Camembert, avocado, papaya and mango.

In the following paper, results of the interlaboratory tests are summarised. Detailed reports are published by this institute.[27-29]

2 METHODS

2.1 Interlaboratory Test To Detect Radiation-Induced Hydrocarbons

Each of the 22 participating laboratories received 6 Camembert samples which were either non-irradiated or irradiated with about 0.5 or 1 kGy and 5 avocado, 5 mango and 5 papaya samples which were either non-irradiated or treated with about 0.35, 0.5 or 1 kGy. The results of one analysis per sample had to be reported. In brief the following instructions were given:

2.1.1 Fat Extraction. From each sample, at least 1 g of fat has to be extracted. This amount is sufficient for one analysis. Homogenise one whole Camembert, the fruit pulp of one avocado, the seeds of 2 papayas or the seeds of 3 mangos and mix thoroughly with anhydrous Na_2SO_4. Homogenise the mixtures with 100 ml (Camembert, avocado) or 150 ml n-hexane (papaya, mango). After centrifugation (900 g, 5 min) concentrate the extracts by vacuum rotary evaporation and transfer the fat to a glass vial. Evaporate solvent residue completely by means of a nitrogen gas flow and store fat at 4°C.

2.1.2 Isolation of Hydrocarbons by Florisil Column Chromatography. Heat Florisil to 550°C overnight and deactivate by addition of 3% double-distilled water (v/w).[13] Fill a glass column (diameter 20 mm; with a frit) with deactivated Florisil in hexane to a height of about 13 cm. Mix 1 g of fat with 1 ml of internal standard (about 1 µg/ml 20:0 in hexane [IS]) and introduce into the column. Elute the HC with 60 ml of hexane at a rate of 3 ml min^{-1}. Add 1 ml isooctane and concentrate by vacuum rotary evaporation (about 25 kPa) to a volume of about 3 ml. Transfer the sample to a graduated glass tube for further concentration under nitrogen to 1 ml and put it in a GC vial.

2.1.3 Gas Chromatography and Detection. It was recommended to separate HC on a non-polar capillary column (100% dimethyl polysiloxane or 5% diphenyl/95% dimethyl polisiloxane; 30 m). For GC analysis, the following recommendations were given: injection: 1 µl splitless at 200°C; temperature programme: 50°C for 2 min. 10°C min^{-1} to 130°C, 5°C min^{-1} to 200°C. HC were detected by flame ionisation (FID) or mass spectrometry (MS).

2.1.4 Identification. On the basis of the fatty acid composition of samples (Table 1) and the availability of standard substances, it was decided that the participants had to quantify the HC listed in Table 2 via IS. A sample had to be identified as irradiated if all the HC to be determined could be clearly identified.

2.2 ESR Interlaboratory Test

Each of the 20 participating laboratories had to examine 8 samples of Norway lobsters *(Nephrops norvegicus)*, 8 samples of brown shrimp *(Crangon crangon)* and 8 samples

Table 1 *Fatty Acid Composition of Interlaboratory Test Samples*

Fatty Acid		Camembert*	Avocado	Mango	Papaya
		Proportion of Total Fat [%]			
Capric	C10:0	2.5			
Lauric	C12:0	3.5			
Myristic	C14:0	11.8			
Palmitic	C16:0	36.3	11.9	7.2	19.5
Palmitoleic	C16:1		3.5		
Stearic	C18:0	13.4		33.1	5.8
Oleic	C18:1	23.4	60.1	48.8	66.8
Linoleic	C18:2	2.0	13.0	9.0	5.7

*Camembert from 2 different manufacturers was used. Mean concentrations are given.

Table 2 *HC to be Determined in Interlaboratory Samples*

Product	$C_{n-2:1}$ and C_{n-1} of Oleic Acid		$C_{n-2:1}$ and C_{n-1} of Palmitic Acid		$C_{n-2:1}$ and C_{n-1} of Stearic Acid	
Camembert	1,7-16:2	8-17:1	1-14:1	15:0		
Avocado	1,7-16:2	8-17:1				
Papaya	1,7-16:2	8-17:1				
Mango	1,7-16:2	8-17:1			1-16:1	17:0

of paprika powder. Shrimp samples were either non-irradiated or treated with 2, 4 or 6 kGy whereas in the case of paprika, doses of 5 or 10 kGy were applied. Sixteen participants examined the samples by means of the Bruker EMS104 ESR spectrometer, 2 used the Bruker ESP300E and one the Bruker ECS106. One laboratory carried out the measurements on two different instruments (EMS104 and ESP300E). The ESR spectrometer settings are given in Table 3.

Shells of Crustaceans were taken from the tail region and dried either at about 40°C in a vacuum oven or at about 60°C without vacuum. Most laboratories fragmented the shells for measurement, although grinding was also allowed. Paprika was examined without any preparation.

For identification, laboratories received examples of typical spectra of non-irradiated and irradiated samples, respectively (Figure 1).

2.3 TL Interlaboratory Test on Shrimps

Each of the 23 participating laboratories received 6 shrimp samples (3 Vietnam Cat Tiger, 3 China Reds) which were either non-irradiated or treated with 1 or 2 kGy. Each sample had to be analysed twice. TL analysis was performed as described elsewhere.[24] In

Table 3 *ESR Spectrometer Settings*

Parameter	Crustacea	Paprika
Magnetic field		
Centre field [mT]	347.8	347.8
Sweep width [mT]	20.0	20.0
Microwave radiation		
Frequency [GHz]	9.76	9.76
Power [mW]	10.0	0.4 - 0.8
Signal channel		
Modulation frequency [kHz]	50 (EMS104) or 100	50 (EMS104) or 100
Modulation amplitude [mT]	0.30	0.80
Time constant [ms]	20.48	20.48
Sweep rate [mT min^{-1}]	14.3	14.3
Number of scans	5	5

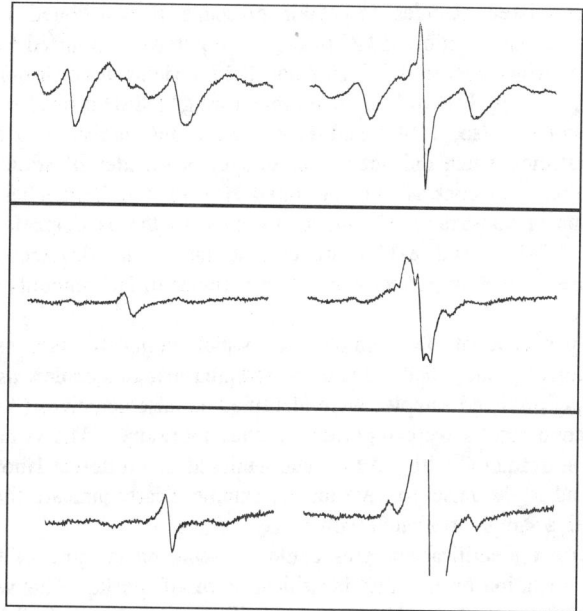

Figure 1 *ESR spectra of non-irradiated (left) and irradiated (right) shells of Norway lobster (top), shells of brown shrimp (middle) and paprika powder (bottom)*

brief, minerals were isolated from the intestines of the animals and a first TL measurement (1st glow) was performed. After irradiation of mineral samples with about 1 kGy, a second glow (2nd glow) was carried out in order to calculate TL ratios (1st/2nd glow of integration area I). Reference light and background intensities of the TL readers had to be recorded regularly. Full process blanks were done in parallel to the analyses of coded samples. Minimal intensities were calculated (50 times of the background intensity of the respective TL reader in integration area I) which had to be exceeded in 2nd glow readings. Otherwise the sample was withdrawn. If TL ratios exceeded 0.50 in two independent analyses, a sample had to be classified as irradiated.

3 RESULTS AND DISCUSSION

3.1 Radiation-Induced Hydrocarbons in Camembert, Avocado, Papaya and Mango

With the exception of 5 samples, no 1,7-16:2 could be detected in non-irradiated samples. Also, 8-17:1 was not found in non-irradiated Camembert, papaya and mango samples with the exception of 2 laboratories, whereas in non-irradiated avocado samples, 8-17:1 was found in 8 samples by 7 laboratories. Both markers are not expected to be present in non-irradiated samples and their presence is considered to be due to contamination or misclassification of GC peaks. The latter is supported by the fact that all the laboratories which reported 1,7-16:2 and 8-17:1 yields in non-irradiated samples identified the GC peaks by FID or "single ion monitoring" (SIM) instead of full scan MS (with one exception). Also, 1-14:1 and 1-16:1 were only detected in non-irradiated samples by laboratories which did not use mass spectra for identification of GC peaks (with one exception). In contrast, the saturated HC, 15:0 in Camembert and 17:0 in mango were found in non-irradiated samples by most of the participants. Mean dose dependencies of 1,7-16:2 and 8-17:1 amounts in fat of the different products are displayed in Figure 2. Nearly perfect linear dependencies of HC amounts on irradiation dose can be stated.

Ninety-eight per cent of 431 samples for which complete data sets had been submitted were correctly identified. One non-irradiated mango sample was misclassified as irradiated and 7 irradiated samples were identified as false negatives (Table 4). Four of the false negative results were reported by one laboratory. The remaining 3 false negatives had been irradiated with 0.3 kGy and analysed in 3 different laboratories. The HC amounts found in the false positive mango sample clearly indicate that the sample was mixed up with a sample irradiated with 1 kGy.

Very high correct identification rates could be achieved in spite of the described problems of contamination or incorrect identification of GC peaks. This was a result of the requirement that a sample could only be identified as irradiated if all markers to be measured were present (Table 2). The probability that a sample is contaminated with all markers or that various GC peaks are misidentified is very low. It becomes even lower if on the basis of a given fatty acid composition, all C_{n-1} and $C_{n-2:1}$ markers to be expected are the prerequisite of positive identifications. The latter condition is demanded for positive identifications in routine food control. Also, in food control applications, the identification of peak identities by MS is a necessity since in addition to the described problems with FID or SIM, some of the radiation-induced HC, e.g. 1,7,10-16:3 and 6,9-17:2 derived from linoleic acid, are not available as standard substances.

Table 4 *Samples Identified as False-Positive or False-Negative*

Sample Code	Product	Dose [kGy]	1,7-16:2	8-17:1	1-14:1	15:0	1-16:1	17:0
				[$\mu g\ g^{-1}$ Fatty acid]				
False positive:								
T17	Mango	0	1.51	0.82			1.76	1.24
False negative:								
B16	Mango	0.32	0.65	0			4.51	1.30
L01	Camembert	1.0	0	0	0	0.08		
L09	Avocado	0.36	0	0				
L17	Mango	0.32	0	0			0.18	0
L18	Mango	1.1	0	0			0.15	0.76
N16	Mango	0.32	0	0			1.91	0.76
R18	Mango	0.32	0	0.04			0.24	0.33

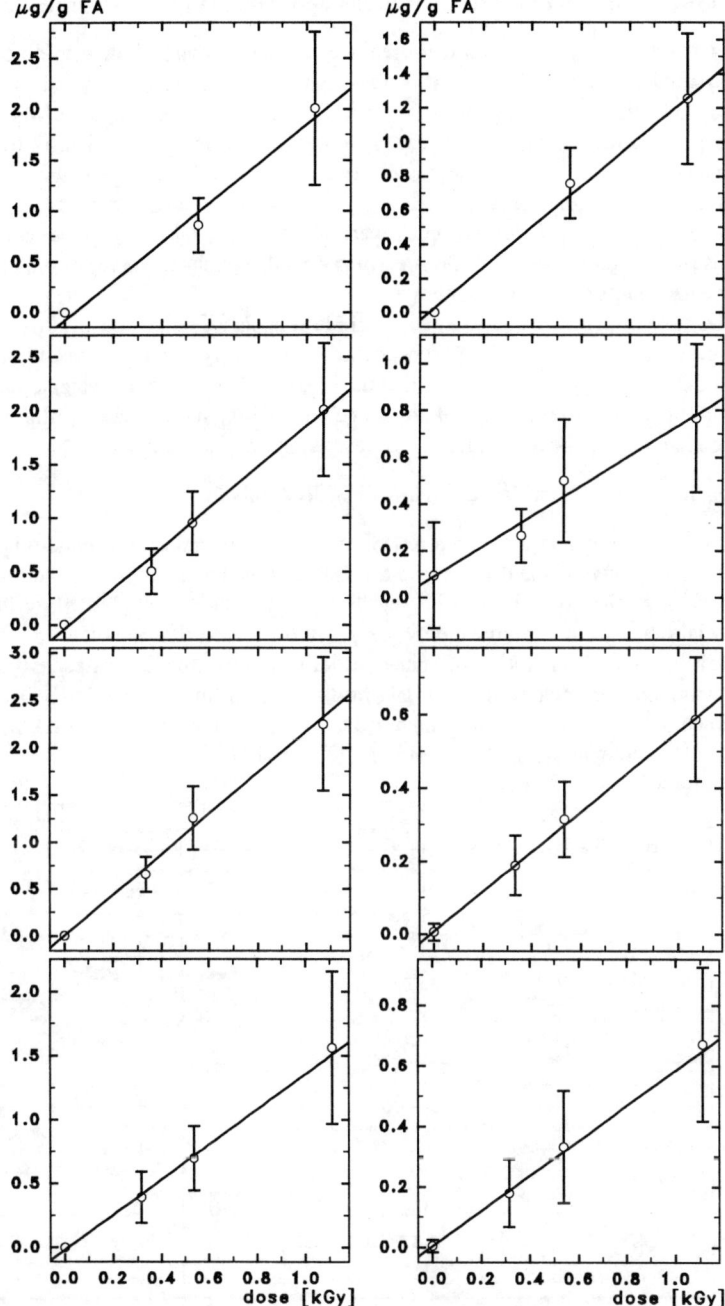

Figure 2 *Dose dependency of 1,7-hexadecadiene (left) and 8-heptadecene (right) mean yields per precursor fatty acid in coded interlaboratory samples*
Linear regressions were calculated from single data. From top to bottom, data are shown in the order Camembert, avocado, papaya and mango.

3.2 Identification of Irradiated Crustacean and Paprika by ESR Spectroscopy

For Norway lobster, 168 results were reported. One irradiated Norway lobster sample was classified as non-irradiated and 3 non-irradiated ones were classified as irradiated whereas 98% were identified correctly. The false positive identifications were done by an unexperienced laboratory. These spectra were sent independently to 3 other participants who were asked to interpret them. They all identified the spectra as being derived from non-irradiated samples.

Two of 168 brown shrimp samples were identified as false negatives and one as a false positive. In the latter case, the spectrum was also misinterpreted, as was proven by 3 other participants as mentioned above.

In the case of paprika, one irradiated sample was classified as non-irradiated whereas all the other 167 samples (99%) were identified correctly. These measurements were done about 3 to 6 weeks after irradiation. To evaluate signal stability, half of the participating laboratories examined the same samples again 3 months after irradiation (the code had already been revealed). All results reported were correct.

3.3 Identification of Irradiated Shrimps by TL Analysis

Nearly all participants were able to isolate enough minerals from the shrimp samples to perform both analyses and to exceed the 2nd glow minimal intensity. All the samples from only one laboratory had to be rejected due to insufficient 2nd glow intensities. Another laboratory could perform only one analysis on each of 3 samples.

Clearly separated TL ratios of minerals derived from non-irradiated and irradiated shrimp samples were achieved by all laboratories which fulfilled the 2nd glow intensity requirements (Figure 3). Slightly higher TL ratios were obtained for 2 kGy samples (Mean: 1.4) in comparison to 1 kGy samples (Mean: 1.0).

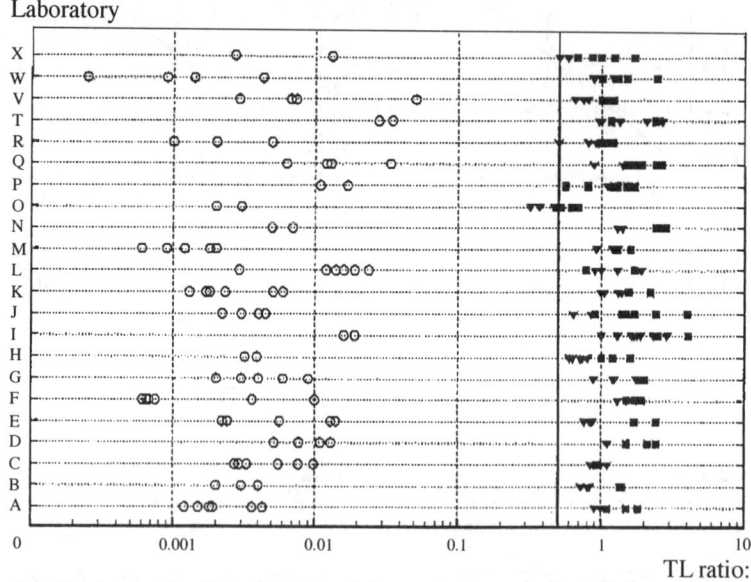

Figure 3 *TL ratios of minerals isolated from shrimp samples which were either non-irradiated (○) or irradiated with about 1 kGy (▼) or 2 kGy (■)*

Using a TL ratio threshold value of 0.50 for identification, 3 irradiated samples were excluded from data analysis because the TL ratios were distributed around a value of 0.50 which meant that the samples had been classified in one determination as irradiated and in the other as non-irradiated. Two samples irradiated with 1 kGy had to be classified as non-irradiated (TL ratios for sample 0.30: 0.37 and 0.47; for sample 0.31: 0.32 and 0.49). All of the other 124 samples (98%) were identified correctly. Even for the 3 samples which were excluded and for the 2 samples which were false negatives, the TL curves of 1st glowings as well as the TL ratios, indicated that these samples had been irradiated. In practice, the analysis would have been repeated until a clear result was achieved.

4 CONCLUSION

The three interlaboratory tests described produced very high correct identification rates (97-99%). These results show that the respective methods can be successfully used to identify irradiation treatment of foods tested within commercially applied dose ranges. The methods will be included in the Official Collection of Methods according to article 35 of the German Foods Act (LMBG) for the applications tested. The data has also be used to extend the draft CEN European Standards to the products tested.

Acknowledgements

We are very grateful to the participants of the interlaboratory trials. We thank Mr J. Ammon (Karlsruhe), Dr U. Ballin (Cuxhaven), Dr U. Bänziger (Solothurn, Switzerland), Mr P. Baumann (Dortmund), Dr R. Brockmann (Bielefeld), Dr J. Brunner (Chemnitz), Mr H. -V. Buchholtz (Berlin), Ms B. Butz (Erlangen), Dr S. Dahmen (Berlin), Dr H. Delincée (Karlsruhe), Mr Ch. Droz (St. Gallen, Switzerland), Mr D. Erning (Hamm), Dr H. Eschelbach (Erlangen), Dr P. Fey (Saarbrucken), Dr B. Fienitz (Berlin), Dr Frieß (Berlin), Mr G. Frohmuth (Kassel), Mr K. Fuchs (Mainz), Dr C. Gemperle (Aargau, Switzerland), Dr T. Göllner (Erlangen), Dr K. Hammerton (Menai, Australia), Ms Ch. Hees (Duisburg), Ms A. Holstein (Oberschleißheim), Dr D. Jahr (Oberschleißheim), Dr K. Jonas (Stendal), Dr E. Kaltwasser (Siegen), Dr J. Kispéter (Szeged, Hungary), Ms H. Klein (Mettmann), Mr W. Krölls (Köln), Mr W. Kruspe (Meiningen), Dr T. Kuhn (Hamburg), Dr M. Langer (Berlin), Mr H. Lohse (Bremen), Ms K. Mainczyk (Düsseldorf), Dr W. Meier (Zürich, Switzerland), Ms H. Meyer (Duisburg), Dr G. Mildau (Karlsruhe), Dr H. Münz (Flensburg), Mr H. Nootenboom (AG Nymegen, The Netherlands), Mr F. Parsch (Berlin), Dr J. Pfordt (Oldenburg), Ms S. Pinnioja (Helsinki, Finland), Dr P. Roberts (Lower Hutt, New Zealand), Mr B. Rönnefahrt (Hamm), Mr J. Rümenapp (Hagen), Dr W. Ruge (Karlsruhe), Dr D.C.W. Sanderson (Glasgow, UK), Dr C. Schleich (Mainz), Ms H. Stemmer (Wuppertal-Barmen), Dr E. Stewart (Belfast, UK), Ms B. Studer (Solothurn, Switzerland), Ms C. Trapp (Rostock), Dr N. Vater (Kassel), Mr N. Vreden (Duisburg), Dr R. Wohlfarth (Wiesbaden), Ms U. Zachäus (Berlin) and Mr H. J. Zehnder (Wädenswill, Switzerland).

References

1. M. Leonardi, J. J. Raffi and J.-J. Belliardo, "Recent Advances on the Detection of Irradiated Food," Commission of the European Communities (BCR), Brussels, Luxembourg, EUR 14315 EN, 1993.
2. G. A. Schreiber, N. Helle and K. W. Bögl, *Int. J. Radiat. Biol.*, 1993, **63**, 105.
3. H. Delincée, *Radiat. Phys. Chem.*, 1993, **42**, 351.
4. H. M. Stevenson, "New Developments in Food, Feed and Waste Irradiation," G. A. Schreiber, N. Helle and K. W. Bögl, Eds., BGA, Berlin, SozEp-Heft 16, 1993, p. 3.
5. M. F. Desrosiers, W. L. McLaughlin, L. A. Sheahen, N. J. F. Dodd, J. S. Lea, J. C. Evans, C. C. Rowlands, J. J. Raffi and J.-P. L. Agnel, *Int. J. Food Sci. Technol.*, 1990, **25**, 682.
6. S. L. Scotter, P. Holley and R. Wood, *Int. J. Food Sci. Technol.*, 1990, **25**, 512.
7. J. J. Raffi, "Electron Spin Resonance Intercomparison Studies on Irradiated Foodstuffs," Commission of the European Communities (BCR), Brussels, Luxembourg, EUR 13630 EN, 1992.
8. J. J. Raffi, M. H. Stevenson, M. Kent, J. M. Thiery and J. J. Belliardo, *Int. J. Food Sci. Technol.*, 1992, **27**, 111.
9. W. Meier and M. H. Stevenson, "Recent Advances on the Detection of Irradiated Food," M. Leonardi, J. J. Raffi and J.-J. Belliardo, Eds., Commission of the European Communities (BCR), Brussels, Luxembourg, EUR 14315 EN, 1993, p. 211.
10. D. C. W. Sanderson, G. A. Schreiber and L. A. Carmichael, "A European Interlaboratory Trial of TL Detection of Irradiated Herbs and Spices. Phase 1: Preliminary Study," Commission of the European Communities (BCR), Brussels, Luxembourg, 1994, SURRC report to MAFF.
11. G. A. Schreiber, N. Helle, G. Schulzki, A. Spiegelberg, B. Linke, U. Wagner and K. W. Bögl, *Radiat. Phys. Chem.*, 1993, **42**, 391.
12. M. F. Desrosiers, "Report on Activities for the Coordinated Research Program on Analytical Detection Methods for Irradiation Treatment of Foods (ADMIT)," Third Research Coordination Meeting, Belfast, Northern Ireland, June 20-24, 1994.
13. G. A. Schreiber, G. Schulzki, A. Spiegelberg, N. Helle and K. W. Bögl, *JAOAC*, 1994, **77**, 1202.
14. G. A. Schreiber, N. Helle and K. W. Bögl, *JAOAC*, 1995, **78**, 88.
15. E. M. Stewart, M. H. Stevenson and R. Gray, *J. Sci. Food Agric.*, 1992, **55**, 653.
16. E. M. Stewart, M. H. Stevenson and R. Gray, *Int. J. Food Sci. Technol.*, 1992, **27**, 125.
17. N. Helle, B. Linke, O. Popoola, K. W. Bögl and G. A. Schreiber, "New Developments in Food, Feed and Waste Irradiation," G. A. Schreiber, N. Helle and K. W. Bögl, Eds., BGA, Berlin, SozEp-Heft 16, 1993, p. 23
18. K. M. Morehouse and M. F. Desrosiers, *Appl. Radiat. Isotop.*, 1993, **44**, 429.
19. E. M. Stewart, M. H. Stevenson and R. Gray, *Appl. Radiat. Isotop.*, 1993, **44**, 433.
20. K. M. Morehouse and Y. Ku, *Radiat. Phys. Chem.*, 1993, **42**, 359.
21. G. Schulzki, A. Spiegelberg, N. Helle, K. W. Bögl and G. A. Schreiber, "New Developments in Food, Feed and Waste Irradiation," G. A. Schreiber, N. Helle and K. W. Bögl, Eds., BGA, Berlin, SozEp-Heft 16, 1993, p. 55.

22. A. Spiegelberg, G. Schulzki, N. Helle, K. W. Bögl and G. A. Schreiber, "New Developments in Food, Feed and Waste Irradiation," G. A. Schreiber, N. Helle and K. W. Bögl, Eds., BGA, Berlin, SozEp-Heft 16, 1993, p. 39.
23. S. Pinnioja, *Radiat. Phys. Chem.*, 1993, **42**, 397.
24. G. A. Schreiber, A. Hoffmann, N. Helle and K. W. Bögl, *Radiat. Phys. Chem.*, 1994, **43**, 533.
25. G. A. Schreiber, B. Ziegelmann, G. Quitzsch, N. Helle and K. W. Bögl, *Food Struct.*, 1993, **12**, 385.
26. G. Schulzki, A. Spiegelberg, K. W. Bögl and G. A. Schreiber, *J. Agric. Food Chem.*, 1995, **43**, 372.
27. G. A. Schreiber, G. Schulzki, A. Spiegelberg, J. Ammon, U. Bänziger, P. Baumann, R. Brockmann, Ch. Droz, P. Fey, K. Fuchs, C. Gemperle, T. Göllner, Ch. Hees, D. Jahr, K. Jonas, W. Krölls, M. Langer, H. Lohse, G. Mildau, F. Parsch, J. Pfordt, B. Rönnefahrt, J. Rümenapp, W. Ruge, B. Studer, C. Trapp, N. Vreden, R. Wohlfarth and K.W. Bögl, "An Interlaboratory Study on the Detection of Irradiated Camembert, Avocados, Papayas and Mangos by Gas Chromatographic Analysis of Radiation Induced Hydrocarbons," Bundesinstitut für gesundheitlichen Verbraucherschutz und Veternärmedizin (BgVV), Berlin, BgVV-Heft 6, 1995.
28. B. Linke, N. Helle, J. Ammon, U. Ballin, R. Brockmann, J. Brunner, S. Dahmen, H. Delincée, J. Erning, H. Eschelbach, A. Holstein, D. Jahr, E. Kaltwasser, W. Krölls, W. Kruspe, T. Kühn, W. Meier, J. Pfordt, C. Schleich, E. Stewart, N. Vater, N. Vreden, K. W. Bögl and G. A. Schreiber, "Elektronenspinresonanz-Spektroskopische Untersuchungen zur Identifizierung Bestrahlter Krebs und Gewürze: Durchführung eines Ringversuches an Nordseekrabben, Kaisergranat und Paprikapulver" Bundesinstitut für gesundheitlichen Verbraucherschutz und Veternärmedizin (BgVV), Berlin, BgVV-Heft 9, 1995.
29. G. A. Schreiber, M. Mager, J. Ammon, J. Brunner, H.-V. Buchholtz, B. Butz, H. Delincée, B. Fienitz, G. Frohmuth, K. Hammerton, D. Jahr, J. Kispéter, H. Klein, W. Kruspe, T. Kühn, K. Mainczyk, H. Meyer, H. Münz, H. Nootenboom, J. Pfordt, S. Pinnioja, P. Roberts, D. C. W. Sanderson, C. Schleich, N. Vreden, U. Zachäus, H. J. Zehnder and K. W. Bögl, "An Interlaboratory Study on the Detection of Irradiated Shrimps by Thermoluminescence Analysis," Bundesinstitut für gesundheitlichen Verbraucherschutz und Veternärmedizin (BgVV), Berlin, BgVV-Heft 3, 1995.

INTERLABORATORY TRIALS OF THE EPR METHOD FOR THE DETECTION OF IRRADIATED SPICES, NUTSHELL AND EGGSHELL

M. F. Desrosiers, D. M. Yaczko, A. Basi and W. L. McLaughlin

Ionizing Radiation Division, Physics Laboratory
National Institute of Standards and Technology
Technology Administration, U.S. Department of Commerce
Gaithersburg, MD 20899
U.S.A.

1 INTRODUCTION

The Electron Paramagnetic Resonance (EPR) spectrometry detection method for irradiated foods is based on the measurement of free radicals produced in the rigid matrices of foods (*e.g.* bones, shell, seeds). For certain foods, these free radicals are stable over part or all of the shelf-life of the food. As part of the International Atomic Energy Agency coordinated research programme ADMIT (Analytical Detection Methods for Irradiation Treatment of Foods), two trials of an EPR method for spices and nutshell were undertaken. In the second of these trials eggshell was included for testing.

This work describes the two ADMIT trials and identifies some essential elements of a protocol for implementation of the EPR method.

2 EXPERIMENTAL DESIGN

The foods to be tested were prepared at the National Institute of Standards and Technology (NIST), and irradiated with a ^{60}Co source (Gammacell 220, Nordion International Inc., $D_{H20} \simeq 15$ kGy h^{-1}, 23°C, calibrated with alanine dosimetry). Food samples were purchased locally and irradiated in groups of individually packaged aliquots. A Bruker ECS106 EPR spectrometer equipped with a TMH microwave resonator was used for the NIST measurements.

2.1 Sample Preparation

Black pepper and paprika were obtained in ground form. Pistachio nuts were removed from their shells. The shells were ground in a high-speed mill and passed through a 0.5 mm sieve. Fresh whole eggs (for Trial II only) were purchased from a local market. The egg was cracked and the yolk and white discarded along with the thin

---

*The mention of commercial products throughout this paper does not imply recommendation or endorsement by the National Institute of Standards and Technology, nor does it imply that the products identified are necessarily the best available for the purpose.

membrane attached to the inside of the shell. The shells were allowed to air-dry overnight, then ground in a high-speed mill and sieved to <0.5 mm.

2.2 Trial Design

In both trials, 6 samples of each type of spice/shell were sent to each laboratory. For a given sample type, 2 of the 6 were labelled (irradiated and non-irradiated) samples intended for use as reference samples by the testing laboratory. The remaining 4 were coded samples, of which at least one, and no more than 2, were non-irradiated samples. For the coded samples a range of doses were selected (Table 1).

2.3 Protocols

A detailed protocol was provided to the trial participants for the analysis of eggshell, black pepper, paprika, and pistachio nut shell in Trials I and II. The protocols are nearly identical to the final version *(vide infra)* and will only be summarised here. The protocols contained the following information: sample tube loading procedures; sample tube positioning within the microwave resonator; centre magnetic field calculation; EPR parameter settings; EPR signal assessment criteria; and examples of EPR spectra (Figures 1-4).

2.4 Test Response

A response form for recording the results of the test was provided to each participant. Each form had an area to identify the participant followed by a response matrix with columns to enter the: sample type; sample code; result of determination (I = irradiated, N = non-irradiated); special comments. For some samples the participant felt the determination was ambiguous. In this instance, the participant choose neither I or N and responded in an appropriate manner with a question mark or special comment. These entries were recorded as uncertain to distinguish a recognized ambiguity from a misidentification.

Table 1 *Absorbed Dose Levels in kGy for Spice / Shell Samples*

Sample Type		Absorbed Dose	
		Trial I	Trial II
Pistachio nut shell	reference	10	10
	unknowns	1, 5	1, 3, 5, 7
Black pepper	reference	10	10
	unknowns	5, 10	3, 5, 7, 10
Paprika	reference	10	10
	unknowns	5, 10	2, 5, 7, 10
Egg shell	reference	not tested	1.8
	unknowns	not tested	0.3, 0.9, 1.8

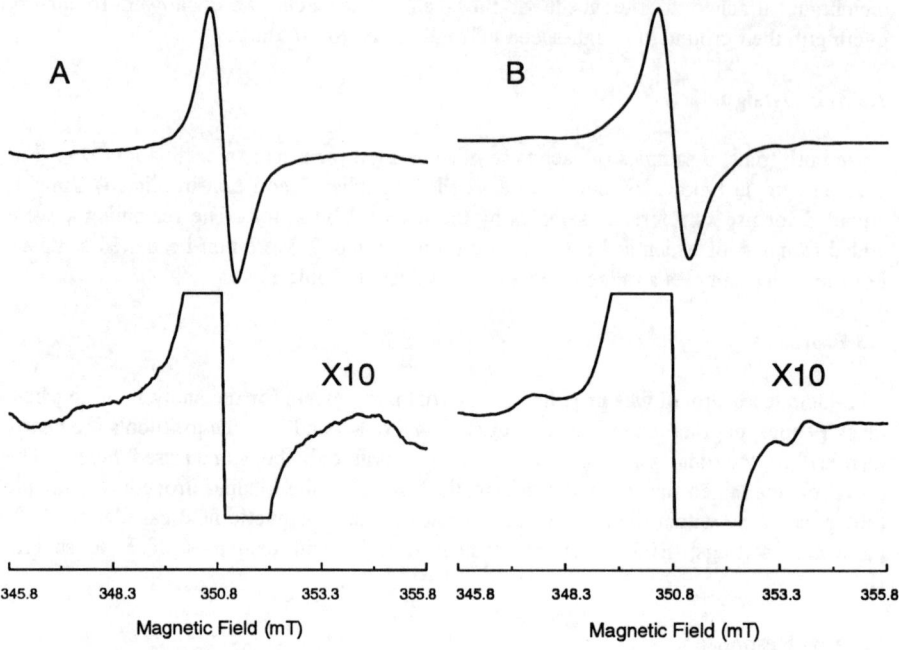

Figure 1 *EPR spectra of black pepper unirradiated (A) and irradiated to 10 kGy (B)*

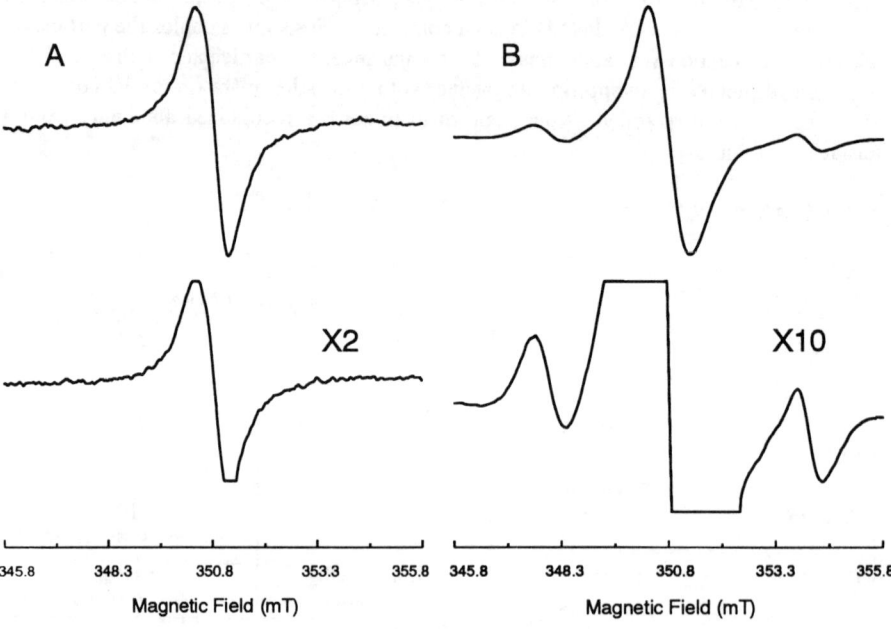

Figure 2 *EPR spectra of paprika unirradiated (A) and irradiated to 10 kGy (B)*

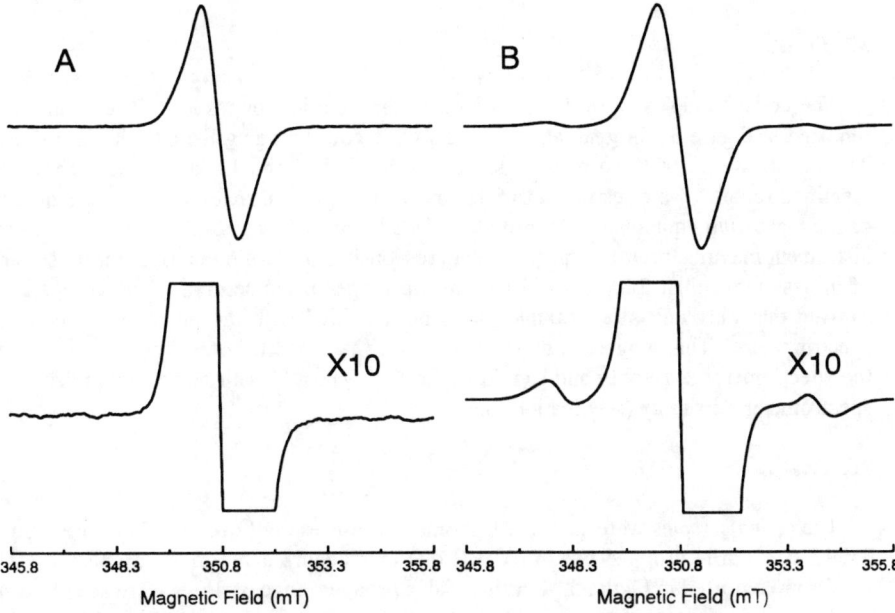

Figure 3 *EPR spectra of pistachio nut shell unirradiated (A) and irradiated to 10 kGy (B)*

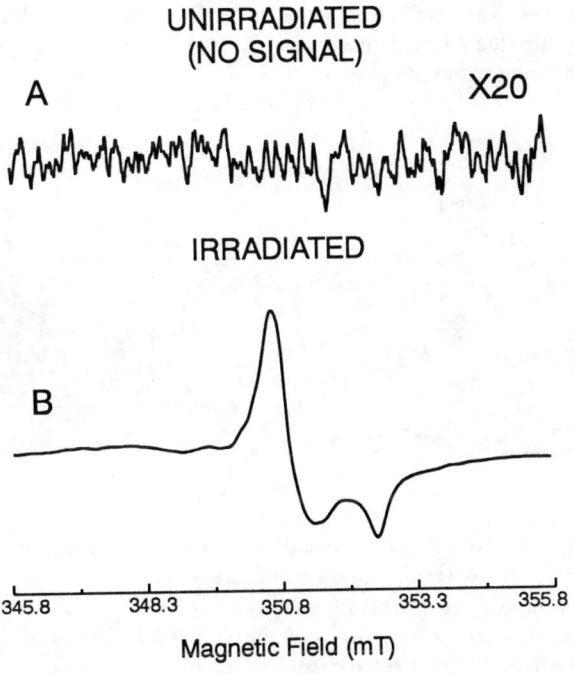

Figure 4 *EPR spectra of eggshell unirradiated (A) and irradiated to 1.8 kGy (B)*

3 RESULTS

3.1 Trial I

The coded samples were distributed to 12 laboratories for testing. The results are reported in Figure 5. In general, the results were good, ranging from 88-100% correct. The results were 100% correct for samples irradiated to the 10 kGy dose. The misidentified samples were restricted to 3 laboratories. One explanation may be the quality of the measuring equipment. After polling the laboratories, the results were grouped by instrument manufacturer: the spectrometers responsible for 5 of 6 incorrect responses are of a class more than 20 years old and are no longer manufactured. The comparison showed that only one of 96 samples were misidentified with the other, more modern, spectrometers. This suggests a dependence not on the spectrometer type *per se*, but on the spectrometer sensitivity and should be further evaluated with direct comparisons of spectrometer sensitivity (see Section 4).

3.2 Trial II

The coded samples were sent to 10 laboratories for testing (2 of the laboratories from Trial I voluntarily dropped out of Trial II due to recognized sensitivity problems). The results are reported in Figure 5. For this trial, a broader range of doses were used to test the sensitivity of the method. For black pepper, misidentification occurred only for the lower dose samples and in the non-irradiated samples. For paprika, the identification failures increased as the dose decreased, and there was one misidentification of an non-irradiated sample. The results were much better for pistachio nut shell; identification was difficult only for the longest times (see Section 5) and lowest dose. No incorrect identifications were reported for eggshell.

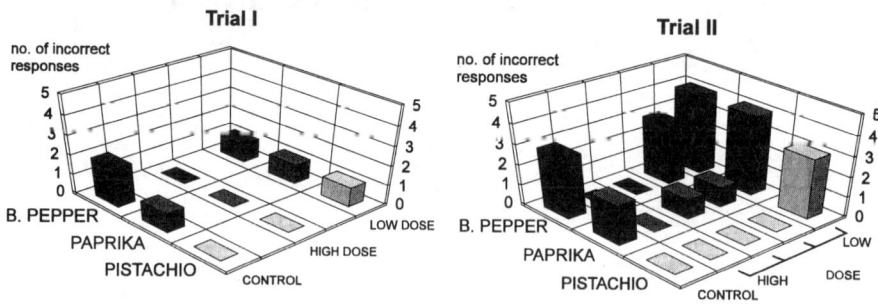

Figure 5 *Bar chart showing the number of incorrect responses for Trial I and Trial II. Here, an uncertain response is counted as incorrect for purposes of clarity. For actual doses in each category see Table 1. The sample size for each type/dose is: 15-17 (A) and for (B), 14-15 (control), 6 (low dose), 7-8 (medium doses), 5 (high dose).*

4 SUMMARY

A graphical summary of the data for both trials are combined in Figures 6 and 7. The results demonstrate that for black pepper and paprika, the application of the EPR method for absorbed doses below 10 kGy would be limited. For spices processed at 10 kGy, detection should be possible if measured within 2-3 months after irradiation. The EPR method is more sensitive for pistachio shell; as no misidentifications were made down to the 3 kGy range. These results are significantly better than those obtained in the only other blind trial (non-ADMIT) carried out on pistachio nut shell.[1] The eggshell trial was 100% correct at the doses used here. This result was anticipated since eggshell has been used to measure much lower doses over very long time periods.[2]

5 CONCLUSIONS

It is foreseen that the EPR method for black pepper would be best used for rapid screening, or as a secondary check in combination with another method. Due to the presence of Mn^{2+} ion in some samples (Figure 8), the radiation-induced signal is sometimes difficult to detect. Furthermore, the Mn^{2+} ion spectral features can be misinterpreted as a false positive by the uninitiated investigator. EPR detection of irradiated black pepper may be reserved for experienced technicians.

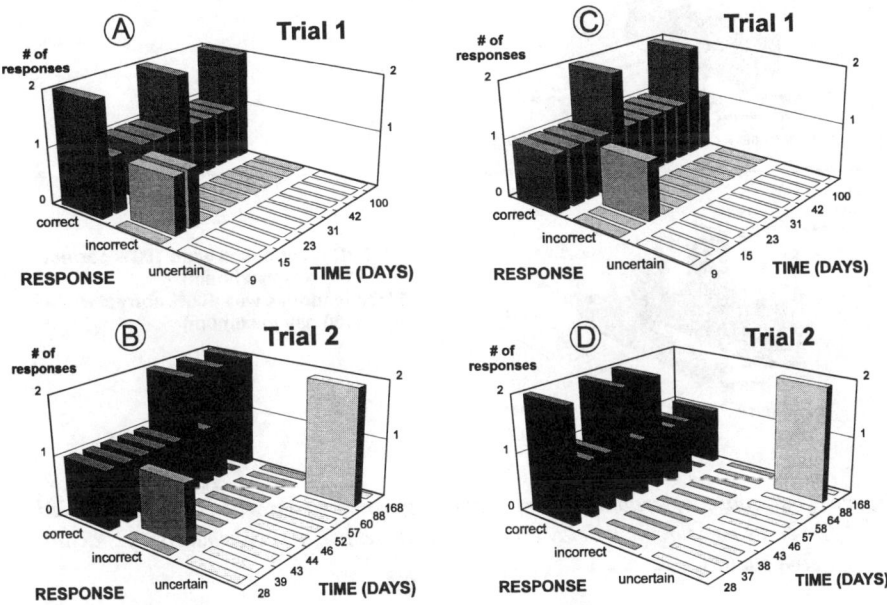

Figure 6 *Bar charts of black pepper (A and B) and paprika (C and D) non-irradiated sample data*

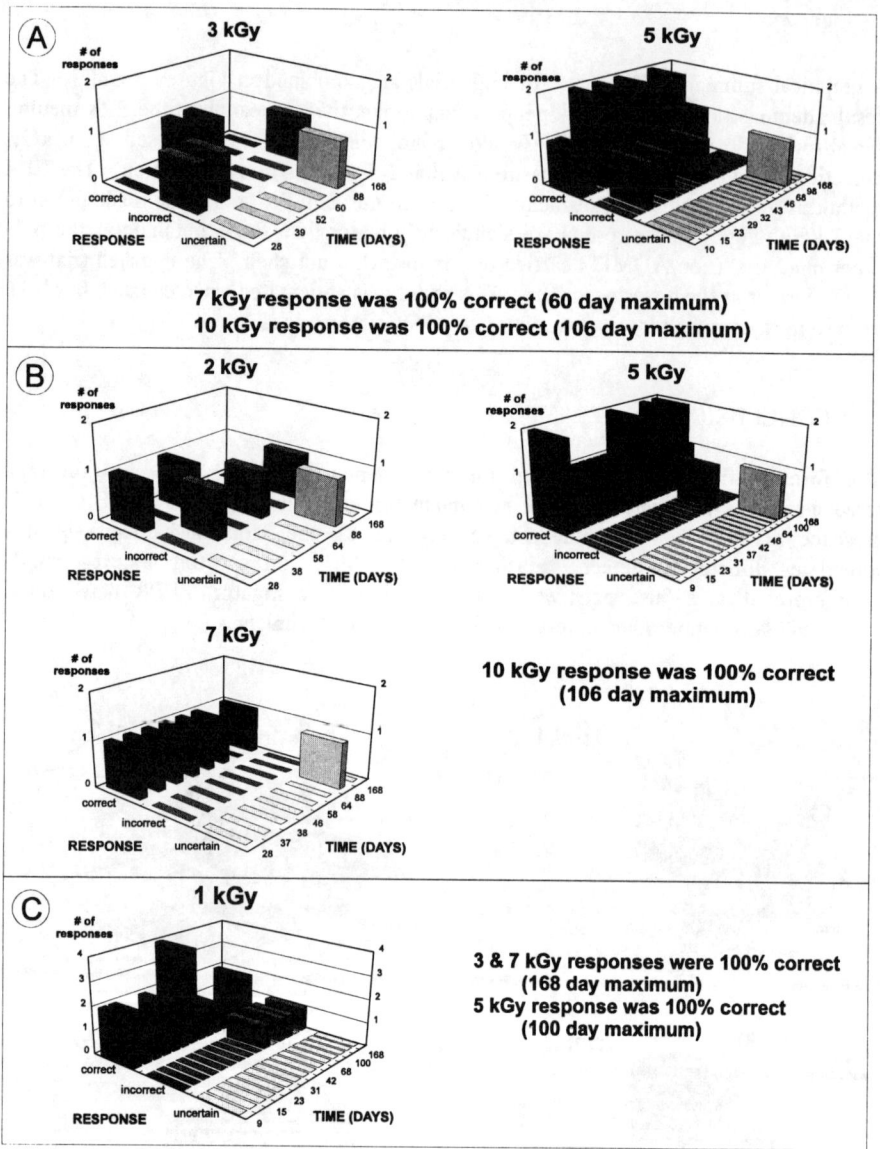

Figure 7 *Bar charts of irradiated sample data separated by absorbed dose and plotted versus measurement delay time for: black pepper (A); paprika (B); pistachio nut shell (C)*

For paprika and pistachio nut shell, no interfering signals (*e.g.* Mn^{2+}) were observed which resulted in improved test reliability. The false positives reported for paprika are attributed to the overall low signal strength of the spectral features. There were no false positives in the relatively intense pistachio spectrum. EPR detection of irradiated eggshell is very straightforward and the probability for successful application in testing laboratories is high.

Application of a second evaluation criterion for black pepper, paprika, and pistachio nut shell based on radiation-induced broadening of the centre resonance was invoked as an attempt to rectify incorrect responses, but had marginal success. Practical application of this criterion to distinguish irradiated foodstuffs would be very limited. The observed line width changes are so small (0.09 to 0.17 mT) that this approach requires that either control samples from the same lot be available, or that extensive studies be undertaken to fully characterise controls from different lots.

The EPR method for detection of irradiated black pepper, paprika, and pistachio nut shell is most limited by the time of measurement after irradiation. It is recommended that systematic studies be done and statistically evaluated to define the useful period after irradiation that the EPR method can be applied.

Finally, there are two general recommendations. Firstly, since results can vary depending on the EPR spectrometer, recommended settings should be regarded as suggested values from which the investigator can begin an acquisition and further adjust to the spectrometer in use. Secondly, there is a need for an EPR spectrometer standard (e.g. weak pitch) to determine if the spectrometer being used has the sufficient sensitivity to use the method, and an agreed standard practice for the application of the sensitivity check.

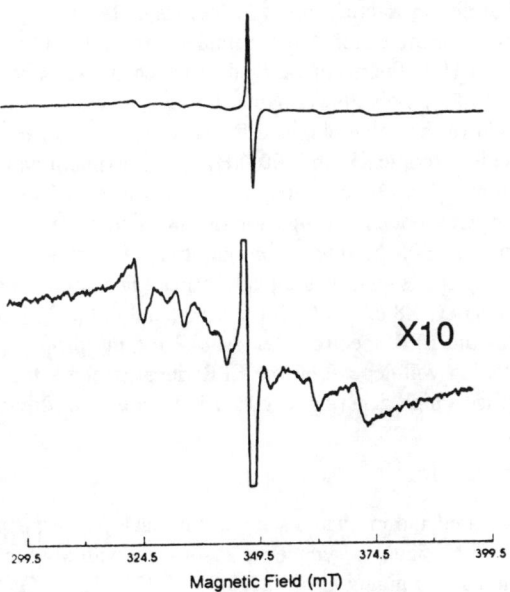

Figure 8 *Wide magnetic field swept EPR spectrum of non-irradiated black pepper showing the 6 line Mn^{2+} spectrum flanking the centre resonance*

6 (DRAFT) PROTOCOL FOR EPR DETECTION OF IRRADIATED SPICES

6.1 Sample Preparation

The samples should be ground and sieved (< 1 mm). Use an amount that will occupy 2-3 cm of a 3-4 mm diameter quartz EPR tube.

6.2 Measurement

The following protocol assumes a complete working knowledge of the EPR spectrometer used according to the particular manufacturer's specifications. It is written for an X-band spectrometer operating at ambient temperature.

Insert the sample tube into the microwave cavity ensuring that the tube is centred in the cavity. If the sample mass fills less than 3 cm of the tube, the best result will come from a sample in the absolute centre (x,y,z directions) of the microwave cavity. Tune the microwave bridge according to the manufacturer's suggested procedure. Using the microwave field value for the tuned bridge, calculate the proper magnetic field strength at which the field sweep should be centred from the relation:

$$H_r = 714.5(v/g)$$

The spectroscopic splitting factor, g, is a constant that is characteristic of the paramagnetic centre measured (for these samples, use 2.004), v is the microwave frequency in GHz, and H_r is the magnetic field at which the resonance condition is met (in gauss, G[10 G = 1 mT]). Solve the equation for H_r.

Set the centre field to the value obtained for H_r with a field sweep range of ± 10 mT (100 G); a modulation frequency of 100 kHz (or maximum value); a modulation amplitude of 0.5-0.6 mT (5-6 G); and a microwave power of 0.5 mW. The modulation amplitude and microwave power settings for the maximum signal intensity should be optimized for the specific EPR hardware configuration. Use a time constant and sweep time appropriate for an EPR signal with a peak-to-peak line width of approximately: 0.7 mT (7 G) for black pepper; 0.8 mT (8 G) for paprika; and 0.9 mT (9 G) for pistachio nut shell. Please consult the EPR spectrometer manual for the proper time constant. The spectrometer gain setting will depend on the EPR signal strength for each sample. It is advisable to initially use a high gain (*e.g.* 5×10^5) and scale up or down as needed.

6.3 Assessment

Both the non-irradiated and irradiated samples for black pepper, paprika and pistachio nut shell will have an EPR signal; a symmetric absorption with a g-factor of about 2.004 with a peak-to-peak width ranging from 0.7-0.9 mT (7-9 G.). The g = 2.004 signal generally increases and slightly broadens upon irradiation. However, the most distinctive markers for ionising radiation are two relatively weak EPR signals approximately 3.0 mT (30 G) on either side of the centre peak (g = 2.004).

Adjust the gain such that the centre signal is at or near full scale (minimum and maximum of the display); record the spectrum. Increase the gain by a factor of 10 (or maximum attainable) and record the spectrum again; the g = 2.004 EPR signal should now be off-scale and the radiation-induced satellite peaks, if present, should be more visible. The non-irradiated sample should only have a single resonance at g = 2.004 (no satellite peaks). Please note, however, that the black pepper control does not have a flat baseline on either side of the g = 2.004 resonance. This is due to the presence of the paramagnetic ion Mn^{2+}.

If necessary, increase the gain further. For pistachio nut shell in particular, it may be necessary to increase the gain 50 or 100 times since the centre resonance is very intense relative to the satellite peaks.

7 (DRAFT) PROTOCOL FOR EPR DETECTION OF IRRADIATED EGGSHELL

7.1 Sample Preparation

Yolk, white and inner membrane (attached to shell) should be discarded. The shell should be air dried at ambient temperature, ground and sieved (<1 mm). Use an amount that will occupy 2-3 cm of a 3-4 mm diameter quartz EPR tube.

7.2 Measurement

The following protocol assumes a complete working knowledge of the EPR spectrometer used according to the particular manufacturer's specifications. It is written for an X-band spectrometer operating at ambient temperature.

Insert the sample tube into the microwave cavity ensuring that the tube is centred in the cavity. If the sample mass fills less than 3 cm of the tube, the best result will come from a sample in the absolute centre (x,y,z directions) of the microwave cavity. Tune the microwave bridge according to the manufacturer's suggested procedure. Using the microwave field value for the tuned bridge, calculate the proper magnetic field strength at which the field sweep should be centred from the relation:

$$H_r = 714.5(v/g)$$

The spectroscopic splitting factor, g, is a constant that is characteristic of the paramagnetic centre measured (for these samples, use 2.002), v is the microwave frequency in GHz, and H_r is the magnetic field at which the resonance condition is met (in gauss, G[10 G=1 mT]). Solve the equation for H_r.

Set the centre field to the value obtained for H_r with a field sweep range of ±5 mT (50 G); a modulation frequency of 100 kHz (or maximum value); a modulation amplitude of 0.10 mT (1.0 G); and a microwave power of 10 mW. Use a time constant and sweep time appropriate for an EPR signal with a peak-to-peak line width of approximately 4 G.

Please consult the EPR spectrometer manual for the proper time constant. The spectrometer gain setting will depend on the EPR signal strength for each sample. It is advisable to use a relatively high gain (5 x 10^4) and scale up or down as needed.

7.3 Assessment

There will be no detectable EPR signal for the non-irradiated eggshell samples. The spectrum of the irradiated samples will be dominated by a major resonance at $g = 2.002$ ($\Delta G_{p-p} = 4$ G) and a minor resonance up-field at $g = 1.997$.

Acknowledgments

The authors acknowledge the sponsorship of the Joint Food and Agriculture Organization/International Atomic Energy Agency under the leadership of the ADMIT secretary, L. Ladomery. The authors wish to thank the following participants and their staff who co-operated in these trials: N. J. F. Dodd, Patterson Institute for Cancer Research, Manchester, UK; E. Marchioni, AERIAL, Strasbourg, France; C. H. McMurray, Department of Agriculture for Northern Ireland, Belfast, UK; K. M. Morehouse, U.S. FDA, Washington, DC, USA; P. Fattibene, Istituto Superiore di Sanita, Rome, Italy; J. R. Pilbrow, Monash University, Melbourne, Australia; J. J. Raffi, CEN, Cadarache, France; R. Roberts, Institute of Geological and Nuclear Sciences, Lower Hutt, New Zealand; G. A. Schreiber, BGA, Berlin, Germany; W. Stachowicz, Institute of Nuclear Chemistry and Technology, Warsaw, Poland; S. Uchiyama, National Institute of Hygienic Sciences, Tokyo, Japan.

References

1 J. Raffi, M. H. Stevenson, M. Kent, J. M. Thiery, and J.-J. Belliardo, *Int. J. Food Sci. Technol.*, 1992, **27**, 111.
2 M. F. Desrosiers, F. G. Le, P. M. Harewood, E. S. Josephson and M. Montesalvo, *J. Agric. Food Chem.*, 1993, **41**, 1471.

Physical Methods
Thermoluminescence

THERMOLUMINESCENCE AND PHOTOSTIMULATED LUMINESCENCE TECHNIQUES TO IDENTIFY IRRADIATED FOODS

G. A. Schreiber

BgVV - Federal Institute for Health Protection of Consumers and Veterinary Medicine
FG 212 - Food Irradiation Laboratory
Postfach 33 00 13
D-14191 Berlin
Germany

Since the publication of the first report about increased thermoluminescence (TL) of irradiated spices, 10 years have passed. At that time, this effect was observed when spices were heated in a TL reader. Meanwhile, the light sources within the foods have been identified and TL methods applied to various foods for identification of irradiation treatment. The methods have been approved by the Ministry of Agriculture, Fisheries and Food (MAFF) of the United Kingdom and included in the Official Collection of Methods according to article 35 of the German Foods Act (LMBG). A draft European Standard has been formulated for approval by the member states of the European Committee for Standardization (CEN) and most importantly, the method is already routinely used by food control authorities to identify irradiation treatment of foods. The strength of TL is its radiation specificity. It counts among the most reliable methods and there is potential for becoming the most sensitive physical method for detection of irradiated foods. Research is still going on to speed up performance, to find new applications and to decrease detection limits. Most recently, a new luminescence technique - photostimulated luminescence - has been introduced which can be rapidly performed and which seems to have the potential to be applied to a similar broad range of foods as TL.

By heating whole spices and herbs in a TL reader, it could be shown that luminescence increases after irradiation and, therefore, this technique was proposed for identification of irradiation treatment of spices.[1] However, examination of a broad range of different spices and herbs has revealed large intersample variation of luminescence intensities.[2] Thus, the irradiation treatment of some samples could be clearly detected whereas in others, this was impossible.[3,4] The reason for these variations has been revealed by Sanderson et al.[5,6] By separation of organic components and mineral contaminants of spices and herbs, it was shown that minerals emitted radiation-induced TL whereas the organic components produced non-specific signals. Therefore, the TL intensity of whole samples depends largely on the degree of mineral contamination. However, even minerals isolated from irradiated products exhibit different TL intensities since various kinds of minerals (quartz, feldspar, *etc.*) emit intensities after irradiation which may differ by several log counts.[7] Therefore, for a clear identification of irradiation, TL intensities have to be normalised and this is achieved by re-irradiation of the already glowed samples with a dose of at least 1 kGy.[5] The TL ratio of first (1st) glow intensity and second (2nd) glow intensity is approximately 1 if the sample has been irradiated prior to examination at or above doses of 1 kGy.

In solid dielectric material, energy is stored during irradiation as trapped charge carriers. By thermal stimulation, excess energy can be released as luminescence emission. In TL, isolated silicate minerals are heated under controlled conditions which give rise to measurable glow curves. Depending on the 'depth of the trap', excess energy is released as light emission at a certain temperature. The stability of the excited electron increases with the 'depth of the trap' or with the temperature at which light is released. Vice versa, fading increases with decreasing release temperature. Therefore, TL which is induced by natural radiation can only be measured at very high temperatures (above ca. 300°C) since at lower temperatures, the induction rate is smaller than the fading rate. TL induced at high doses of artificial radiation, on the other hand, is also released at intermediate (about 200 - 300°C) and low temperatures (< about 200°C). Therefore, the differences of TL intensities between non-irradiated and irradiated samples is most pronounced within the low and intermediate temperature ranges.

Instead of thermal stimulation for release of trapped energy, light can be used as a stimulus. This process is known as optically stimulated or photostimulated luminescence (OSL or PSL). Infrared light is most efficient for stimulation. Emissions from irradiated foods show Anti-Stokes behaviour since their wavelengths are shorter than the wavelengths used for excitation, whereas non-irradiated samples are unable to participate in these transitions.[8,9]

In interlaboratory studies, including blind trials, it was established that the TL mineral method is suitable for clear identification of irradiated spices and herbs even after prolonged storage.[4,10] Already at an early stage it was observed that increased TL could also be measured on surfaces of irradiated fruit.[11] After the source of light had been identified, it was concluded that irradiation treatment of all foods contaminated by silicate minerals might be identified by TL analysis. Following isolation of minerals from strawberries and mushrooms, clearly separate TL ratios from non-irradiated and irradiated samples could be achieved.[12] The TL mineral technique can even be applied to identify irradiated shrimps and prawns by extracting minerals from the intestines of animals. From mussels, minerals can be obtained by rinsing with water.[13-16] Recent studies indicate that even the very low doses used for inhibition of sprouting of potatoes or onions can be detected by TL.[17] A fast screening method has been proposed using shells of seafood instead of isolating minerals.[18] Using this technique, increased luminescence levels could be measured on all shells examined even after a 1 - 3 month storage period.[19]

Spices, herbs and shellfish were also examined by the new PSL technique. Besides isolated minerals, whole products were examined without any sample preparation. Although many irradiated samples could be identified by the whole sample technique, minerals have to be isolated if the food item is contaminated by low amounts of minerals. For fast screening applications in food control laboratories, a low-cost PSL instrument has been designed which can be used for whole product measurements.[9]

References

1. L. Heide and K.W. Bögl, "Die Messung der Thermolumineszenz - Ein neues Verfahren zur Identifizierung strahlenbehandelter Gewurze," BGA, Berlin, ISH-Heft 58, 1984.
2. L. Heide, G. A. Schreiber and K.W. Bögl, "Umfassende Zusammenstellung von Daten zur Identifizierung strahlenbehandelter Gewurze und Trockenlebensmittel mittels Chemi- und Thermolumineszenz-Analyse am Gesamtprodukt," BGA, Berlin, SozEp-Heft 2, 1992.
3. L. Heide, R. Guggenberger and K. W. Bögl, *Radiat. Phys. Chem.*, 1989, **34**, 903.
4. G. A. Schreiber, N. Helle and K. W. Bögl, *JAOAC*, 1995, **78**, 88.
5. D. C. W. Sanderson, C. Slater and K. J. Cairns, *Radiat. Phys. Chem.*, 1989, **34**, 915.
6. D. C. W. Sanderson, C. Slater and K. J. Cairns, *Nature*, 1989, **340**, 23.
7. T. Autio and S. Pinnioja, *Z. Lebens. Unters. Forsch.*, 1990, **191**, 177.
8. D. C. W. Sanderson, "Potential New Methods of Detection of Irradiated Food," J. J. Raffi and J.-J. Belliardo, Eds., Commission of the European Communities (BCR), Brussels, Luxembourg, EUR 13331 EN, 1991, p. 159.
9. D. C. W. Sanderson, L. A. Carmichael, S. Ni Riain, J. Naylor and J. Q. Spencer, *Food Sci. Technol. Today*, 1994, **8**, 93.
10. D. C. W. Sanderson, G. A. Schreiber and L. A. Carmichael, "A European Interlaboratory Trial of TL detection of Irradiated Herbs and Spices," SURRC report to the Commission of the European Communities (BCR), Brussels, Luxembourg, 1994.
11. L. Heide, R. Guggenberger and K. W. Bögl, *J. Agric. Food Chem.*, 1990, **38**, 2160.
12. U. Wagner, M. Jakob, A. Leffke, N. Helle and K. W. Bögl, G. A. Schreiber, "Recent Advances on the Detection of Irradiated Food," M. Leonardi, J. J. Raffi and J.-J. Belliardo, Eds., Commission of the European Communities (BCR), Brussels, Luxembourg, EUR 14315 EN, 1993, p. 152.
13. T. Autio and S. Pinnioja, "Recent Advances on the Detection of Irradiated Food," M. Leonardi, J. J. Raffi and J.-J. Belliardo, Eds., Commission of the European Communities (BCR), Brussels, Luxembourg, EUR 14315 EN, 1993, p. 177.
14. S. Pinnioja, T. Autio, E. Niemi and O. Pensala, *Z. Lebens. Unters. Forsch.*, 1993, **191**, 111.
15. S. Pinnioja, *Radiat. Phys. Chem.*, 1993, **42**, 397.
16. G. A. Schreiber, A. Hoffmann, N. Helle and K. W. Bögl, *Radiat. Phys. Chem.*, 1994, **43**, 533.
17. G. A. Schreiber, B. Ziegelmann, G. Quitzsch, N. Helle and K. W. Bögl, *Food Struct.*, 1993, **12**, 385.
18. J.-P. Agnel, M. S. Dutraive, I. Vaux, I. Rustan and J. J. Raffi, "Recent Advances on the Detection of Irradiated Food," M. Leonardi, J. J. Raffi and J.-J. Belliardo, Eds., Commission of the European Communities (BCR), Brussels, Luxembourg, EUR 14315 EN, 1993, p. 186.
19. L. A. Carmichael, D. C. W. Sanderson and S. Ni Riain, *Radiat. Meas.*, 1994, **23**, 455.

RECENT ADVANCES IN THERMOLUMINESCENCE AND PHOTOSTIMULATED LUMINESCENCE DETECTION METHODS FOR IRRADIATED FOODS

D. C. W. Sanderson, L. A. Carmichael and J. D. Naylor

Scottish Universities Research and Reactor Centre (SURRC)
East Kilbride
Glasgow G75 0QU
U.K.

1 INTRODUCTION

Thermoluminescence (TL) and photostimulated luminescence (PSL) are radiation-specific phenomena resulting from energy storage by trapped charge carriers in dielectric materials following irradiation. Releasing such stored energy by thermal or optical stimulation can result in detectable luminescence emission during the relaxation processes which follow. These approaches can be applied to inorganic components present either as inherent parts of foods or as adhering contaminants, and to bio-inorganic systems. The strengths of these techniques lies in their radiation-specificity, and the wide range of sample types which may be analysed. The Scottish Universities Research and Reactor Centre (SURRC) has been involved in the development and application of luminescence methods since 1986, during which time over 4000 analyses of more than 800 different food samples have been performed for research purposes, or in support of UK food labelling regulations.

This paper discusses the present scope of luminescence techniques, and identifies areas where recent work has extended the range of applications, and indicates areas where further investigations may be worthwhile.

2 TL ANALYSES USING EXTRACTED MINERALS

TL was first applied to detecting irradiated foods using an approach whereby whole samples of herbs and spices were heated directly for TL measurement.[1] However, although such measurements are simple to conduct, this approach is limited in scope by the nature of the signals, which originate in minor mineral contamination,[2-4] and the damage which is caused to the organic material in the sample by heating. Reproducibility of whole sample measurements from aliquot to aliquot is poor, due in part to the sub-sampling variability of the minor (0.01-5%) mineral contents in aliquots of a few mg in mass. The organic matter itself produces a poor and variable thermal contact between TL reader and the luminescence phase, and is damaged by the measurement. This results in a raised instrumental background due to light released by pyrolysis and auto-oxidation, and prevents single aliquot calibration to measure the luminescence sensitivity. Whole sample measurements in uncalibrated (screening) mode are thus prone to false negative results from low-sensitivity samples, and although performance can be improved by measuring a paired aliquot irradiated to a known dose before TL readout[5] the limitations of this approach are increasingly being appreciated.

The use of pure mineral extracts for analysis has the advantage of presenting the sample for measurement in a more appropriate manner. Thermal contact between sample and equipment is improved, resulting in better formed TL glow curves. Instrumental background levels are not artificially raised by spurious, non-TL, signals from the organic matter. The samples can be re-irradiated for calibration purposes, to allow sensitivity variations to be taken into account. These features are necessary to overcome the inherent sensitivity variations of natural polymineral, silicate assemblages, although of course the analytical demands are greater.

The original SURRC development of mineral extraction and measurement procedures, was based on over 200 different samples of herbs, spices and seasonings irradiated using both ^{60}Co and electron-beam sources. The essential features were rapidly confirmed overseas,[6,7] and validated by interlaboratory trials conducted under the European Commission (BCR) concerted action[8] and subsequently by the German Federal Health Office (BGA).[9] The full SURRC procedure was adopted in the UK in 1992[10] as the first Ministry of Agriculture, Fisheries and Food (MAFF) validated technique for the detection of irradiated foods. A similar approach has also been applied in Finland[6] in support of import regulations, and in Germany for routine food control purposes.[9] An international standard is currently being prepared by the European Standardization Committee (CEN) to cover the use of silicate methodologies for analysis of irradiated foods, and it is therefore appropriate to examine the methodology and the present range of sample types and dose ranges which can be detected reliably.

2.1 Analysis of Herbs, Spices and Seasonings

Under UK food labelling regulations[11] all irradiated foods, or food ingredients, must be clearly labelled at all points in the supply chain, including warehouses or catering outlets. This has resulted in a steady demand for routine analyses in support of labelling regulations, providing an opportunity to assess the performance of the validated method under practical conditions. Some 333 samples of herbs, spices and dried products have been submitted to SURRC by food suppliers and retailers for TL analysis since 1992, and analysed using the full validated MAFF protocol.

The procedure involves extraction of the silicates from two parallel aliquots of each sample, using wet sieving to pre-concentrate the minerals, and density separation in sodium polytungstate solution to remove organic components. The minerals are deposited on 0.25 mm thick stainless steel discs, after hydrochloric acid treatment to remove carbonates, using an acetone suspension which is dried at 50°C overnight before measurement. TL glow curves are recorded immediately after preparation (First Glow:G1) and following a standard calibrating radiation dose of 1 kGy delivered with a ^{60}Co source at a dose rate of 1-2 kGy h^{-1} (Second Glow:G2). The sample preparation, handling and TL measurements are all conducted under safelight conditions to minimise optical bleaching of TL. TL curves are recorded at 6°C s^{-1} from room temperature to 400°C. TL glow ratios (G1/G2) are evaluated from 220-240°C and used to form the basis of judgements of the outcome of the analysis.

Quality assurance steps include the positive verification that all glassware and sample discs are free from luminescence materials, and measuring full process blanks with every batch of samples. The process blank levels are used to define minimum detectable levels (MDL) for the analysis, in this case as the mean value plus 3 standard deviations using long term data. Both upper and lower acceptance limits are defined for TL sensitivity (measured from G2) to prompt re-analysis of samples with signal strength above the saturation limits of the TL reader, and of samples with insufficient sensitivity to permit irradiated and non-irradiated samples to be discriminated with confidence. The TL glow ratios (G1/G2) are used, together with glow shape evidence and a check for concordance between the paired analyses to judge whether the sample is irradiated. Two critical values are used in assessing G1/G2. Samples below an investigation level of 0.1, on both aliquots are considered non-irradiated, while those with G1/G2 >1 are considered irradiated. Intermediate values are investigated in detail with critical attention to glow curve shape and are usually remeasured. Table 1 shows the quality assurance parameters achieved in 1993. Considerable efforts were required over a number of years to achieve a consistently low process blank with a standard deviation less than 100 counts. Figure 1 demonstrates the importance of a low blank by comparison with the second glow (G2) of samples analysed in the same period.

Samples with sensitivity less than 10 times MDL are considered too low to make a reliable assessment of glow ratio (G1/G2) with respect to the investigation limit of 0.1, while those with sensitivity greater than the upper limit of linearity are reviewed to assess whether the glow ratios depend critically on photon counting saturation, and are re-analysed if necessary. The sensitivity range spans 5-6 orders of magnitude due to the inherent variations in mineralogy, TL sensitivity per unit mass, and mineral yields. In the face of such variations it is essential to adopt good laboratory procedures to control cross contamination, and to incorporate performance checks on every stage of the preparation and handling process.

The importance of concordance testing as a laboratory quality assurance step was recognised at an early stage, and has been adopted at SURRC principally to check for handling problems. However, it has recently become apparent that this test is more important than that. Figures 2 and 3 show the G1 versus G2 plots and glow ratio histograms of concordant results from commercially submitted samples analysed from 1993 to 1994 showing the distribution of results from samples identified as irradiated and

Table 1 *SURRC TL Quality Control Parameters During 1993*

	Mean Integral Counts: 250-260°C
Instrument Blank	-1.5 ± 5
Glassware Blank	57 ± 300
Full Process Blank	49 ± 257
Minimum Detection Level (MDL)	820
Upper Limit of Linearity (ULL)	7.5×10^5
Upper Saturation Limit (USL)	1.5×10^7

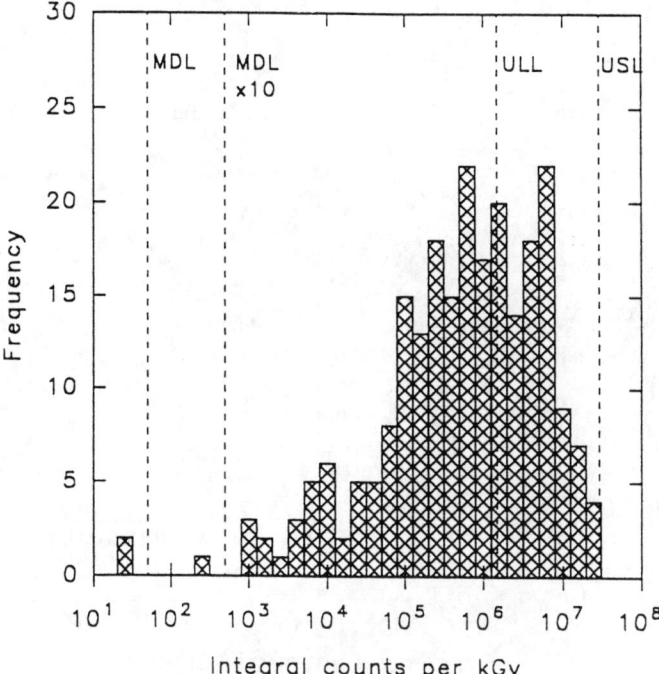

Figure 1 *Second glow sensitivity histogram for TL samples examined in 1993*

non-irradiated using standard criteria. Again the range of sensitivities and G1 signals spans many orders of magnitude. The need for calibration to measure sensitivity is self evident from Figure 2, which clearly shows the correlations between the G1 signal and TL sensitivity for both irradiated and non-irradiated samples. Attempts to use G1 intensity alone would result in rather poor performance for samples analysed under market conditions where the majority of specimens are non-irradiated.

It is notable that concordant samples fall into the same two distinct distributions originally defined by experiments with paired irradiated and non-irradiated samples.[3,5,10] Also, the variability of glow ratios from non-irradiated samples is greater than that of irradiated samples, as was originally observed. Of the 333 samples analysed between 1992 and 1994, some 23 have been positively identified as irradiated, satisfying all the criteria listed above. However, the concordance test was not satisfied by some 33 pairs of samples, a rather greater proportion than expected on the basis of experience under controlled conditions. Figure 4 shows glow ratios from discordant samples

In each case the samples were examined in detail, and in the majority of cases repeat analyses were performed. Discounting the few data sets where discordance could be associated with specific analytical problems, and resolved by repeat analysis, it was concluded that many of these samples contained a mixture of non-irradiated and irradiated components. In some instances the materials examined were themselves blends of separate ingredients, or had been handled in the same production as known irradiated materials. However, there were other cases, *e.g.* black pepper, where the consolidation of

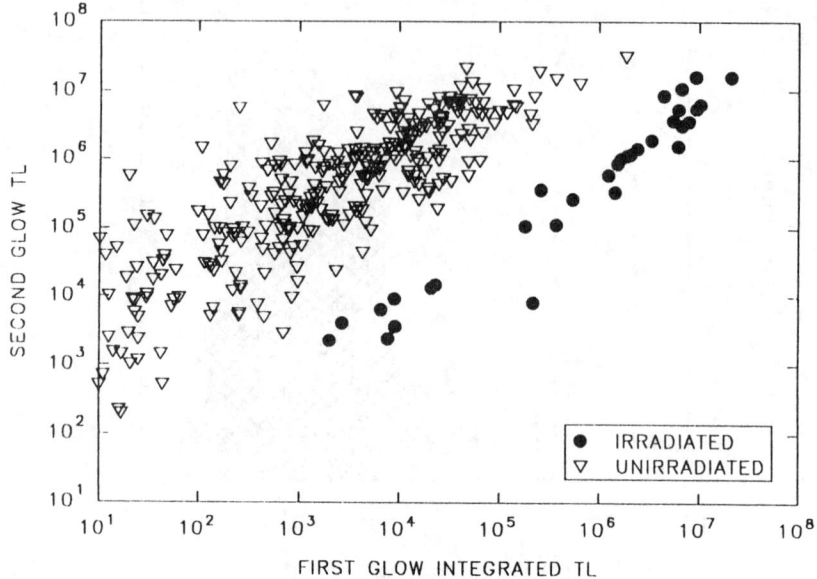

Figure 2 *Glow 1 versus Glow 2 plot for concordant samples analysed from 1993-94*

large batches of material from separate sources takes place commonly, and it is possible that small proportions of irradiated material are introduced at this stage. Finally, it must also be remembered that the present prejudice against consumption of labelled irradiated foods provides a commercial incentive for blending to take place. The recognition that blended products are in circulation presents a considerable analytical challenge, which might be met using imaging luminescence techniques. It is certainly necessary to conduct replicated analyses to identify discordant specimens, and for a representative set of samples to be analysed if large consolidated batches of material are being examined.

The results have confirmed the effectiveness of the TL method for herbs and spices in practice, and the necessity of applying the quality control measures described above. Simplification of the procedures would unfortunately lead to a loss of reliability. Another aspect of methodology which still attracts debate is the use of laboratory safelights. It has been pointed out[9] that both the BCR and BGA interlaboratory trials produced qualitatively acceptable results from laboratories which did not use safelights, from which it has been inferred that this is not necessary. A short experiment was conducted taking paired discs from 30 samples of foods, irradiating them to 5 kGy and comparing the effects of dark and light storage for 24 h. Figure 5 shows the results of these tests, expressed as glow ratios. The samples stored in the dark show glow ratios generally >1, with a similar distribution to the irradiated samples in Figure 3.

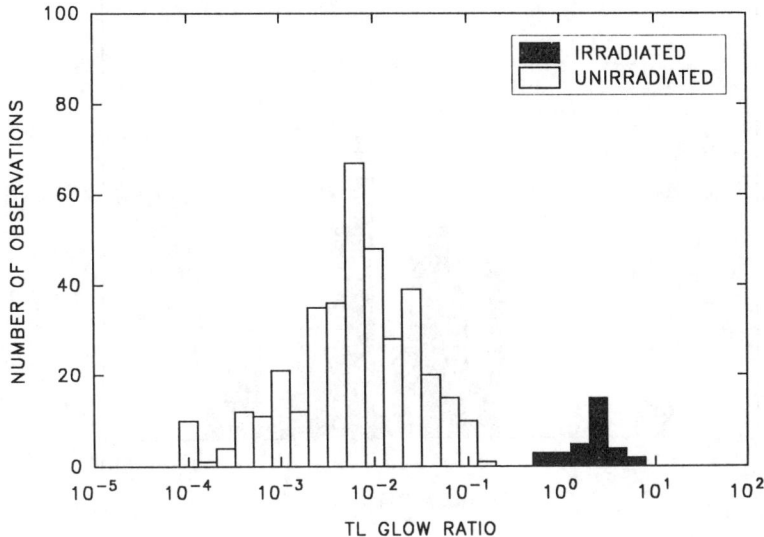

Figure 3 *Glow ratio histogram for concordant samples examined from 1993-94*

By contrast samples exposed to light show glow ratios generally between 0.5 and 1, with some examples down to 0.16. Clearly, the exposure of prepared samples to light will reduce the differences between irradiated and non-irradiated samples, and will adversely effect minimum detectable doses.

2.2 TL Detection of Irradiated Fruits and Vegetables

A similar approach can be applied to fruits and vegetables, irradiated to doses between 100 Gy and 2-3 kGy. A MAFF funded study of 44 varieties conducted from 1990-92[12] confirmed that TL measurements could distinguish between all irradiated and non-irradiated samples under controlled conditions providing that full laboratory quality assurance criteria were satisfied. However, it was recognised that the minerals are more vulnerable to optical bleaching than those from herbs and spices. Systematic investigations of the effects of optical exposure on residual TL signals were therefore undertaken using a set of 134 mango samples, exposed to energy densities of up to 100 J cm^{-2} in two well characterised light boxes. This again showed clearly that prolonged exposure to light prior to analysis reduces the magnitude of TL signals from irradiated fruits and vegetables. The rate of reduction is greatest at the start of bleaching, and rapidly reduces until glow ratios are almost invariant with further exposure.

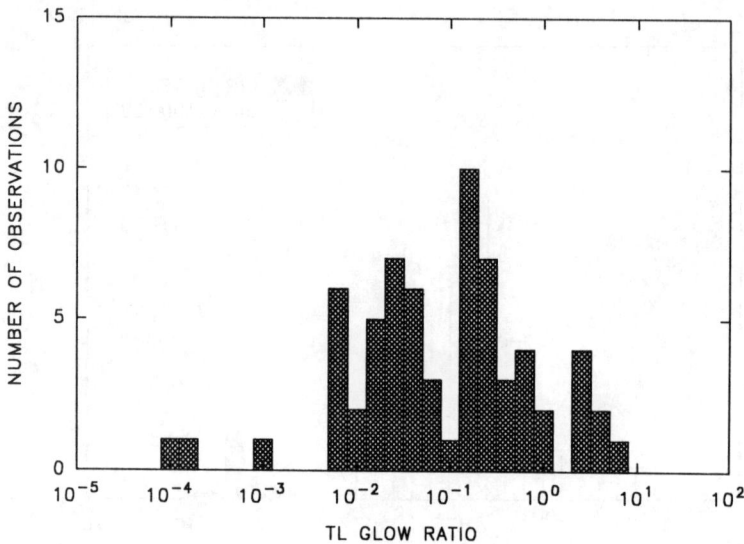

Figure 4 *TL glow ratios from discordant samples analysed from 1993-94*

Therefore, optical bleaching will not remove all detectable TL from an irradiated fruit. However, glow ratios from irradiated samples could be reduced to levels of 0.3 or lower under practical conditions. When the possibility of exposed surfaces acquiring additional non-irradiated grains through post-irradiation handling is considered, it is unclear whether the compounded effects will result in false negative analyses. Although positive TL results are a reliable indicator of irradiation for these food classes, analytical procedures must take account of optical bleaching to minimise the possibility of misclassifying low sensitivity irradiated samples.

A brief interlaboratory examination of 5 varieties of fruits and vegetables in Germany[13] produced a disappointingly low detection rate (approximately 50%), based on classification criteria for herbs and spices. Further international trials are needed to determine detection rates for TL analysis of fruits and vegetables, using methodologies which take the additional effects of optical bleaching, and consequently reduced TL signals, into account. The detection rates for low dose materials would be greatly enhanced by further studies aimed at reducing the residual background from non-irradiated samples, and by imaging luminescence detection.

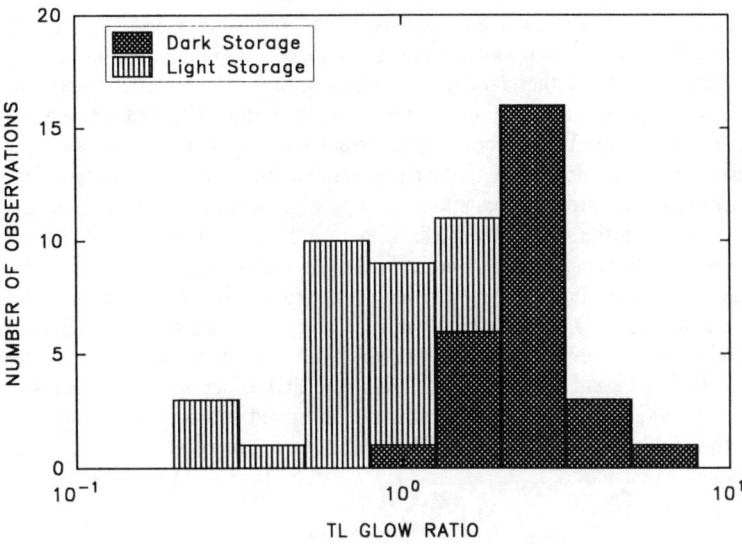

Figure 5 *Glow ratios from irradiated samples stored under dark and light conditions*

3 TL AND PSL ANALYSIS OF SHELLFISH

Since 1992 MAFF funded research at SURRC has focused on the extension of luminescence methods to shellfish. Two complementary approaches have been identified, based on intestinal grits and shells. Many Crustaceans contain inorganic intestinal grits, which can be used for TL analysis following standard procedures, providing that sufficient quantities of suitable minerals can be recovered.[14-16] Physical recovery of grits for TL analysis is laborious, and work has been undertaken to explore the use of tissue solubilisation, and direct PSL measurements to simplify analysis. Acid hydrolysis methods have been applied successfully to release trapped silicates, and may provide a useful complimentary approach to physical separation techniques. Perhaps more importantly it has been possible to stimulate PSL, from trapped grains, through the membranes of dissected intestinal material, and even in some cases through whole shrimps. This avoids the need for laborious sample preparation in analysis of shellfish. Shells themselves, where available, provide a bio-inorganic material, comprising organic membranes infiltrated with minerals, which can also be used for luminescence detection. Preliminary studies in France demonstrated that TL from calcite shells could be detected shortly after irradiation from well mineralised Mediterranean species.[17] This work has been followed up at SURRC using a range of 9 types of shellfish, including warm water species of immediate regulatory interest, and poorly mineralised examples. TL measurement conditions have been examined in detail.[18] It is necessary to limit the measurement temperature to 250°C to minimise problems with elevated backgrounds induced by thermal damage to the organic phases. The main TL emission from calcite shells has been shown to occur in a band associated with Mn^{2+} (550-600 nm), which, by analogy with published studies of geological calcites, suggests defect agglomeration in the biogenic materials. Replacement of the standard TL filters with KG3 or BG39 filters increases sensitivity by two orders of magnitude. Under these conditions TL can be detected from all irradiated samples examined, shortly after irradiation. Extensive stability tests have been undertaken to examine the extent of post-irradiation fading. In most cases signal losses are less than one order of magnitude over storage times of 1-3 months at -20°C or 5°C, which go beyond the normal shelf-life of these products (Figure 2). However, there are still some uncertainties as to why individual species show apparently different fading behaviour, which require further consideration. It is known from geological literature that fully mineralised calcite has higher temperature, stable, TL signals originating from deep traps. Luminescence stimulation methods which avoid heating the sample, based on photo-transfer (PTTL) and direct photostimulation (PSL) to access stable signals without elevated backgrounds thus offer opportunities for improved detection conditions. Results from TL and PSL studies of irradiated shellfish are presented elsewhere in these proceedings.[19]

4 PULSED INFRA RED PSL

Finally, a new method for rapid detection of irradiated foods has been developed at SURRC based on pulsed photostimulated luminescence (PSL). PSL uses light rather than heat to stimulate the release of trapped charge carriers, and is thus able to examine inorganic systems in the presence of organic matter.[20,21,22] Since the sample is not damaged by this, non-contact measurement it is possible to calibrate the result by re-irradiation of a single aliquot. SURRC has developed two research spectrometers for studying PSL excitation characteristics and for time domain studies. Preliminary work using both instruments confirmed the viability of direct detection of silicate containing food without the need for sample preparation, and has provided important information about excitation mechanisms in feldspars, micas, and biogenic calcite. Either infra-red (IR) stimulation or certain visible bands can be used to detect irradiated silicates; the IR band being associated with a thermally assisted transition from a trap which is meta-stable over quaternary timescales. The IR transitions are thus sufficiently stable to store signals over the shelf-life of irradiated food, but are insufficiently stable to accumulate geologically induced background signals equivalent to kGy doses. The research spectrometer was able to demonstrate rapid detection of the most sensitive irradiated herbs and spices without mineral separation, and also to verify the radiation specificity of the Anti-Stokes luminescence resulting from IR stimulation. However, it was not sufficiently sensitive to detect signals from unseparated materials of intermediate or low sensitivity foods. A prototype pulsed IR system was assembled at SURRC in 1991 to increase sensitivity, and this was able to distinguish approximately 80% of irradiated herb and spice samples using 1 cm diameter specimens. However, again, the least sensitive samples were still not detected, despite adoption of computer based signal recovery procedures based on multichannel scaling of single photon counts.

The data from these experiments, however, were used to design a new low-cost instrument based on digital lock-in photon counting, and pulsed IR stimulation. This instrument is sufficiently sensitive to allow direct measurements of herbs and spices, for screening purposes, without the need for sample preparation or re-irradiation. Samples are introduced in disposable Petri dishes, with no other preparation, and the instrument produces a qualitative screening measurement in 15 seconds, either through a front panel display, or via a computer interface. The lock-in system automatically suppresses photo-multiplier dark signals, improving the signal to background ratio by some two orders of magnitude compared with earlier IR systems. Figure 6 shows a typical signal, recorded from irradiated and non-irradiated parsley presented in a 50 mm diameter Petri dish. The accumulated photon count from the instrument rapidly outpaces the instrumental background, allowing discrimination within a few seconds between irradiated and non-irradiated specimens. The instrument is capable of detecting significant signals from pure IAEA F1 feldspar samples irradiated with doses down to a few mGy, and therefore will respond to kGy materials present in minute quantities. Figure 7 shows the distribution of results obtained from duplicate analyses of 45 varieties of herbs, spices and seasonings deliberately selected to span the full range of sensitivities observed in TL analysis.

The irradiated samples show the wide range of sensitivities expected on the basis of TL experience (>5 orders of magnitude), however, even the least sensitive sample examined produces approximately 10^3 counts, which is more than one order of magnitude higher than the instrumental baseline of $\sim \pm 100$ counts. The majority of non-irradiated samples produce signals below 10^3 counts, however, a small proportion show greater

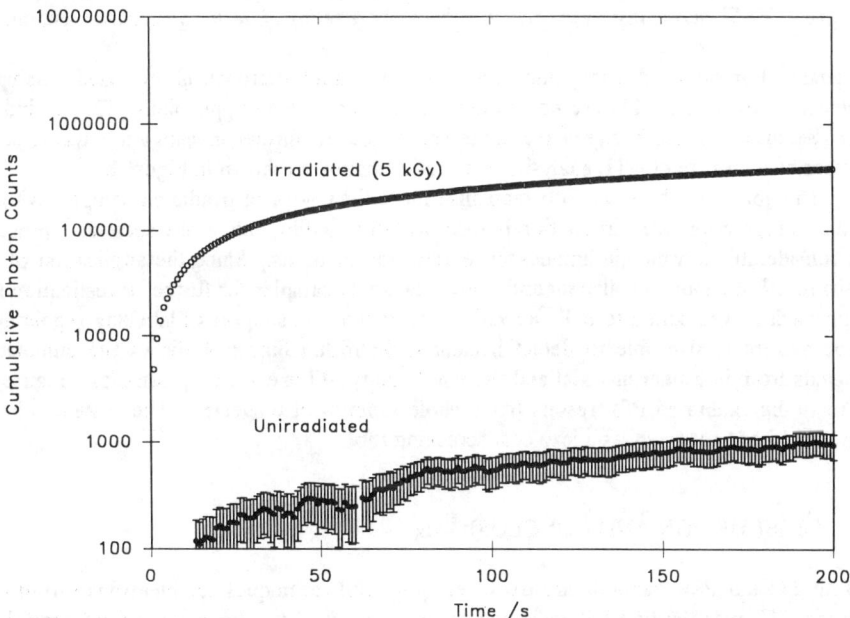

Figure 6 *Pulsed PSL signals from irradiated and non-irradiated parsley* quantities.

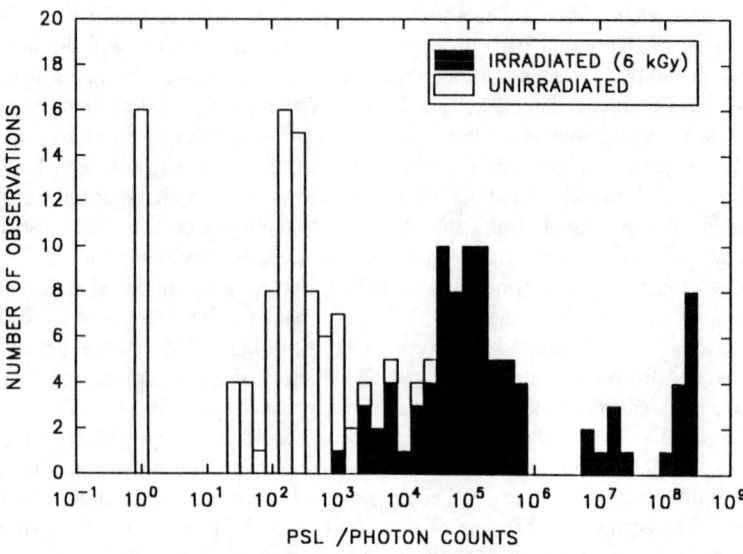

Figure 7 *PSL screening measurements from 45 varieties of herbs, spices and seasonings*

signals. For rapid screening purposes a two threshold approach is proposed. Samples below a lower limit (T1) are non-irradiated; those above an upper limit (T2) are judged irradiated, and those between the limits are subject to further investigation which could either be conducted by TL analysis, or by re-irradiation as shown in Figure 8.

The ability of the system to recognise more than 90% of irradiated samples without either preparation or re-irradiation in measurements lasting only a few seconds represents a considerable advance in luminescence detection methods. Since the application of the two threshold approach also identifies low sensitivity samples for further investigation this approach is well suited to bulk screening programmes in support of labelling regulations. The system is also able to detect irradiated shellfish. Figure 9 shows the cumulative signals from intestinal material and the whole body of brown shrimp samples. Figure 10 shows the calibrated PSL results from whole samples of 6 species. The system is being produced commercially as a low-cost screening tool.

5 DISCUSSION AND CONCLUSIONS

Both TL and PSL methods are extremely powerful techniques for identifying irradiated foods. TL was the first UK validated method, and is in use for routine food analysis in the UK and in several other countries. Practical experience has confirmed the need for

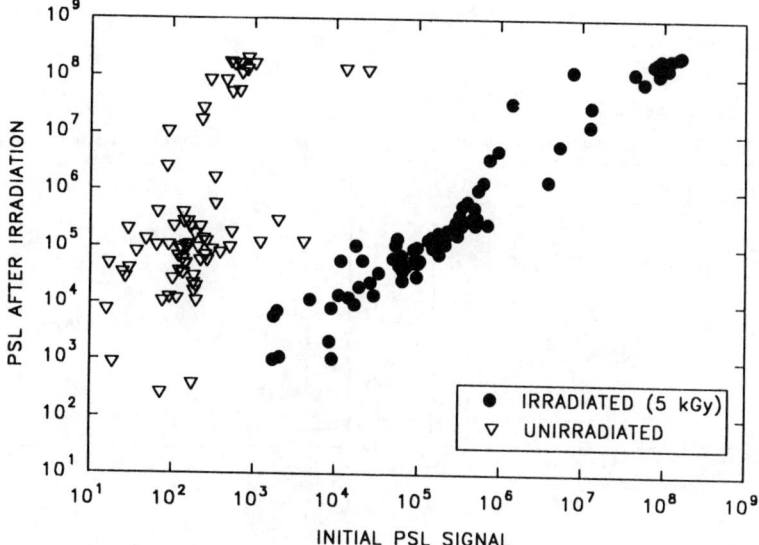

Figure 8 *Calibrated PSL results for the samples of Figure 7*

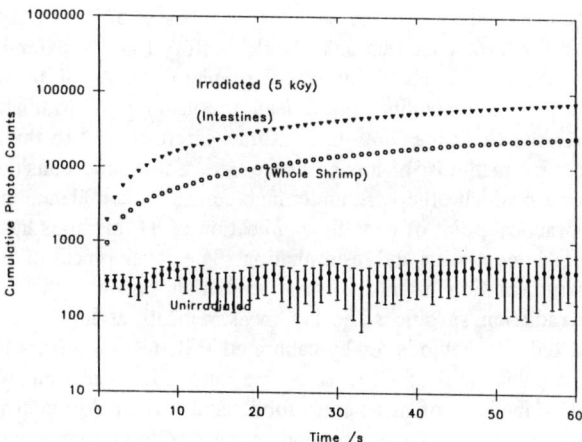

Figure 9 *Pulsed PSL signals from irradiated and non-irradiated brown shrimp*

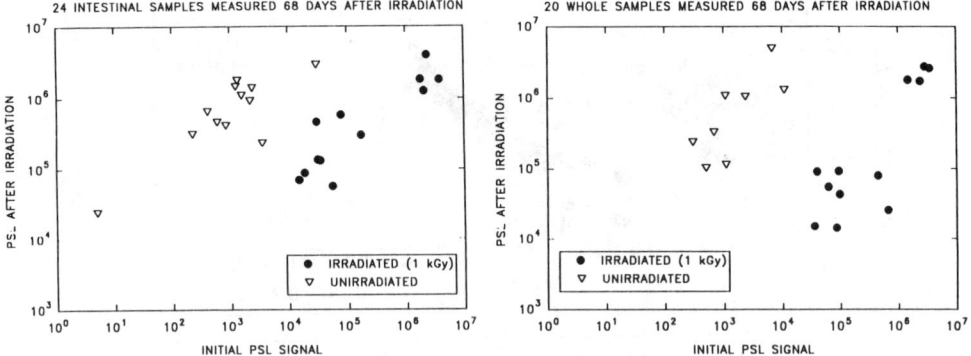

Figure 10 *Calibrated PSL results from 6 species of shellfish irradiated 68 days before measurement*

full laboratory quality assurance procedures; however, it is gratifying to note that the method holds up well to commercial conditions when these are used. The range of sample varieties which can be analysed is wide, and is continually being extended. Further work is needed to understand the underlying nature of variation in residual levels from non-irradiated samples with a view to increasing power for low dose materials, and for detection of highly bleached or blended samples, such as some fruits and vegetables. There is also a need for further international interlaboratory trials to extend the range of validated food classes, and to clarify outstanding minor details of procedure. The observation that commercial conditions can lead to blending of irradiated and non-irradiated foods presents challenges both to consumer interests, and to the analyst. The development of imaging methods for luminescence detection would considerably aid the detection of such cases, which otherwise undermine consumer confidence in the meaning of labels. From a practical point of view the application of TL methods has been limited by the need for sample preparation and re-irradiation. The development of high sensitivity PSL screening instrumentation offers industrial and public analysts the opportunity of conducting highly radiation specific screening measurements at low cost, and can be supported by validated TL methods, or by calibrated PSL measurements in a reference laboratory. The combination of PSL screening and TL validation will hopefully encourage accurate labelling of irradiated food, and eventually contribute to the acceptance of radiation technology when properly applied to food processing.

References

1. L. Heide and W. Bögl, "Die Messung der Thermoluminesenz - Ein Neues Verfahren zur Identifizierun strahlenbehandler," Institute fur Strahlenhygiene, Neuherberg, ISH-Heft 58, 1984.
2. D. C. W Sanderson, C. Slater and K. J. Cairns, *Nature*, 1989, **340**, 23.
3. D. C. W Sanderson, C. Slater and K. J. Cairns, *Rad. Phys. Chem.*, 1989, **34**, 915.
4. D. C. W Sanderson, C. Slater and K. J. Cairns, *Int. J. Rad. Biol.*, 1989, **55**, 5.
5. D. C. W. Sanderson, "Food Irradiation and the Chemist," D. E. Johnston and M. H. Stevenson, Eds., Royal Society of Chemistry, Cambridge, 1990, p. 25.
6. T. Autio, S. Pinnioja, *Z. Lebens. Unters. Forsch.*, 1990, **191**, 177.
7. H. Y. Goksu, D. F. Regulla, B. Heitel and G. Popp, *Rad. Protect. Dosim.*, 1990, **34**, 319.
8. D. C. W Sanderson, G. A. Schreiber and L. A. Carmichael, "A European Interlaboratory Trial of TL Detection of Irradiated Herbs and Spices. Phase 1," SURRC report to the Commission of the European Communities (BCR), Brussels, Luxembourg, 1991.
9. G. A. Schreiber, U. Wagner, A. Leffke, N. Helle, J. Ammon, H. V. Buchholtz, H. Delincée, S. Estendorfer, H. U. von Grabowski, W. Kruspe, K. Mainczyk, H. Munz, C. Schleich, N. Vreden, C. Weizorek and K.W Bögl, "Thermoluminescence Analysis to Detect Irradiated Spices, Herbs and Spice-and-Herb Mixtures - An Intercomparison Study," BGA, Berlin, SozEp-Heft 2, 1993.
10. MAFF, "Thermoluminescence detection of Irradiated Herbs and Spices," MAFF Validated Methods for Analysis of Feedstuffs, No. V27, MAFF Food Science Laboratory, Norwich, UK, 1992.
11. Anon, The Food Labelling (Amendment) (Irradiated Food) Regulations 1990, No. 2489, Her Majesty's Stationery Office, London, 1990.
12. D. C. W. Sanderson, L. A. Carmichael, P. A. Clark and R. J. Clark, "Development of Luminescence Tests to Identify Irradiated Foods," Final Report N1701: Identification of Irradiated Fruits and Vegetables, MAFF N1701, 1992.
13. G. A. Schreiber, U. Wagner, N. Helle, J. Ammon, H. V. Buchholtz, H. Delincée, S. Estendorfer, H. U. von Grabowski, W. Kruspe, K. Mainczyk, H. Munz, C. Schleich, N. Vreden, C. Weizorek and K. W. Bögl, "Thermoluminescence analysis to detect irradiated Fruit and Vegetables - An Intercomparison Study," BGA, Berlin, SozEp-Heft 3, 1993.
14. T. Autio and S. Pinnioja, "Recent Advances on Detection of Irradiated Food," M. Leonardi, J. J. Raffi and J.-J. Belliardo, Eds., Commission of the European Communities (BCR), Brussels, Luxembourg, EUR 14315 EN, 1990, p. 177.
15. S. Pinnioja, T. Autio, E. Niemi and O. Pensala, *Z. Lebens. Unters. Forsch.*, 1993, **196**, 111.
16. S. Pinnioja, *Rad. Phys. Chem.*, 1991, **42**, 397.
17. J.-P. Agnel, M. S. Dutraive, I. Vaux, I. Rustan, J. Raffi, "Recent Advances on Detection of Irradiated Food", M. Leonardi, J. Raffi and J.-J. Belliardo, Eds., Commission of the European Communities (BCR), Brussels, Luxembourg, EUR 14315 EN, 1992, p. 186.
18. L. A. Carmichael, D. C. W. Sanderson and S. Ni Riain, *Radiat. Meas.*, 1994, **23**, 455.
19. D. C. W. Sanderson, L. A. Carmichael, J. D. Naylor, These proceedings.

20. D. C. W. Sanderson, "Potential New Methods of Detection of Irradiated Food," J. J. Raffi and J.-J. Belliardo, Eds., Commission of the European Communities (BCR), Brussels, Luxembourg, EUR 13331 EN, 1991, p. 159.
21. R. J. Clark and D. C. W. Sanderson, *Radiat. Meas.*, 1994, **23**, 641.
22. D. C. W. Sanderson and R. J. Clark, *Radiat. Meas.*, 1994, **23**, 633.

LUMINESCENCE DETECTION OF SHELLFISH

D. C. W. Sanderson, L. A. Carmichael, J. Q. Spencer and J. D. Naylor

Scottish Universities Research and Reactor Centre (SURRC)
East Kilbride, Glasgow
G75 0QU
U.K.

1 INTRODUCTION

The Scottish Universities Research and Reactor Centre (SURRC) has been active in the development and application of luminescence techniques in the detection of irradiated foods,[1-7] in support of UK legislation. Thermoluminescence (TL), photostimulated luminescence (PSL) and photo-transfer luminescence (PTTL) are radiation-specific phenomena which arise due to energy stored by trapped charge carriers following irradiation. The energy released following stimulation is accompanied by detectable luminescence. The TL method involves preparation of pure silicate extracts from the sample and subsequent TL analysis, whereas PSL uses stimulation by electromagnetic radiation (visible, or near visible wavelengths) thus avoiding heating the sample.

The research has recently focused on the use of luminescence as a detection method for irradiated shellfish.[8] Two approaches have been identified for TL, based on intestinal grits and shells.[9-11] Crustaceans contain many inorganic intestinal grits which are suitable for TL analysis, however, recovery of sufficient quantities of the material can prove problematic and the recovery is an extremely laborious task. To simplify and/or compliment this physical method, SURRC have developed an acid hydrolysis method to release trapped silicates for TL measurements. The shells provide a bio-inorganic material comprising of organic membranes infiltrated with minerals, and therefore, can also be used for TL detection.[12-16] However, there are limitations due to the presence of organic material which is damaged by heat, producing interfering light emission, and darkening the sample thus precluding calibration. This is a serious drawback when extending TL methods from herbs, spices, fruits and vegetables to bio-inorganics.

An alternative to TL is the use of PSL techniques[17,18] and PTTL techniques, where the energy to release trapped charge carriers is provided optically. Direct PSL measurements record Anti-Stokes luminescence emitted during and following stimulation, where the output wavelengths are shorter than the stimulating wavelengths. This is highly radiation-specific since energy conservation principles demand that the quantum energy differences between stimulation and luminescence are balanced by energy stored in the form of trapped charge carriers. Non-irradiated samples, therefore, are unable to participate in these transitions. PTTL uses a process of transfer of charge carriers from a deep trap, associated with high temperature TL emissions, to a shallow trap which is capable of producing a low temperature TL peak. This approach is useful in (i) limiting the temperatures of TL heating of bio-inorganic systems to avoid damage to the organic components, and (ii) providing a means of detecting luminescence at wavelengths which are close to the emission band wavelengths. Cryogenic PTTL has been used here to investigate the excitation spectra of biogenic calcite to identify promising stimulation and detection schemes for PSL.[19]

2 TL OF SHELLS

A series of experiments were conducted on a suite of irradiated and non-irradiated powdered shells in order to establish whether sufficient TL signals could be detected. The background signals were found to be highly variable, and that the results on re-heating were not always lower than the first glow. Comparison of these results with instrumental backgrounds clearly showed that the elevated signal originated from the sample. These elevated backgrounds are associated with the release of trapped moisture, pyrolysis, auto-oxidation of organic material, and the use of freeze-drying. Restricting glow temperatures (to 250°C), substantially reduced these spurious signals.

Further experiments were conducted to determine the filter or filter combination most suitable to the emission spectrum of shells to optimise signal intensity. This clearly showed that the major TL signal was emitted in the 500-600 nm region and by using a KG3 filter, signal levels were increased by two orders of magnitude compared with the blue filter system originally used.[8] On this basis the KG3 filter, or a broad band filter (*e.g.* BG18, BG39) is required for TL analysis of calcite shells.

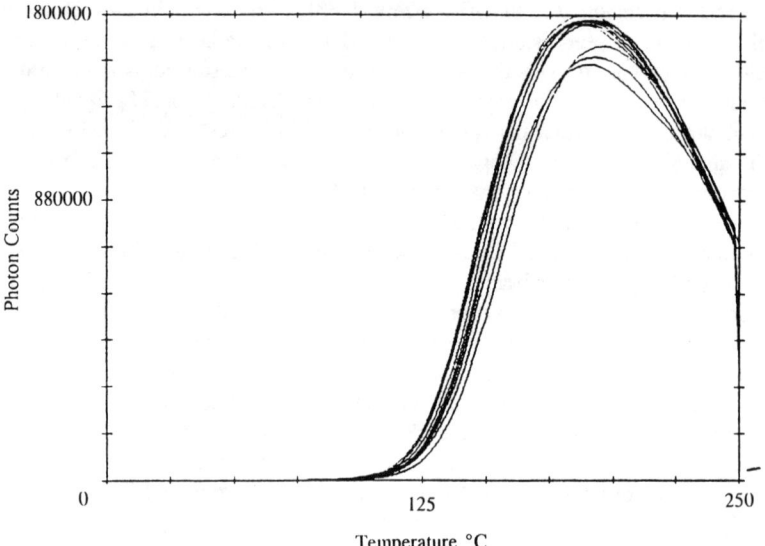

Figure 1 *TL glow curves of powdered calcite shell from king scallop irradiated with a 5 kGy dose*

Direct TL measurements were then carried out on lobster, warm water shrimps, black tiger prawns, *Nephrops norvegicus* (Norway lobster), king and queen scallops, mussels, crabs and oysters using these optimised instrumental conditions. The samples were dried, powdered and sieved through 250 mm mesh. The samples were freeze-dried and split into two aliquots, one of which was irradiated to a 5 kGy dose in a ^{60}Co source, and the other of which was retained as a control. Ten discs of each pair of samples were prepared and the glow curves were recorded using the KG3 filter system and heating from room temperature to 250°C. Freshly irradiated samples of all species are readily distinguishable from the non-irradiated controls. Examples of glow curves from king scallop are shown in Figure 1; other examples can be found elsewhere.[8]

Having established that TL from freshly irradiated powdered shells can be detected, signal stability becomes an important consideration. It is not possible to make any detailed prediction of stability characteristics of an unknown material and, even with known trap parameters from kinetic analyses, there is a need to supplement models with empirical results. The TL glow curves comprise a pseudo trap-depth spectrum and in general the thermal lifetime increases with glow temperature, and depends strongly on storage temperatures. Observations are required over the full potential lifetime and storage conditions that the foodstuffs may be exposed to. Aliquots of each shellfish species were stored at refrigerator and freezer temperatures. The samples were measured directly after irradiation and at intervals over a period of 3 months. Results for the 230-250°C temperature region in the glow curve for both storage temperatures are shown in Figures 2 and 3.

The results show that TL from all the 130-150°C, 150-170°C, 170-190°C, 190-210°C and 230-250°C temperature regions are subject to post irradiation fading, by up to one order of magnitude, over storage times of 1-3 months at -20°C or 5°C. For well mineralised species (lobster, mussel, oyster and scallop) these losses are unlikely to lead

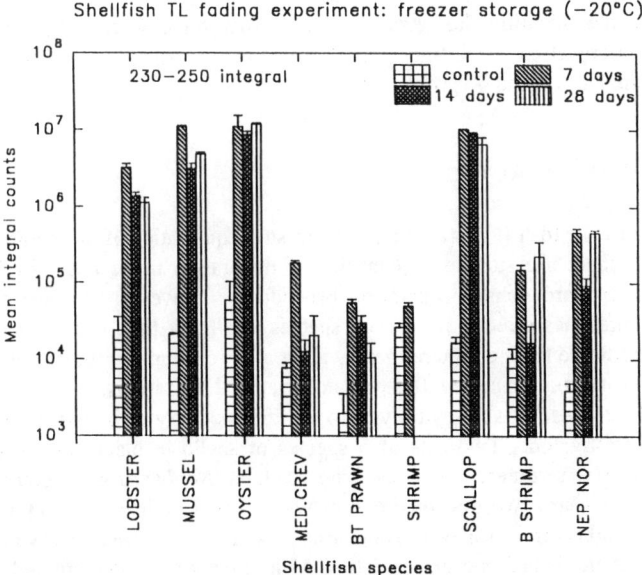

Figure 2 *Stability of TL signals from irradiated shells stored at -20°C*

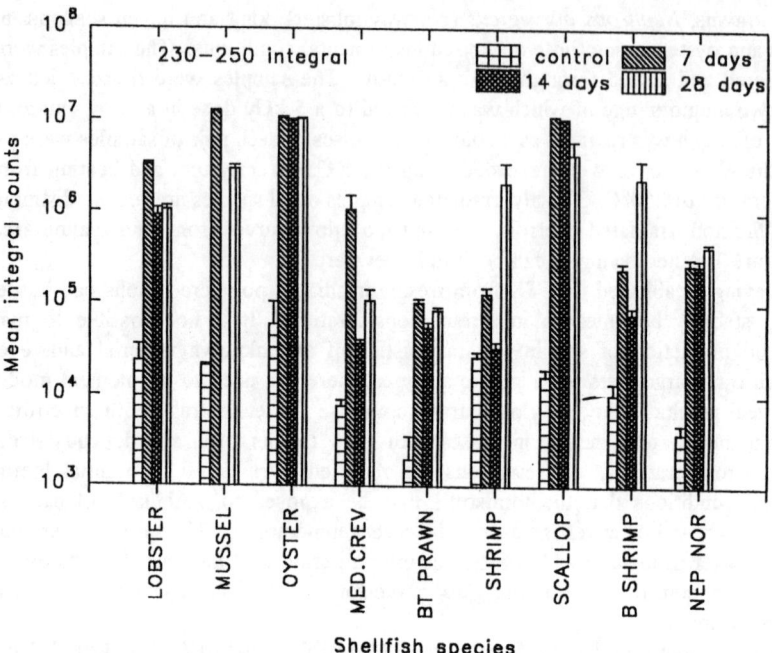

Figure 3 *Stability of TL signals from irradiated shells stored at 5°C*

to ambiguous identification, however, for the poorly mineralised warm water species (Mediterranean crevettes, black tiger prawn, warm water shrimp and brown shrimps) the worst case signal loss could be problematic.

3 TL OF INTESTINAL GRITS

The intestines of shellfish (Crustaceans) contain small quantities of inorganic grits which, providing that there are sufficient quantities of minerals, can be used for TL analyses following the standard density separation procedure. However, physical extraction of minerals and intestines directly from each species is a most time-consuming part of the analyses. For routine testing, where there will be a large sample turnover and also a high possibility of cross-contamination, this presents practical limitations.

At SURRC the use of acid hydrolysis to aid the recovery of intestinal grits has been successfully investigated. Two sets of 6 species of shellfish, black tiger prawns, warm water shrimps, Mediterranean crevettes, king scallops, *Nephrops norvegicus* and brown shrimps, were split into two portions, one of which was irradiated to 1 kGy in the ^{60}Co source and the other of which was retained as a control. The samples were left for 68 days in the freezer, shelled and placed into 6 M hydrochloric acid, refluxed for 2-3 h at 110-120°C, then left to cool. They were then diluted with deionised water to give an approximately 2 M solution and left to settle. The solution was decanted off and the remaining minerals were washed several times with deionised water. The minerals were rinsed in acetone and dispensed onto stainless steel discs.

TL measurements were recorded in duplicate for each species of shellfish and every irradiated sample could readily be distinguished from the non-irradiated controls as can be seen from Figures 4 and 5.

4 PSL OF SHELLFISH

Several photostimulation techniques have been investigated at SURRC as a tool for detecting irradiated shellfish. Fully mineralised geological calcite has stable TL signals, originating from deep traps at glow temperatures above 250°C.[14] The need to limit TL glow temperatures to 250°C in modern biogenic calcite prevents access to these signals. The aim of the photostimulation research was to develop methods for measuring luminescence from such deep traps, without the need for heating above 250°C.

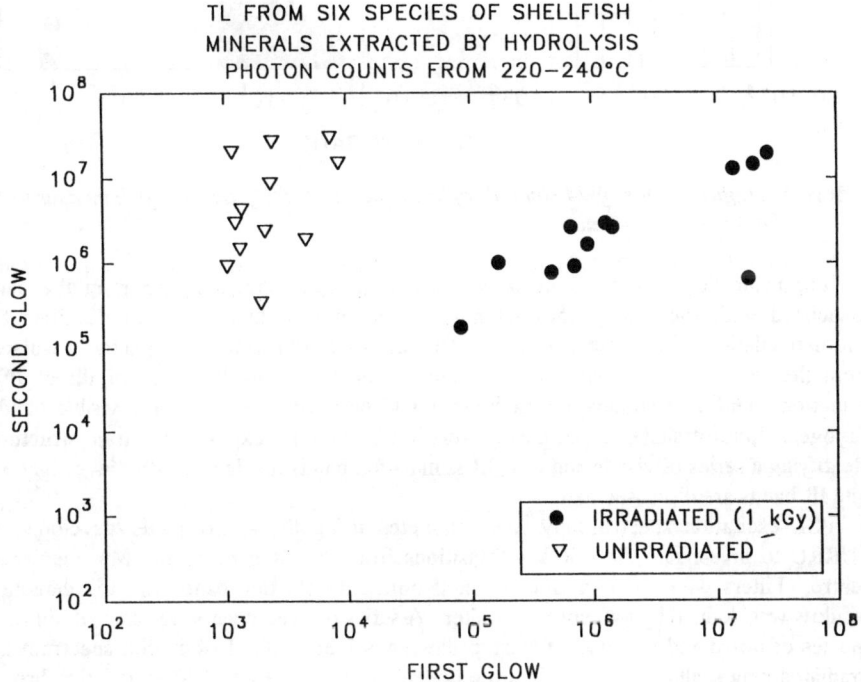

Figure 4 *First glow versus second glow plot of 24 samples of minerals extracted from shellfish measured 68 days after irradiation*

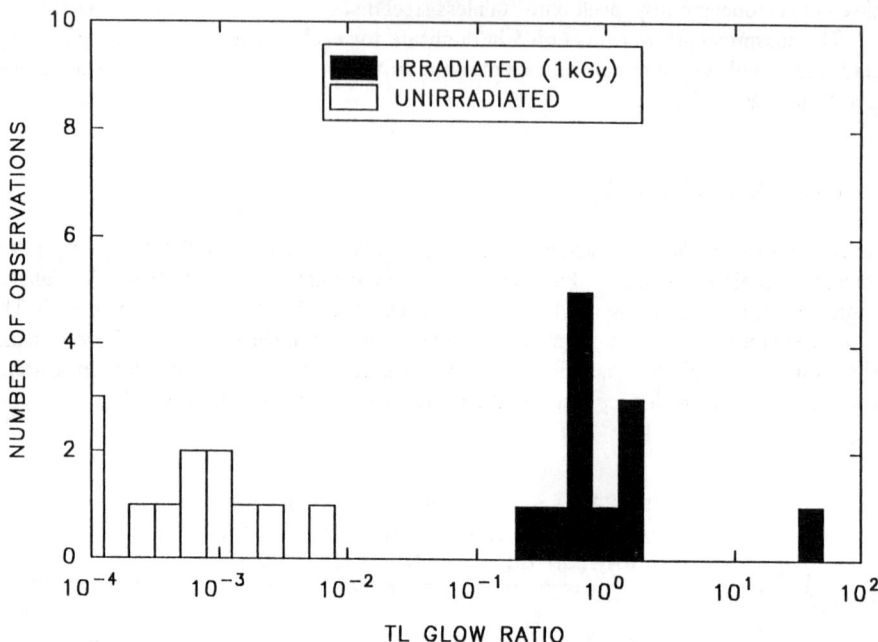

Figure 5 *TL glow ratios of 24 samples of minerals extracted from shellfish measured 68 days after irradiation*

The main TL emission from calcitic shells has been shown to occur in the band associated with the Mn^{2+} (550-600 nm), which in turn limits the opportunities for photostimulation with direct detection. The infra-red (IR) bands are spectrally isolated from the Mn^{2+} emission band, and therefore, can potentially be used for direct PSL detection. Different stimulation methods would be needed to utilise the visible bands. Cryogenic phototransfer experiments were conducted to explore the trap structure, identifying a series of visible and near IR stimulation bands for detailed PSL investigation. The IR bands are discussed here.

PSL excitation spectroscopy was conducted using the spectrometer developed at SURRC to investigate Anti-Stokes transitions from the IR band to the Mn^{2+} emission centre. Filters were used to isolate stimulation and detection bands and the detection window was defined by a 4 mm BG39 filter. A series of experiments was carried out on 7 species of powdered shellfish. Figure 6 illustrates a normalised excitation spectrum for irradiated king scallop, showing the presence of a complex series of IR stimulation bands. Similar bands have been observed from 9 other samples of shellfish with the main variations observed are the sensitivity and ratios of the main spectral components.

IR stimulation of shells was further investigated at different times after irradiation, using the high sensitivity pulsed PSL system[20] developed for screening herbs and spices, in conjunction with a TL reader to vary sample temperature, to examine the stability and

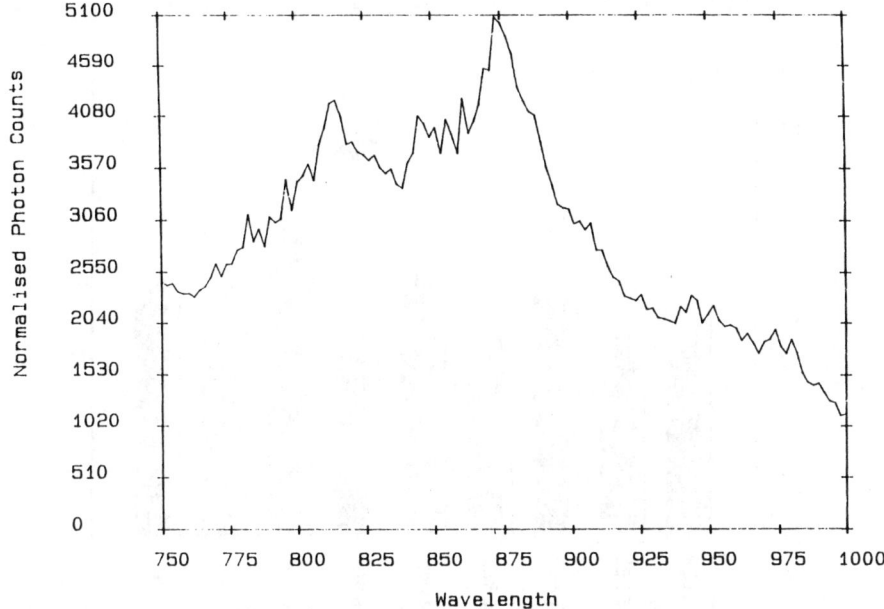

Figure 6 *Power normalised IR excitation spectra for king scallop shells irradiated to 2 kGy one week before measurement*

sensitivity of this band. Freshly irradiated samples of shell show a pronounced IR signal, Figure 7, which fades over the course of the sample shelf-life, to an extent comparable with TL signal below 250°C. In all cases strong initial signals were detected. However, losses during 14 days storage at -20°C were typically between 5 and 10 times. In the case of mussels and Mediterranean crevettes this instability presents practical limitations, which apply to a lesser extent to the other species.

This leads to the conclusion that the IR components are associated with a similar trap depth distribution, and that stable PSL signals will require stimulation of the visible bands identified by PTTL studies. However, the possibility of accessing the long wavelength tail of the 700 nm band using the pulsed IR system at elevated sample temperatures was explored. Small samples (5 mg) of both "old" irradiated powdered shell (irradiated 6 months prior to experiment) and non-irradiated controls were measured for 60 sec periods at the increased temperatures of 20, 40, 60, 80, 100, 140 and 180°C. The results for king scallop are shown in Figure 8, demonstrating the dramatic "recovery" of IR PSL from an "extinct" sample when stimulated at temperatures above approximately 100°C. This is attributed to thermal broadening of the 700 nm absorption band at elevated temperatures, and provides one possible means of accessing stable signals from calcite shell.

Pulsed PSL measurements have also been conducted using silicate containing parts of shellfish. It has been possible to stimulate PSL both from dissected intestinal material, which contains silicates, and from whole samples. Figure 9 shows examples of the cumulative signals from irradiated and non-irradiated intestines. More detailed analyses of a larger number of examples are in progress.

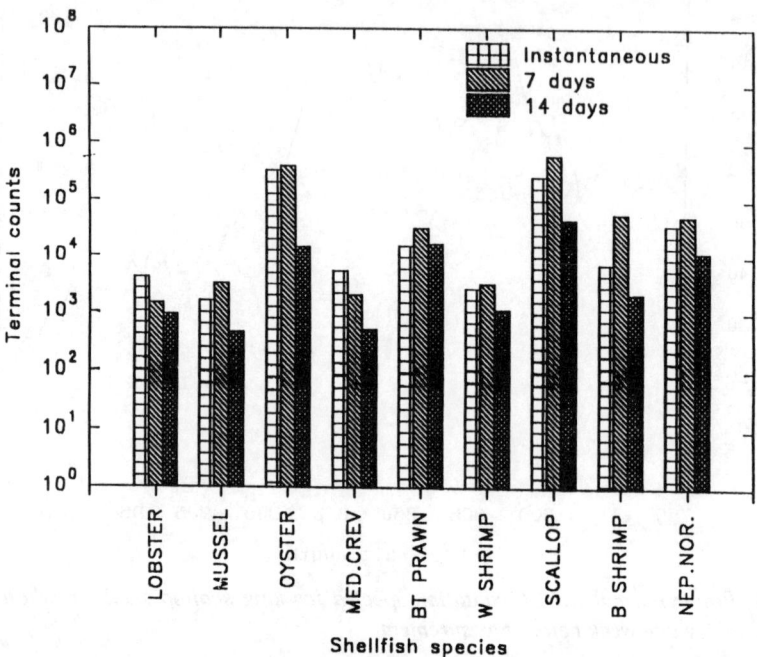

Figure 7 *Pulsed IR PSL results from 9 species of powdered shellfish irradiated to 2 kGy*

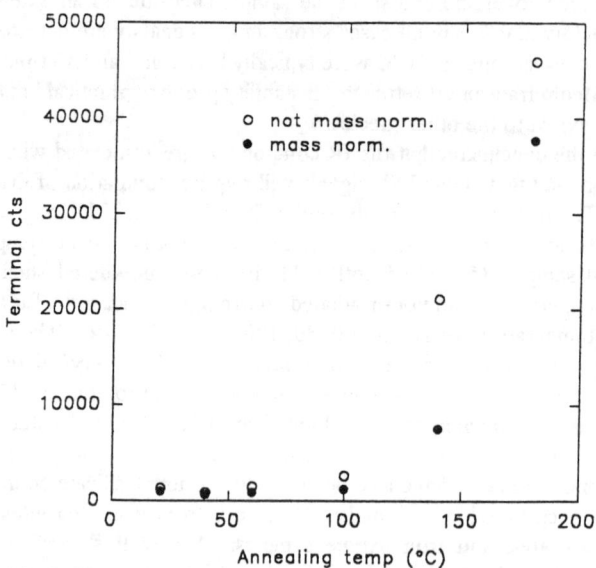

Figure 8 *IR PSL of irradiated calcite shell (6 month fading) measured at elevated temperatures*

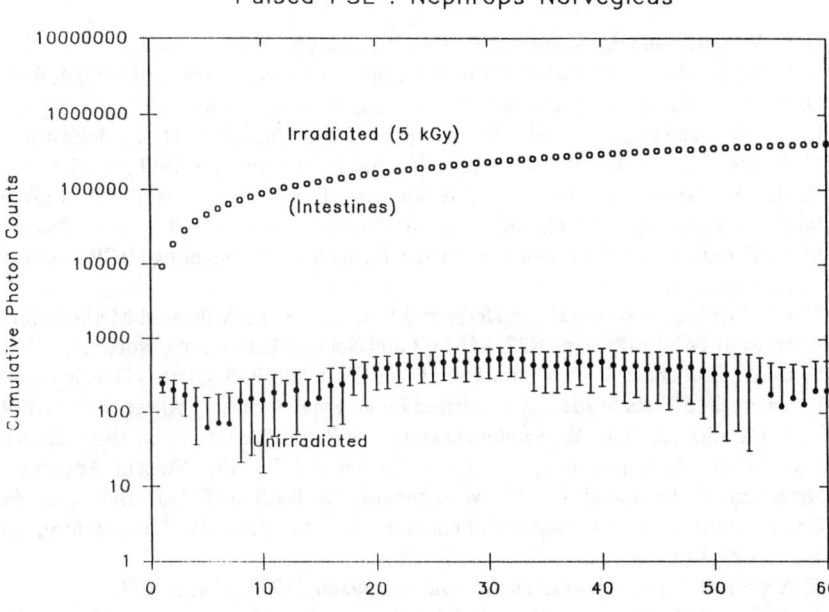

Figure 9 *Pulsed PSL measurements from the intestines removed from shellfish. The signal is attributed to trapped minerals*

5 DISCUSSION AND CONCLUSION

Both TL and PSL methods provide possible means of detecting irradiated shellfish, both using the shell, and using trapped intestinal grits. For TL the need to restrict glow curve measurements to 250°C limits the stability of the signals, however, well mineralised species are detectable over reasonable product shelf-lives, and signals can be detected for over 14 days in poorly mineralised species. Extraction of intestinal grits provides a more stable signal, and it has been shown that acid hydrolysis provides an effective alternative to mechanical extraction procedures coupled to heavy liquid separation. PSL signals from shells are detectable in the IR band for freshly irradiated specimens; for aged samples the IR response can be enhanced by raising sample temperatures, alternatively there are visible stimulation schemes. The demonstration that pulsed PSL can be detected from intestinal grits, both through the membranes of the intestine, and in many case through the body of the specimen, provides an important additional tool for screening commercial samples.

The range of luminescence methods available will therefore deal with situations where either shell or intestinal grits are present, and both TL laboratories or those with PSL equipment will be able to make measurements. A small scale international trial of TL analysis of silicates is in progress, and there is a need for further trials of the additional approaches identified here.

References

1. D. C. W Sanderson, C. Slater and K. J. Cairns, *Nature,* 1989a, **340,** 23.
2. D. C. W. Sanderson, C. Slater and K. J. Cairns, *Rad. Phys. Chem.*, 1989c, **34,** 915.
3. D. C. W. Sanderson, C. Slater and K. J. Cairns, *Int. J. Rad. Biol.*, 1989, **55,** 5.
4. D. C. W Sanderson, "Food Irradiation and the Chemist," D. E. Johnston and M. H. Stevenson, Eds., Royal Society of Chemistry, Cambridge, 1990, p. 25.
5. D. C. W Sanderson, G. A. Schreiber and L. A. Carmichael, "A European Interlaboratory Trial of TL Detection of Irradiated Herbs and Spices. Phase 1," SURRC report to the Commission of the European Communities (BCR), Brussels, Luxembourg, 1991.
6. MAFF, "Detection of Irradiated Herbs and Spices," MAFF Validated Methods for the Analysis of Foodstuffs, No. V27, MAFF Food Science Laboratory, Norwich, 1992.
7. D. C. W. Sanderson, L. A. Carmichael, P. A. Clark and R. J. Clark, "Development of Luminescence Tests to Identify Irradiated Food. Final Report," Project N1701, 1992.
8. L. A. Carmichael, D. C. W. Sanderson and S. Ni Riain, *Radiat. Meas,* 1994, **23,** 455.
9. J.-P. Agnel, M. S. Dutraive, I. Vaux, I. Rustan and J. Raffi, "Recent Advances on Detection of Irradiated Food," M. Leonardi, J. Raffi and J.-J. Belliardo, Eds., Commission of the European Communities (BCR), Brussels, Luxembourg, EUR 14315 EN, 1993, p. 186.
10. T. Autio and S. Pinnioja, *Z. Lebens. Unters. Forsch.*, 1990, 191, p. 177.
11. T. Autio and S. Pinnioja, "Recent Advances on Detection of Irradiated Food," M. Leonardi, J. Raffi and J.-J. Belliardo, Eds., Commission of the European Communities (BCR), Brussels, Luxembourg, EUR 14315 EN, 1993, p. 177.
12. C. Christodoulides and J. H. Fremlin, *Nature,* 1971, **232,** 257.
13. H. S. T. Driver, PACT, 1979, **3,** 290.
14. J. S. Down, R. Flower, J. A. Strain and P. D. Townsend, *Nucl. Tracks,* 1985, **10,** 581.
15. N. M. Johnson, *Sediment. Pet.*, 1960, **30,** 305.
16. N. M. Johnson and R. L. Blanchard, *Americ. Min.*, 1967, **52,** 1297.
17. D. C. W Sanderson, "Potential New Methods of Detection of Irradiated Foods," J. J. Raffi and J.-J. Belliardo, Eds., Commission of the European Communities (BCR), Brussels, Luxembourg, EUR 13331 EN, 1991, p. 159.
18. D. C. W. Sanderson and R. J. Clark, *Radiat. Meas.*, 1994, **23,** 633.
19. J. F. Lima, P. Trzesniak, E. M. Yoshimura and E. Okuno, *Radiat. Protec. Dosim.*, 1990, **33,** 143.
20. D. C. W. Sanderson, L. A. Carmichael and J. D. Naylor, "Identification of Irradiated Food: Current Status," ADMIT, see these proceedings.

COMPARISON OF THERMOLUMINESCENCE DETECTION METHODS FOR IRRADIATED SPICES

Y. Kawamura, M. Murayama, S. Uchiyama[*] and Y. Saito

National Institute of Health Sciences, Setagaya, Tokyo 158, Japan
([*]Present address: Food and Drug Safety Center, Hatano Research Institute, Hatano, Kanagawa 257, Japan)

1 INTRODUCTION

Thermoluminescence (TL) analysis has been shown to be one of the most applicable methods for the detection of γ-irradiated spices. This analysis was introduced by Heide et al. as a detection technique for irradiated spices using the whole sample.[1,2] Sanderson et al. then revealed the origin of the TL response to be mineral dust adhering to the spices.[3,4] TL measurement on separated minerals and the normalised TL measurement by re-irradiation was then established.[5,6]

This paper details investigations on TL measurements carried out using clean powdered spices stored for one year after being irradiated with doses of 1, 5, 10 and 30 kGy in order to clarify their applicable dose range, the effect of storage and mineral content. The effect of the mineral separation was also studied.[7]

2 MATERIALS AND METHODS

2.1 Samples and Irradiation

The samples used were pure and clean powdered spices obtained on the Japanese market. They were irradiated with ^{60}Co γ-rays at doses of 1, 5, 10 and 30 kGy. After irradiation they were stored for one year prior to analysis.

2.2 Sample Preparation

TL analysis was performed using three different methods. The whole sample measurement used 1 mg of spice which was directly measured for TL intensity. For the measurement of separated minerals, samples were sonicated and centrifuged in a sodium tungstate saturated solution (density, 1.6 g ml^{-1}) as the first step. If the separation was not complete, sodium polytungstate solution (density 2.1 g ml^{-1}) was used for the second step. The minerals were washed and dried, then a 1 mg sample was analysed for TL intensity. For renormalised measurements, the TL ratio was used as the parameter to detect irradiated samples. After the initial TL intensity measurement the minerals were re-irradiated with a 1 kGy dose and their TL intensity was re-measured. The TL ratio was calculated from the TL intensity before re-irradiation divided by that after re-irradiation.

2.3 TL Measurements

Samples or minerals were heated from 50°C to 350°C at a heating rate of 10°C sec^{-1}. TL glow curves were measured using a Harshow TLD4000 system.

3 RESULTS AND DISCUSSION

3.1 Whole Sample Measurements

Fifteen spice samples were analysed using the whole sample measurement technique (Table 1). The TL intensity of turmeric increased about 40 times and that of black pepper, celery leaf, red pepper and thyme increased 3 times or more due to irradiation. However, the other spices showed only a small or no increase in TL intensity.

The distribution of TL intensity at each irradiation dose was broad and overlapped each other (Figure 1). The difference in TL intensity during the whole sample measurement was not sufficient to discriminate between non-irradiated and irradiated samples even at a dose level of 30 kGy.

3.2 Measurement of Separated Minerals

For the mineral separation from powdered spices, the sodium polytungustate solution (density 2.1 g ml^{-1}) was most efficient. However, a sodium tungustate saturated solution (density 1.6 g ml^{-1}) was used in the first step, since sodium polytungustate is expensive and is not readily available for purchase in Japan. If minerals could not be separated, a sodium polytungstate solution was then used. Twenty-seven samples were analysed and minerals were obtained from 17 samples using sodium tungstate and from 7 additional samples using sodium polytungstate. The latter were mainly the finely powdered spices. Mineral yields ranged from 0.02% to 5.28%, but were mainly under 0.2%. Minerals obtained at a high percentage still retained much organic matter. Minerals were not obtained from 2 samples of allspice and one sample of cinnamon, and therefore, about 10% of the samples could not be measured.

The TL intensity of minerals separated from spices was significantly increased for all samples with γ-irradiation (Table 2 and Figure 2). Irradiated red pepper and turmeric showed the highest intensity. However, the distribution of the TL intensity at each irradiation dose still overlapped the non-irradiated samples for each dose. If the threshold was determined to be 10 nC mg^{-1}, all samples irradiated at 30 kGy and 70% of the samples irradiated at 10 kGy could be correctly identified, while 2 samples gave a false positive judgment.

From Table 2 it can be seen that the same kind of spice samples showed a similar TL intensity. If a threshold could be set up for each kind of spice, it would allow for better discrimination.

Table 1 Effect of γ-Irradiation on TL Intensity of Spices Measured using Whole Sample[7]

Spices	TL Intensity (nC mg^{-1})				
	0 kGy	1 kGy	5 kGy	10 kGy	30 kGy
Allspice	0.18 ± 0.03 (1.0)	0.20 ± 0.06 (1.1)	0.27 ± 0.05 (1.5)	0.22 ± 0.05 (1.2)	0.22 ± 0.06 (1.2)
Black pepper	0.24 ± 0.03 (1.0)	0.51 ± 0.21 (2.0)	1.49 ± 2.20 (6.0)	1.11 ± 0.42 (5.0)	0.64 ± 0.60 (3.0)
Celery leaf	0.46 ± 0.13 (1.0)	1.06 ± 0.35 (2.0)	2.45 ± 1.54 (5.0)	2.76 ± 0.93 (6.0)	2.61 ± 0.95 (7.0)
Cinnamon (1)	0.07 ± 0.07 (1.0)	0.13 ± 0.05 (1.9)	0.11 ± 0.04 (1.5)	0.17 ± 0.05 (2.4)	0.21 ± 0.07 (2.9)
Cinnamon (2)	0.07 ± 0.01 (1.0)	0.05 ± 0.03 (0.7)	0.07 ± 0.01 (1.0)	0.15 ± 0.11 (2.3)	0.33 ± 0.59 (4.6)
Coriander	0.18 ± 0.03 (1.0)	0.24 ± 0.03 (1.3)	0.26 ± 0.02 (1.4)	0.37 ± 0.10 (2.0)	0.35 ± 0.05 (1.9)
Laurel	0.86 ± 0.17 (1.0)	1.14 ± 0.07 (1.4)	1.14 ± 0.22 (1.4)	1.11 ± 0.25 (1.3)	1.04 ± 0.30 (1.2)
Mustard	0.09 ± 0.01 (1.0)	0.19 ± 0.06 (2.0)	0.13 ± 0.03 (1.4)	0.92 ± 1.49 (10)	0.20 ± 0.06 (2.0)
Nutmeg	0.05 ± 0.01 (1.0)	0.13 ± 0.16 (2.6)	0.05 ± 0.00 (1.0)	0.07 ± 0.03 (1.4)	0.07 ± 0.01 (1.4)
Red-pepper	0.24 ± 0.06 (1.0)	0.81 ± 0.54 (3.0)	1.56 ± 1.35 (6.5)	0.84 ± 0.36 (3.5)	1.47 ± 0.57 (6.0)
Rosemary	0.87 ± 0.11 (1.0)	1.57 ± 0.27 (1.8)	2.32 ± 0.39 (2.7)	2.99 ± 0.97 (3.5)	2.31 ± 0.53 (2.7)
Sage	0.31 ± 0.03 (1.0)	0.56 ± 0.43 (1.8)	0.45 ± 0.05 (1.4)	0.80 ± 0.59 (2.6)	0.71 ± 0.18 (2.3)
Thyme	0.28 ± 0.04 (1.0)	1.09 ± 0.19 (3.9)	0.83 ± 0.18 (2.9)	1.54 ± 0.56 (5.4)	2.19 ± 0.46 (7.7)
Turmeric	0.06 ± 0.02 (1.0)	3.28 ± 2.74 (53)	2.30 ± 1.19 (38)	2.39 ± 1.42 (39)	2.75 ± 1.97 (44)
White pepper	0.29 ± 0.05 (1.0)	0.36 ± 0.05 (1.2)	0.45 ± 0.09 (1.5)	0.41 ± 0.04 (1.4)	0.35 ± 0.08 (1.2)

Values are the mean of TL intensity (nC mg^{-1}) ± S.D. for 5 or more trials and the rate of increase against 0 kGy is in parentheses

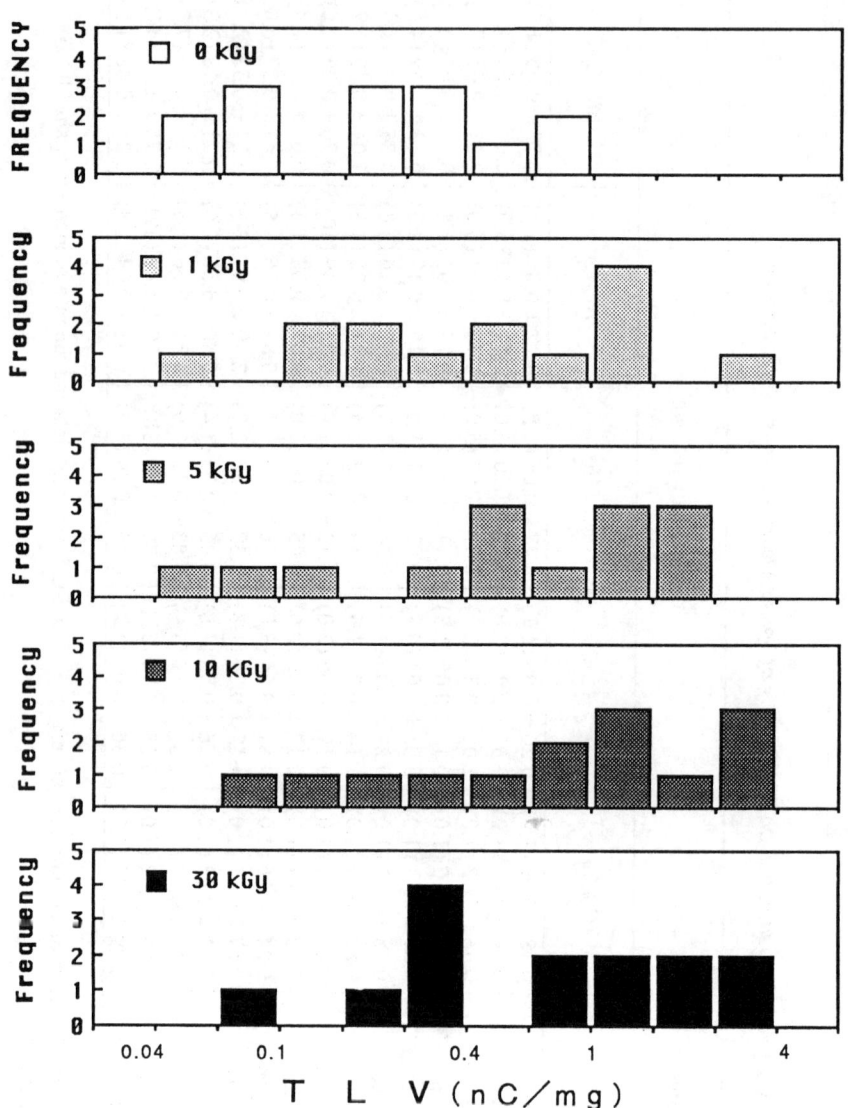

Figure 1 *TL intensity ranges of unirradiated and irradiated spices measured using whole sample* [7]

Table 2 Effect of γ-Irradiation on TL Intensity of Minerals Separated from Spices[7]

Spices	Yield (%)	TL Intensity (nC mg^{-1})				
		0 kGy	1 kGy	5 kGy	10 kGy	30 kGy
Basil	1.18 ± 0.79	0.56 ± 0.17 (1.0)	5.42 ± 2.74 (9.7)	1.79 ± 0.99 (3.2)	9.97 ± 8.52 (5.0)	48.9 ± 3.15 (87)
Black pepper (1)	0.07 ± 0.01	1.86 ± 0.55 (1.0)	67.6 ± 27.7 (36)	165 ± 68 (89)	235 ± 60 (126)	313 ± 79 (168)
Black pepper (2)	0.18 ± 0.04	1.20 ± 0.08 (1.0)	1.93 ± 0.46 (1.6)	2.35 ± 0.75 (2.0)	3.01 ± 1.10 (2.6)	19.5 ± 14.3 (16)
Celery leaf	0.14 ± 0.01	2.14 ± 0.87 (1.0)	567 ± 157 (265)	947 ± 175 (442)	1048 ± 233 (489)	1284 ± 144 (600)
Cinnamon (1)	0.10 ± 0.05	0.27 ± 0.19 (1.0)	2.53 ± 1.09 (9.6)	4.01 ± 4.73 (15)	4.92 ± 1.36 (18)	24.3 ± 22.3 (92)
Cinnamon (2)	0.40 ± 0.09	0.03 ± 0.00 (1.0)	0.21 ± 0.15 (6.7)	1.27 ± 1.03 (40)	4.52 ± 3.65 (141)	12.6 ± 11.5 (393)
Coriander (1)	0.17 ± 0.03	0.60 ± 0.08 (1.0)	1.99 ± 0.82 (3.0)	7.08 ± 5.27 (11)	4.36 ± 2.43 (7.0)	15.6 ± 10.8 (26)
Coriander (2)	0.09 ± 0.03	0.79 ± 0.22 (1.0)	7.12 ± 2.53 (9.0)	3.88 ± 3.19 (4.9)	11.3 ± 8.1 (14)	63.4 ± 60.7 (80)
Cumin	1.29 ± 0.54	3.67 ± 1.44 (1.0)	115 ± 36 (31)	162 ± 55 (53)	537 ± 128 (146)	184 ± 79 (50)
Dill	5.28 ± 2.37	0.53 ± 0.69 (1.0)	7.23 ± 7.72 (14)	16.4 ± 5.7 (31)	17.6 ± 8.7 (33)	25.1 ± 19.9 (47)
Laurel	0.06 ± 0.01	0.17 ± 0.06 (1.0)	24.7 ± 10.4 (145)	3.23 ± 0.09 (19)	23.4 ± 11.6 (138)	19.8 ± 2.0 (117)
Marjoram	4.13 ± 0.62	0.05 ± 0.03 (1.0)	7.85 ± 4.42 (157)	51.6 ± 26.5 (1032)	29.4 ± 4.6 (588)	114 ± 38 (2272)
Mustard	1.29 ± 0.45	0.02 ± 0.01 (1.0)	3.14 ± 5.13 (157)	1.59 ± 0.63 (80)	5.11 ± 5.58 (256)	11.4 ± 10.9 (570)
Nutmeg (1)	0.03 ± 0.01	0.10 ± 0.04 (1.0)	2.33 ± 0.96 (25)	29.8 ± 42.0 (314)	16.1 ± 13.5 (167)	53.6 ± 17.3 (563)
Nutmeg (2)	0.11 ± 0.12	0.05 ± 0.02 (1.0)	1.59 ± 0.96 (32)	5.04 ± 0.12 (101)	2.37 ± 2.58 (47)	64.8 ± 66.1 (1296)
Red pepper	0.04 ± 0.01	38.3 ± 35.3 (1.0)	790 ± 330 (21)	2179 ± 766 (57)	2090 ± 1136 (54)	2112 ± 184 (55)
Rosemary (1)	0.31 ± 0.29	1.96 ± 1.37 (1.0)	112 ± 47 (57)	67.0 ± 40.4 (34)	91.4 ± 45.9 (47)	153 ± 63 (78)
Rosemary (2)	0.58 ± 0.33	1.13 ± 0.21 (1.0)	37.4 ± 16.8 (33)	53.6 ± 17.3 (47)	39.7 ± 26.6 (35)	24.0 ± 12.8 (21)
Sage	0.02 ± 0.01	2.89 ± 1.44 (1.0)	117 ± 79 (41)	26.9 ± 4.4 (9.0)	74.6 ± 20.9 (26)	93.1 ± 0.0 (32)
Staranise	0.04 ± 0.03	0.88 ± 0.53 (1.0)	77.1 ± 0.0 (7.6)	126 ± 0 (142)	68.0 ± 28.4 (77)	831 ± 0 (944)
Thyme	0.25 ± 0.08	1.87 ± 0.72 (1.0)	40.9 ± 15.9 (22)	95.0 ± 4.1 (51)	168 ± 66 (106)	242 ± 38 (129)
Turmeric (1)	0.20 ± 0.06	25.2 ± 12.3 (1.0)	873 ± 43 (35)	2327 ± 522 (92)	2802 ± 139 (111)	1810 ± 509 (72)
Turmeric (2)	0.13 ± 0.07	5.33 ± 1.72 (1.0)	282 ± 62 (53)	149 ± 45 (28)	113 ± 64 (21)	118 ± 91 (22)
White pepper	0.02 ± 0.01	1.24 ± 0.00 (1.0)	2.11 ± 0.10 (1.7)	6.21 ± 0.00 (5.0)	14.8 ± 12.0 (12)	30.3 ± 31.9 (24)

Values are the mean S.D. for 5 or more trials and the rate of increase against 0 kGy is in parentheses

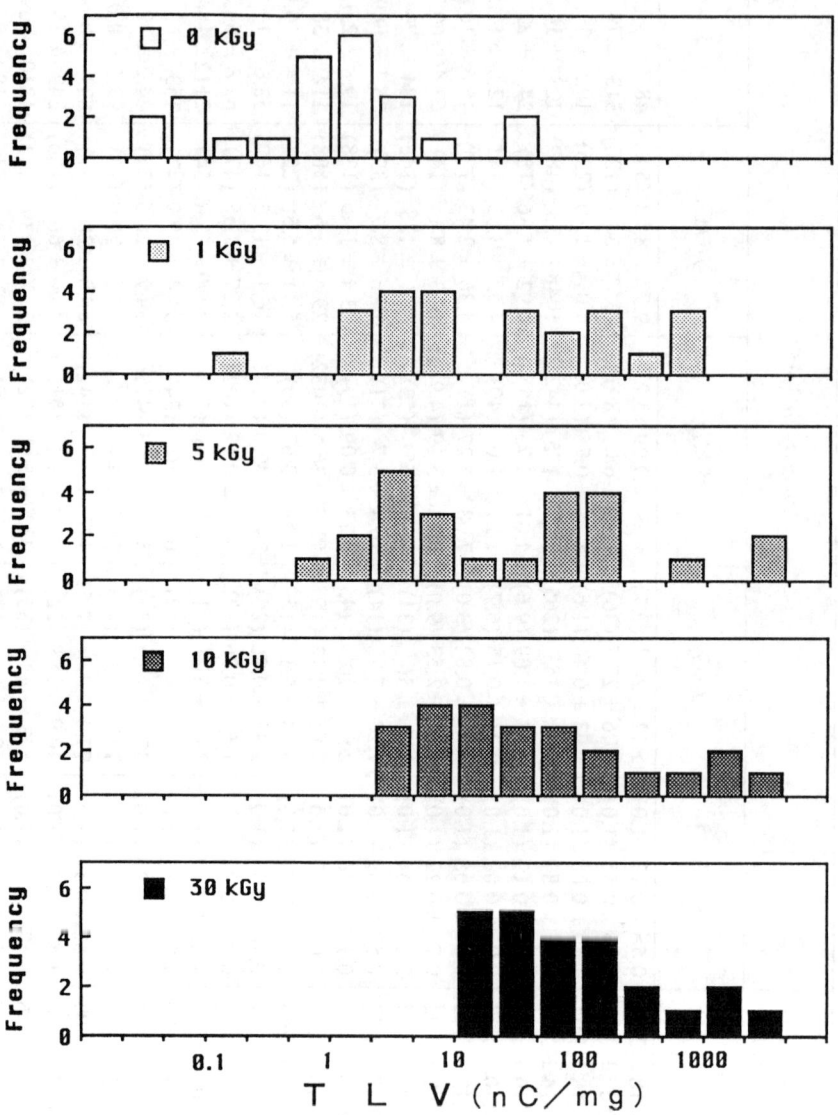

Figure 2 *TL intensity ranges of minerals separated from unirradiated and irradiated spices* [7]

Table 3 Effect of γ-Irradiation on TL Ratio of Spices Using Normalised Measurement[7]

Spices	TL Intensity (nC mg^{-1})				
	0 kGy	1 kGy	5 kGy	10 kGy	30 kGy
Basil	0.06 ± 0.06 (1.0)	0.76 ± 0.31 (12)	1.24 ± 0.97 (21)	1.75 ± 1.27 (29)	2.23 ± 0.96 (37)
Black pepper	0.30 ± 0.14 (1.0)	1.78 ± 0.39 (5.9)	0.84 ± 0.38 (6.1)	1.73 ± 1.34 (5.8)	3.35 ± 1.72 (11)
Cumin	0.03 ± 0.01 (1.0)	0.74 ± 0.22 (25)	0.98 ± 0.36 (33)	1.18 ± 0.30 (39)	1.87 ± 1.49 (62)
Dill	0.01 ± 0.00 (1.0)	2.31 ± 2.91 (231)	1.41 ± 1.03 (141)	2.76 ± 3.48 (276)	3.62 ± 5.12 (362)
Marjoram	0.02 ± 0.03 (1.0)	0.91 ± 0.63 (46)	1.42 ± 0.86 (71)	2.47 ± 1.12 (124)	1.69 ± 0.45 (85)
Nutmeg	0.04 ± 0.06 (1.0)	10.2 ± 15.4 (255)	2.06 ± 0.00 (52)	5.62 ± 5.49 (141)	260 ± 65 (6500)
Rosemary	0.05 ± 0.05 (1.0)	1.16 ± 0.65 (23)	2.12 ± 0.62 (42)	1.53 ± 0.91 (31)	2.38 ± 2.03 (48)
Staranise	0.24 ± 0.13 (1.0)	1.94 ± 0.00 (8.1)	3.86 ± 0.00 (16)	2.16 ± 1.21 (9.0)	3.59 ± 0.00 (15)
Turmeric	0.01 ± 0.00 (1.0)	0.34 ± 0.03 (34)	1.16 ± 0.45 (116)	1.13 ± 0.34 (113)	1.57 ± 0.20 (157)

Values are the mean of TL ratio ± S.D. for 5 or more trials and the rate of increase against 0 kGy is in parentheses

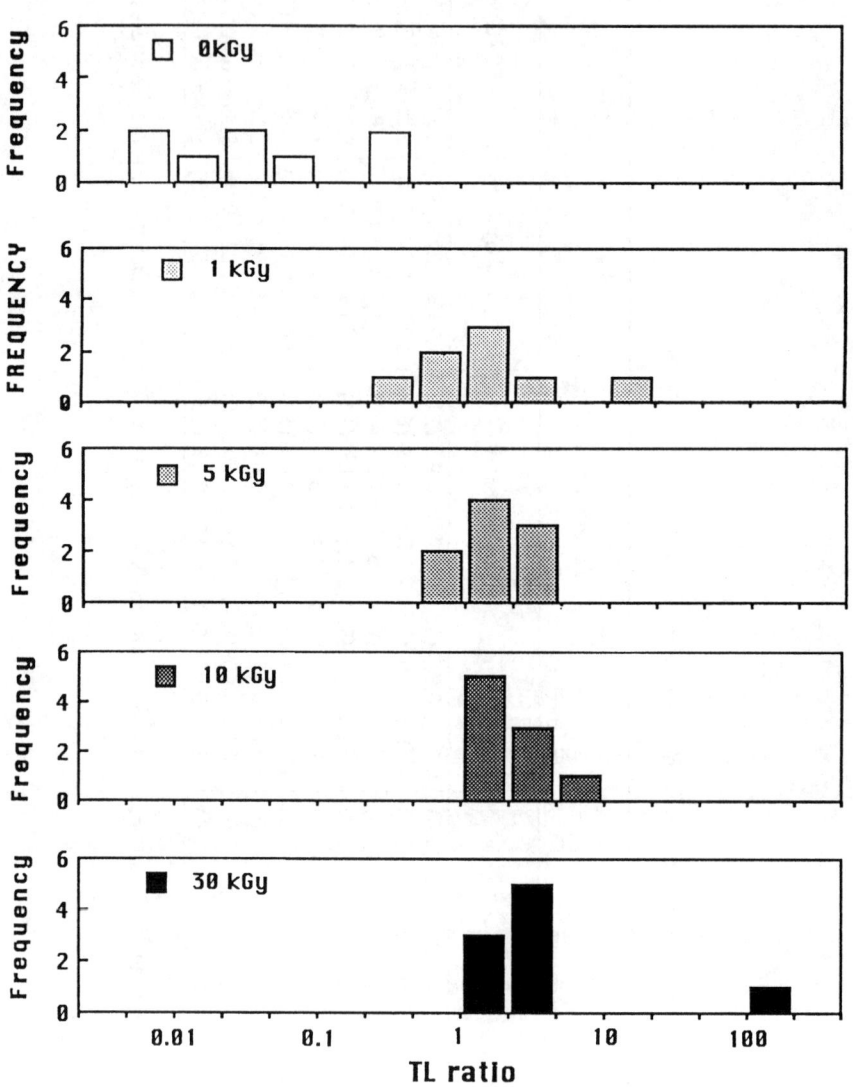

Figure 3 *TL ratio ranges of unirradiated and irradiated spices using normalized measurement* [7]

3.3 Re-normalised Measurement

Nine samples were analysed using a re-normalised measurement (Table 3 and Figure 3). The distribution of the TL ratio at each irradiation dose showed good separation of the non-irradiated and irradiated samples. The TL ratio of non-irradiated samples was under 0.30. On the other hand, the TL ratio of 1 kGy irradiated samples ranged from 0.34 to 10.2, that of 5 kGy from 0.84 to 3.86, the 10 kGy samples from 1.13 to 5.62 and that of 30 kGy was 1.57 to 260. If the threshold of the TL ratio was determined to be 0.5, all samples irradiated with 5 kGy or more, and about 90% of the samples irradiated with 1 kGy, can be correctly identified without any false positive judgments. In this case, the detection limit of the irradiation dose is 1 kGy even if the sample have been irradiated one year before to analysis.

4 CONCLUSIONS

The TL ratio of the re-normalised measurement is the foremost parameter for TL analysis to detect irradiated spices. Samples stored for one year after irradiation could still be sufficiently distinguished from non-irradiated samples and samples stored longer than one year might certainly be detected. One problem, however, is that occasionally no minerals can be extracted from the spices, therefore, they cannot be measured. Another problem is that not all food control laboratories have facilities for re-irradiation in Japan. However, this method is the most suitable as a practical detection method for irradiated spices.

References

1. L. Heide and K. W. Bögl, "Report of the Institute for Radiation Hygiene of the Federal Health Office," Bundesgesundheitsamt (BGA), Berlin, ISH-Heft 58, 1984.
2. L. Heide, R. Guggenberger and K.W. Bögl, *Radiat. Phys. Chem.*, 1989, **34**, 903.
3. H. Y. Gökusu-Ögelman and D. F. Regulla, *Nature*, 1989, **340**, 23.
4. D. C. W. Sanderson, C. Slater and K. J. Cairns, *Nature*, 1989, **340**, 23.
5. D. C. W. Sanderson, C. Slater and K. J. Cairns, *Radiat. Phys. Chem.*, 1989, **34**, 915.
6. T. Autio and S. Pinnioja, *Z. Lebensm. Unters. Forsch.*, 1990, **191**, 177.
7. Y. Kawamura, K. Kojima, T. Sugita, T. Yamada and Y. Saito, *J. Food Hyg. Soc. Japan*, 1995, **36**, 55.

THERMOLUMINESCENCE IDENTIFICATION OF IRRADIATED FOODSTUFFS: LARQUA RESEARCH

G. Lesgards,[+] A. Fakirian[+]* and J. Raffi*

Laboratoire de Chimie des produits Naturels (+) and
Laboratoire de Recherche sur la Qualité des Aliments (*)
Faculté de Saint-Jérôme, Avenue Escadrille Normandie Niémen
F-13397 Marseille cedex 20
France

1 INTRODUCTION

Food irradiation by X- and gamma-rays and by electron-beam has been introduced recently, as a new technological process, mainly to reduce losses and to improve hygienic quality.[1,2] Following many investigations devoted to possible health impacts, a Joint Food and Agriculture Organisation (FAO)/International Atomic Energy Agency (IAEA)/ World Health Organisation (WHO) Expert Committee[3] meeting concluded in 1980 that "the irradiation of any food commodity up to an overall average dose of 10 kGy presents no toxicological hazard; hence, toxicological testing of foods so treated is no longer required." The Commission of the European Communities,[4] concluded as did many Health Authorities, that these aspects were investigated to a degree that no other method of food processing has been examined so far. As a consequence, the radiation treatment of different foods is now legally accepted in at least 37 countries, although it is still prohibited in others. To facilitate trade in irradiated foods, regulatory authorities in all countries appear to be interested in having simple and reliable methods to detect foods treated by irradiations[5,6] and, consequently, to check on compliance with labelling regulations.

The changes that occur in irradiated foodstuffs are very small and generally similar to those produced by classic food treatment processes (heating, freezing) or natural evolution of the foodstuffs (auto-oxidation). In order to develop a test, these changes must be 'characteristic' of the ionising treatment, at least under certain conditions (to be well known), and measurable and stable for at least the commercial shelf-life of the foodstuff.

Several detection methods have been discussed previously.[1,7-10] Considerable progress has been made recently, particularly due to the actions of the Reference Bureau of the Commission of the European Communities (Brussels)[11-13] and of the Joint Division of the FAO/IAEA in Vienna. Consequently the European Committee for Standardization is examining some protocols as possible official methods for identification of irradiated foodstuffs: electron spin resonance (ESR for meat, fish and some fruits), thermoluminescence (TL, for spices) and gas chromatography ('lipid method' for meats).

Despite the fact that the 'official' TL protocol is only intended to demonstrate irradiation treatment of spices, research at LARQUA studied the application of TL to aromatic herbs, with the intention of establishing a screening method rather than a test to prove irradiation treatment has been carried out. Consequently a systematic study of the TL recording conditions was carried out on some aromatic herbs as described, showing the respective advantages of direct TL and TL after extraction of silicate mineral impurities. In the same way, as described in the third part of this paper, the application of TL to shells has also been studied.

2 TL STUDY OF AROMATIC HERBS

When a foodstuff exhibiting TL is exposed to ionising radiation, electron-hole pairs are produced. Some electrons and holes may become trapped at certain sites in the material, where they remain, in particular in solid and dry foodstuffs, until they acquire sufficient thermal energy to escape from the energy levels between the valence and conduction bands. The TL intensity is the amount of light emission which occurs on heating a solid material. The glow curve is obtained by continuously increasing the temperature and recording the amount of light. The first glow curve refers to the TL directly recorded from the sample. The second glow refers to the subsequent measurement of the TL response of the sample following exposure to a known dose of radiation.

2.1 The TL Protocol

Up to the 1980's, TL was only recorded on the whole sample,[14-16] the method is simple, quick and requires only small samples (less than 20 mg). However, the differences between the 'whole sample TL intensity' of irradiated and non-irradiated samples are not always very obvious and consequently only lead to a presumption of irradiation. More recently, Sanderson *et al.*[17,18] found that the TL signal comes mainly from the mineral silicates present with the plant. Consequently, a protocol was developed requiring an extraction of these minerals from the plant before recording of the glow curve of the mineral part. Moreover, in order to avoid any misunderstanding, the final protocol requires comparison of the ratio between the first glow curve and the glow curve recorded after exposure to a fixed known dose.

However, obtaining this ratio, even if it gives proof of irradiation, requires more time than a simple measurement on the whole sample. This measurement requires about 15 min (Figure 1), including the sample grinding, but the other method requires about 3-4 hours, mainly due to the extraction of silicates, usually by a density separation step (using sodium polytungstate),[18] and TL needs 1 h more, if an irradiation facility is required. Thus, at LARQUA it was decided to examine the simplicity and the time required for each method.

2.2 TL Studies on the Whole Plant

2.2.1. Experimental. All the irradiation was carried out, at room temperature, in the Cadarache ^{137}Cs gamma-cell delivering a dose rate of 60 Gy min^{-1}. The TL apparatus was a HARSHAW 4000A connected to a microcomputer. The aromatic plants studied were parsley, thyme, savory, sage, rosemary, origanum and fennel.

The first results from the 1970-1980's showed very large variations between these samples,[1,14] thus, the authors proposed to make 7 measurements and to obtain the minimal and maximal values. In this study, the experimental research methodology has been used to obtain the optimal performance of the system.[19,20] After determination of the main factors (mass and granulometry of the sample, nitrogen flow rate, temperature rate), the statistical analysis has been performed using a DOEHLERT experimental matrix. Study of the response equation, and the canonic equation leads to the values of parameters which optimise the experimental response, and allow relatively reproducible results[21] *i.e.* sample mass: 7 mg, initial temperature; 50°C; final temperature, 350°C; temperature rate, 10°C min^{-1}.

2.2.2. Results. As can be seen in Table 1 and Figure 2, the ratio between irradiated and non-irradiated samples, using the whole sample, is never definitive, ranging from 2.3 (rosemary) to 19.6 (thyme). This is surely not sufficient enough to demonstrate proof of irradiation and does not generally allow even a presumption of irradiation, with perhaps the exception of parsley, thyme and origanum.

2.3.1. Experimental. The protocol which is proposed by the European Committee for Standardization was used, *i.e.* the following steps (Figure 1):

- shake vigorously a suspension of 5 mg of aromatic plant in 50 ml of water together with 5 ml of 1.7 g ml^{-1} sodium polytungstate solution and agitate with ultrasound for about 3 min;
- centrifuge for 2 min at 1000 g. Silicate minerals (density 2.5-2.7 g ml^{-1}) sink whereas organic components float;
- fill up the tube carefully with water and discard the upper water layer and the organic material;
- wash the mineral residue and add an acid solution;
- neutralise the acid by ammonium hydroxide;
- transfer the isolated minerals on stainless steel discs and store the disc overnight at 50°C in a laboratory oven;
- record the TL curve on minerals.

Table 1 *Areas of TL Signals Recorded on the whole Aromatic Herbs Irradiated (10 kGy) or not (Reference).*

Herb	Reference	Irradiated	TL(I)/TL(R)
Parsley	1.6 ± 0.1	23 ± 0.5	14.4
Thyme	2.4 ± 0.1	47 ± 1.8	19.6
Savory	2.0 ± 0.1	12 ± 0.7	5.9
Sage	1.3 ± 0.1	13 ± 0.9	9.9
Rosemary	11.6 ± 0.1	27 ± 0.8	2.3
Origanum	0.6 ± 0.01	11 ± 0.5	18.5
Fennel	1.9 ± 0.2	19 ± 0.6	10.0

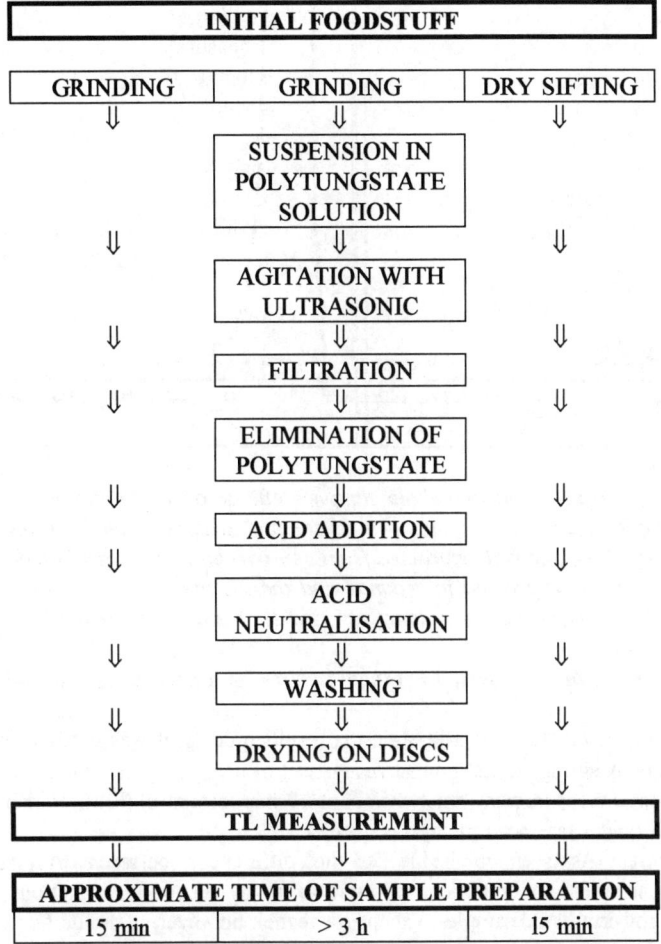

Figure 1 *Different ways of sample preparation for TL measurements*

2.3 TL Studies on Silicate Minerals Extracted from the Plant

2.3.2. Results. The results are completely different (Table 2 and Figure 2), as there is a ratio of about 2 orders of magnitude, at least and this can be sufficient as a presumptive test.

In the same way, studying the influence of time storage, there is no difficulty in distinguishing the irradiated from the non-irradiated samples more than 8 months after irradiation, whatever the aromatic plant studied.

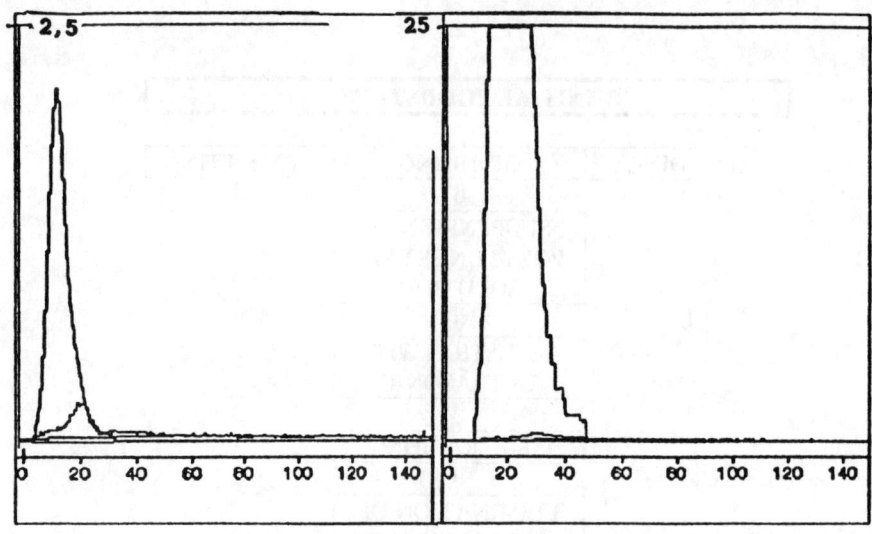

Figure 2 *TL of parsley: On the whole sample (left) or on the minerals (right): the samples masses are 6 mg (whole sample) and 0.25 and 0.1 mg for the minerals respectively extracted from non-irradiated and irradiated plant. The time is expressed in seconds and the TL intensities are respectively 2 and 23 u.a. for the whole sample, 5 and 1068 u.a. for minerals.*

2.4 Improvement of the Extraction Protocol in View of a Screening Method

2. 4.1. Experimental. In order to obtain the results quickly, it was decided to extract the mineral parts by sifting on 125 μm sieves instead of using polytungstate. In this case (Figure 1) the total measurement time was about 20 min instead of 3-4 h, which is much more suitable if only a screening method is needed.

2.4.2. Results. As seen in Table 3, the differences between irradiated and non-irradiated samples are sufficient enough to allow a good distinction between irradiated and non-irradiated samples, except for fennel: however, it should be noted that fennel is the only plant of which seeds are analysed..

Table 2 *Areas of TL Signals Recorded on the Minerals Extracted by Density Separation from Irradiated (10 kGy) or Non-Irradiated Aromatic Herbs*

Herb	Reference	Irradiated	TL(I)/TL(R)
Parsley	20.3 ± 1.2	9600 ± 380	473
Thyme	2.3 ± 0.1	940 ± 53	409
Savory	8.5 ± 0.6	830 ± 66	97.7
Sage	3.3 ± 0.2	1000 ± 16	303
Rosemary	2.6 ± 0.1	1100 ± 49	423
Origanum	1.6 ± 0.1	830 ± 50	519
Fennel	9.5 ± 0.6	900 ± 60	94.7

Table 3 *Areas of TL Signals Recorded on the Minerals Extracted by Filtration from Irradiated (10 kGy) or Non-Irradiated Aromatic Herbs*

Herb	Reference	Irradiated	TL(I)/TL(R)
Parsley	2.5 ± 0.2	976 ± 78	390
Thyme	2.7 ± 0.2	198 ± 11	73
Savory	4.0 ± 0.3	201 ± 13	50.3
Sage	2.3 ± 0.2	484 ± 36	213
Rosemary	5.4 ± 0.3	757 ± 58	140.7
Origanum	2.9 ± 0.2	141 ± 90	48.7
Fennel	3.3 ± 0.2	35 ± 30	10.6

3 TL STUDY OF SHELLS

In the case of shell-fish such as clams, oysters, scallops *etc.*, a typical shelf-life extension of 2 to 3 fold can be achieved, at 0 to 5°C storage temperatures, using radiation

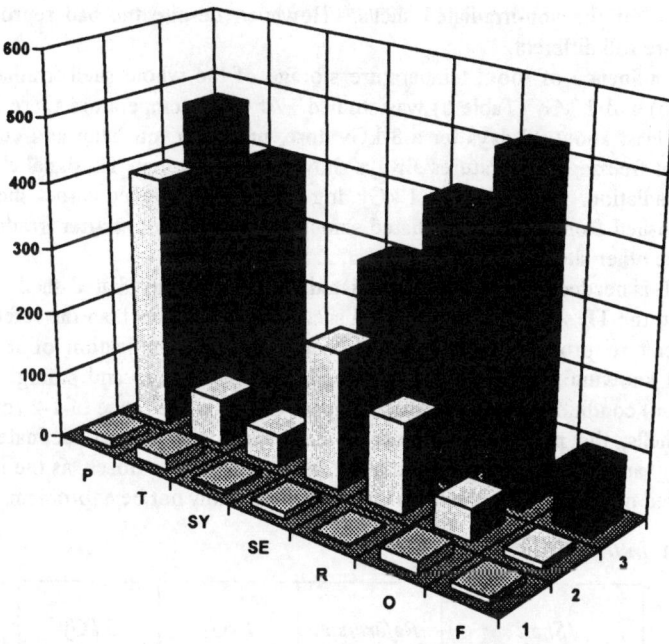

Figure 3 *Comparison on TL(I)/TL(R) ratio between irradiated and non-irradiated samples depending on the measurement (1 = whole sample, 2 = extraction of minerals by sifting, 3 = extraction of minerals by filtration) for the different plants: P = parsley, T = thyme, SY = savory, SE = sage, R = rosemary, O = origanum, F = fennel*

processing[28-30] in the absorbed dose range of 0.5 to 2.5 kGy. The studied shells were quahaug (*Venus mercenaria*, French: praire), common-mussel (*Mytilus edulis*, French: moule), knotted-cockle (*Cardium tuberculatum*, French: coque), dog-cockle (*Pectunculus glycemeris*, French: amande de mer) and grooved carpet shell (*Tapes decussatus*, French: palourde). A crushing mill (type DANGOUMEAU, Prolabo Cie) was used to grind the shells, the granulometry of the powder being controlled using sieves with different hole diameters.

Whatever the choice of sufficient maximal temperature, the glow curve (Figure 4) has the same general shape, *i.e.* a first maximum around 160°C, which is radio-induced, followed by a decrease, and a new and important final increase of the TL signal (maximum 380°C) in all samples. The following recording conditions were chosen:
- pre-heating of 10 sec at 50°C
- heating phase (9°C s^{-1}) for 30 sec
- maximum: 320°C.

It was found that, for a constant sample mass, the sensitivity as well as the reproducibility depend on the granulometry; thus, a small and constant granulometry was chosen.[31]

Immediately after irradiation at 1 or 8 kGy (Table 4), there is no problem in distinguishing the irradiated from reference samples. The case is less favourable for the grooved carpet shells irradiated at 1 kGy which give a TL signal of 3 ± 0.9 instead of 0.9 ± 0.3 for the non-irradiated shells. However, despite the bad reproducibility, the values are still different.

The influence of room temperature storage of the whole shell irradiated at 8 kGy (Table 5) and 1 kGy (Table 6) was studied. At room temperature there is no problem until at least about 30 days for a 8 kGy dose, moreover, quahaug and common-mussel stored at freezing temperatures always show a relatively high TL signal even 1200 days after irradiation. In the case of 1 kGy dog-cockle and grooved carpet shell they can be distinguished from the non-irradiated samples at least 22 days after irradiation, and for the three other shells, at least 29 days.

If TL is needed to identify the irradiated from the non-irradiated shells, one has to be sure that the TL signal of the irradiated shell is always higher than the reference shell. It is difficult to produce a definitive protocol for the identification of irradiated shells without knowing exactly the irradiation (dose, temperature) and storage (temperature, time, *etc.*) conditions. However, as the doses will be in the range of 1-2 (or 2.5) kGy for fresh shells, the protocol proposed in this paper can be used, definately in case of quahaug, and also probably for the other shells. For higher doses, as the irradiation will be carried out in freezing conditions, there will probably not be a problem.

Table 4 *Initial TL Signal Heights*

Shell	Reference	1 kGy	8 kGy
Quahaug	2.0 ± 0.5	246 ± 204	553 ± 260
Common-mussel	2.0 ± 0.3	16 ± 13	106 ± 84
Knotted-cockle	0.6 ± 0.3	8 ± 6	26 ± 14
Dog-cockle	1.0 ± 0.4	47 ± 2	104 ± 18
Grooved carpet	0.9 ± 0.3	3 ± 0.9	11 ± 2

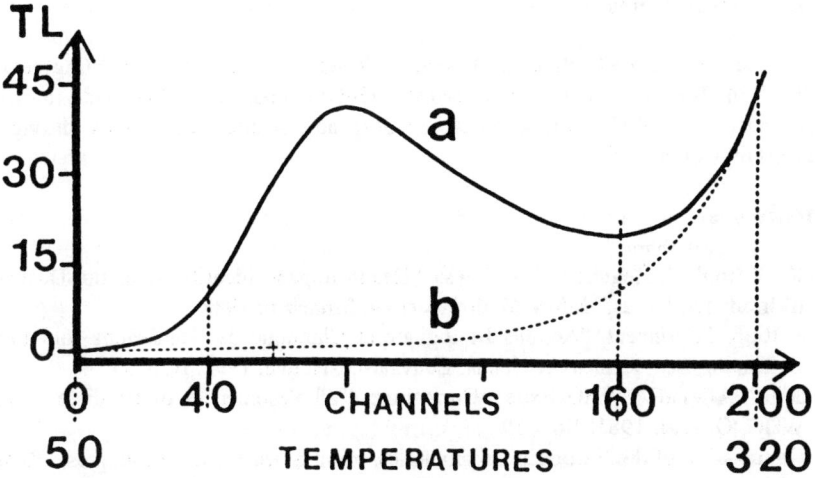

Figure 4 *TL signals of irradiated (a) and non-irradiated (b) Quahaug*

Table 5 *Influence of Storage Time (Days) on TL Signal Height of Shells Irradiated at 8 kGy*

Shell	Storage Time (Days)					
	0	23	28-34	600	1200	Control
Quahaug	553 ± 231		81 ± 4	35 ± 8	27 ± 6	2 ± 0.5
Common-mussel	106 ± 72		27 ± 0.5	19 ± 7	15 ± 6	2 ± 0.3
Knotted-cockle	26 ± 13		3 ± 0.6		1.8 ± 0.6	0.6 ± 0.3
Dog-cockle	104 ± 24	28 ± 3			19 ± 6	1 ± 0.4
Grooved carpet	11 ± 2		4.5 ± 0.6		1.6 ± 0.5	0.9 ± 0.3

Table 6 *Influence of Storage Time (days) on TL Signal Height of Shells Irradiated at 1 kGy*

Shell	Storage Time (Days)			
	0	22	28-29	Control
Quahaug	246 ± 173		24 ± 4	2 ± 0.5
Common-mussel	16 ± 11		6 ± 0.8	2 ± 0.3
Knotted-cockle	8 ± 7		3 ± 0.6	0.6 ± 0.3
Dog-cockle	47 ± 6	10 ± 1.5		1 ± 0.4
Grooved carpet	3 ± 1	3 ± 0.8		0.9 ± 0.3

Acknowledgements

We are indebted to the Community Bureau of Reference (BCR, Brussels) (Agreements 5348/1/5/340/90/4-BCR-F[10] and 5415/1/5/340/90/11-BCR-F[10]) and to IAEA (Agreement N°5154/CF) for financial support and helpful discussions during the meetings they organised.

References

1. K. W Bögl, D. Regulla and M. Suess, "Health Impact, Identification, and Dosimetry of Irradiated Foods," MMV Medizin Verlag, Munchen, 1988.
2. J. Raffi, P. Vincent, "Actions Biologique et Chimique des Radiations Ionisantes," B. Tilquin, Ed., Teknea-Academia, Louvain-La-Neuve, 1992, II, p. 91.
3. Joint FAO/IAEA/WHO Expert Committee, "Wholesomeness of Irradiated Food," WHO, Geneva, 1981, No. 659.
4. Commission of the European Communities, Food Science and Techniques, "Reports of the Scientific Committee for Food," Commission of the European Communities (BCR), Brussels, Luxembourg, 1986, EUR 10840 EN.
5. FAO/IAEA, Research Co-ordination Meeting on Food Irradiation Programme for the Middle East and European Countries, Wageningen, 17-21 April 1989.
6. A. J. Swallow, *Radiat. Phys. Chem.*, 1990, **35**, 311
7. K. W. Bögl, *Radiat. Phys. Chem.*, 1990, **35**, 301.
8. H. Delincée, "Analytical Detection Methods for Irradiated Foods. A Review of the Current Literature," IAEA-TECDOC-587, Vienna, IAEA, 1992.
9. H. Delincée and D. Ehlermann, *Radiat. Phys. Chem.*, 1989, **34**, 877.
10. J. Raffi, *Cahiers de l'ENSBANA*, Dijon, 1992, **8**, 235.
11. J. Raffi and J.-J. Belliardo, "Potential New Methods in Identification of Irradiated Food," Commission of the European Communities (BCR), Brussels, Luxembourg, 1991, EUR 13331 EN.
12. M. Leonardi, J. Raffi and J-J. Belliardo, "Recent Advances on Detection of Irradiated Food," Commission of the European Communities (BCR), Brussels, Luxembourg, 1993, EUR 14315 EN.
13. J. Raffi, H. Delincée, E. Marchioni, C. Hasselmann, A-M. Sjoberg, M. Leonardi, M. Kent, K-W. Bögl, G. Schreiber, M. H. Stevenson and W. Meier, "New Methods for the Detection of Irradiated Food," Commission of the European Communities (BCR), Brussels, Luxembourg, 1993/4, EUR 15261 EN.
14. L. Heide and K. W. Bögl, *Int. J. Food Sci. Technol.*, 1987, **22**, 93.
15. J. Oduko and N. Spyrou, *Radiat. Phys. Chem.*, 1990, **36**, 603.
16. A-M. Sjöberg, M. Manninen, P. Harrnala and S. Pinnioja, *Z. Lebens. Forsch*, 1990, **190**, 99.
17. D. Sanderson, C. Slater and K. Cairns, *Radiat. Phys. Chem.*, 1989, **34**, 915.
18. D. Sanderson, G. A. Schreiber and L. Carmichael, "A European Interlaboratory Trial of TL Detection of Irradiated Herbes and Spices," SURRC Report to BCR, 1991.
19. D. Doehlert, *Appl. Statistics*, 1970, **19**, 231.
20. P. Giamarchi, A. Fakirian, A-A. Chaouch, I. Pouliquen, G. Lesgards, M. Sergent and R. Phan Tan Luu, *Analusis*, 1994, **22**, 127.

21. A. Fakirian, PhD Thesis, University of Aix-Marseille III, 6 April 1994.
22. J. Raffi and S. Benzaria, *J. Radiat. Steril.*, 1994, **September**, 281.
23. M. Desrosiers, W. McLaughlin, L. Shean, N. Dodd, J. Lea, J. Evans, C. Rowlands, J. Raffi and J-P. Agnel, *Int. J. Food Sci. Technol.*, 1990, **25**, 682.
24. A. Rossi, G. Poupeau, O. Chaix, J. Raffi and J-P. Agnel, "ESR Applications in Organic and Bioorganic Materials," B. Catoire Ed., Springer-Verlag, Berlin, 1992, p. 151.
25. J. Raffi, M.H. Stevenson., M. Kent, J.M. Thiery and J-J. Belliardo, 1992, *Int. J. Food Sci. Technol.*, **27**, 111.
26. J. Raffi and J-P. Agnel, 1989, *Radiat. Phys. Chem.*, **34**, 891.
27. J. Raffi, J-P. Agnel and S-H. Ahmed, *Food Technol.*, 1991, 3/4, 26.
28. WHO, Consultation on Microbiological Criteria for Foods to be Further Processed Including by Irradiation, Geneva, 29 May - 2 June 1989.
29. N. Chuaqui-Offermanns, AECL-9062, 1989, p. 9.
30. A-M. Dollar, "Proceedings of the Asian Workshop on Food Irradiation, Jakarta, IDN, Asia COFAF, 1985, p. 31.
31. J-P. Agnel, M-S. Dutraive, I. Vaux, I. Rustan and J. Raffi, "Recent Advances on Detection of Irradiated Food," Commission of the European Communities (BCR), Brussels, Luxembourg, 1993, EUR 14315 EN, p. 193.

DETECTION OF IRRADIATED SPICES BY THERMOLUMINESCENCE ANALYSIS

K. M. Hammerton and C. Banos

Australian Nuclear Science and Technology Organisation
PMB 1
Menai NSW 2234
Australia

1 INTRODUCTION

Spices are used extensively in prepared foods. The high levels of contamination of many spices with microorganisms poses a problem for the food industry. Irradiation treatment is the most effective means of reducing the microbial load to safe levels. Although the process is currently subject to a moratorium in Australia, it is used in several countries for the decontamination of spices. Methods for detecting irradiation treatment of spices are necessary to enforce compliance with labelling requirements or with a prohibition on the sale of irradiated foods.

Thermoluminescence (TL) analysis of spice samples has been shown to be an applicable method for the detection of all irradiated spices.[1,2] It was established that the TL response originates from the adhering mineral dust in the sample. Definitive identification of many irradiated spices requires the separation of a mineral extract from the organic fraction of the spice sample. This separation can be achieved by using density centrifugation with a heavy liquid, sodium polytungstate. Clear discrimination between untreated and irradiated spice samples has been obtained by re-irradiation of the mineral extract after the first TL analysis with an absorbed dose of about 1 kGy (normalisation). The ratio of the first to second TL response was about one for irradiated samples and well below one for untreated samples. These methods have been investigated with a range of spices in order to establish the most suitable method for routine control purposes.

2 METHODS

The established method for separating a mineral extract from spice samples was used.[1,2] Spice samples (5 g) were sonicated in water for 30 min, sieved through 125 or 250 μm nylon mesh, separated by density centrifugation with sodium polytungstate (1.7 g ml^{-1}), acidified with 1 M HCl, and the mineral extract weighed after evaporation of acetone with nitrogen. For TL analysis, 0.5 mg aliquots of the mineral extract in acetone were dispensed onto 1 cm aluminium discs. The total integral of the glow curve over the temperature range 150°C to 350°C was measured in a Victoreen Model 2800 TL reader.

The units of TL intensity are given as counts per 0.5 mg of mineral extract. After the first glow analysis (Glow 1), the discs were irradiated with a dose of 1 kGy and the second glow analysis (Glow 2) undertaken one week after irradiation. The TL ratio (Glow 1/Glow 2) was then determined.

3 RESULTS AND DISCUSSION

TL analysis was performed on the mineral extracts from 45 samples of 20 spices. Untreated samples and samples treated with 5 and/or 10 kGy were analysed. The samples could be distinguished as untreated or irradiated on the basis of first glow analysis (Glow 1) by using the total integral of Glow 1 and the shape of the glow curve.

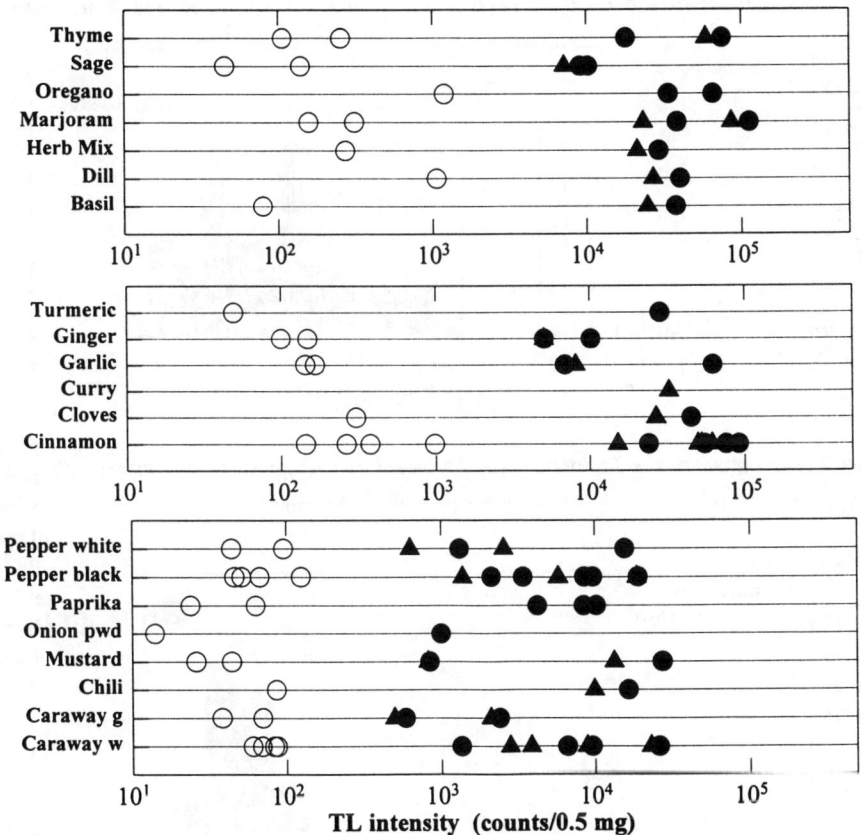

Figure 1 *TL intensities of mineral extracts from spices, either untreated (O) or irradiated with 5 kGy (▲) or 10 kGy (●). The spices have been divided into three groups as described in the text.*

It was found that the spice samples tended to fall within three groups (Figure 1):
(1) herbs, with a high mineral yield (> 5 mg from 5 g sample),
(2) spices with a high mineral yield (> 5 mg from 5 g sample), and
(3) spices with a low mineral yield (< 5 mg from 5 g sample or very low mineral content in extract).

With the latter group, microscopic examination of the mineral extracts revealed that there were very few mineral grains in some extracts (< 10 grains per 0.5 mg) and the presence of a large amount of organic material, especially with ground caraway. Consequently, the TL intensities were relatively lower for the extracts from both the untreated and irradiated samples of these spices including caraway, mustard, onion powder, black and white pepper.

For each group the TL intensities of the untreated and irradiated samples differed by at least one order of magnitude. But taken altogether, the TL intensities of the untreated and irradiated samples overlapped (Figure 2) including 3 untreated and 5 irradiated

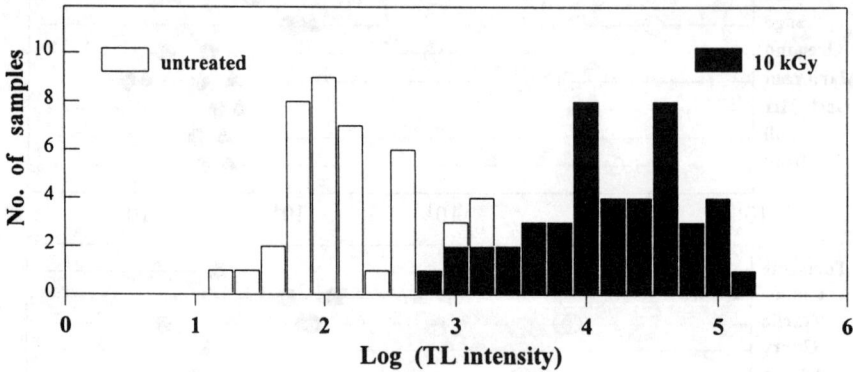

Figure 2 *Histogram of the TL intensities of mineral extracts from spices, either untreated or irradiated with 10 kGy (Glow 1 data given in Figure 1).*

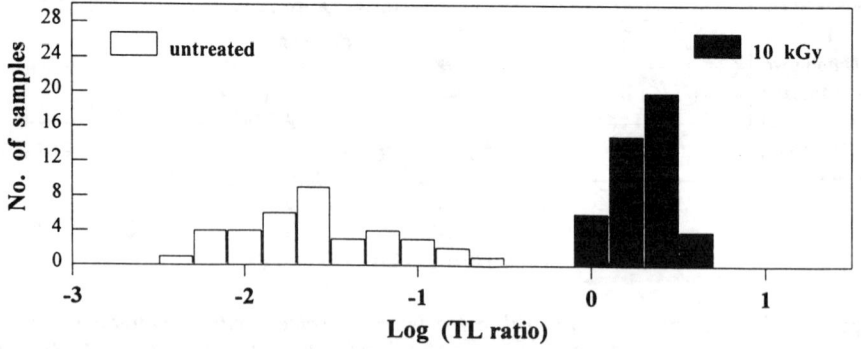

Figure 3 *TL ratios (Glow 1/Glow 2) of mineral extracts from spices, either untreated or irradiated with 10 kGy. The samples (Glow 1 data given in Figure 1) were re-irradiated with a dose of 1 kGy (Glow 2).*

samples. However, these samples could be identified on the basis of the shape of their glow curves. The glow curves for the untreated samples with relatively high TL intensities (cinnamon, dill, oregano) displayed a high temperature peak (in the region 250°C to 300°C) and no peak at 200°C. The low TL intensities observed for the 5 irradiated samples were due to the low mineral content in the extracts obtained from these samples (caraway, mustard, onion powder, black and white pepper) but the glow curves for all these samples displayed a definite peak at 200°C.

Normalisation by re-irradiation with a dose of 1 kGy from a ^{60}Co source, clearly identified all samples as either untreated or irradiated (Figure 3). The TL ratios varied between 0.0039 and 0.19 for untreated samples, between 0.79 and 2.4 for samples irradiated with 5 kGy and between 0.87 and 2.8 for samples irradiated with 10 kGy. During the course of this work it was found that the magnitudes of the TL intensity and the TL ratio were dependent on the mineral composition of the extract, the treatment radiation dose, the re-irradiation dose, the quantity of extract analysed, and the fading of the TL response with time.

4 CONCLUSION

From our investigations on the identification of irradiated spices with TL analysis it was concluded that for routine control purposes: (a) the separation of a mineral extract from the spice sample was necessary, (b) for the majority of spices identification of irradiation treatment could be determined on the basis of the TL response of the mineral extract, and (c) for some spice samples, such as caraway, mustard, black and white pepper, for which the yield of minerals was low, clear identification required the normalisation of the TL response by re-irradiation of the mineral extract.

References

1. D. C. W. Sanderson, "Food Irradiation and the Chemist," D. E. Johnston and M. H. Stevenson, Eds., Royal Society of Chemistry, Cambridge, 1990, p. 25.
2. G .A. Schreiber, U. Wagner, A. Leffke, N. Helle, J. Ammon, H. V. Buchholtz, H. Delincée, S. Estendorfer, K. Fuchs, H. U. von Grabowski, W. Kruspe, K. Mainozyk, H. Munz, H. Nootenboom, C. Schleich, N. Vreden, C. Weizorek and K. W. Bögl, "Thermoluminescence Analysis to Detect Irradiated Spices, Herbs and Spice-and-Herb Mixtures - An Intercomparison Study," Federal Health Office (BGA), Berlin, SozEp Heft 2, 1993

THE DETECTION OF IRRADIATION OF FOODSTUFFS BY A THERMOLUMINESCENCE METHOD, AND THE LIMITS OF ITS PRACTICAL APPLICABILITY

J. Kispéter and L. I. Kiss

College of Food Industry
Mars tér 7
H-6724 Szeged
Hungary

1 INTRODUCTION

The analysis of thermoluminescence (TL) curves is one of the physical methods widely used for the detection of irradiated foodstuffs and radiation-induced changes. The phenomenon of TL was discovered by Boyle in 1663. He observed that diamond and calcium fluoride emitted weak light in response to human body warmth. The analysis of TL curves has been and is successfully applied to determine the parameters of electron traps in solids and inorganic semi-conductors on the basis of the band model. In simpler cases, an exact mathematical description has been given,[1] which promoted the development of semi-conductor devices.

From the early 1950s, it became clear that TL could be advantageously applied for the measurement of doses of ionising radiation (TL dosimetry).[1]

During the past decade, this method was successfully used primarily for the detection of irradiation treatment of spices, dried vegetables and herbs. Heide and Bögl reported on the first detailed investigation of whole samples.[2] Later international interlaboratory studies proved the applicability of this method.[3] It was found that the intensity of radiation-induced TL signals varied considerably for different samples within a given product. From a practical point of view, the primary purpose was to demonstrate, after treatment and during storage, that the samples had been irradiated.

It follows from an interpretation of the phenomenon of TL that the intensity of the TL signal and the area under the TL curve scarcely change during storage if γ-radiation fills those traps for which the temperatures of thermal energy needed for the release of trapped electrons are considerably higher (by more than 100°C) than that of storage (usually room temperature).

For whole samples, the rapid decrease in TL intensity can be attributed to the fact that, besides the absorption of γ-radiation in electron traps, other energy absorptions and transformations take place as well. Their rearrangement may proceed at room temperature (*e.g.* within some weeks), decreasing the TL intensity.

It seemed obvious to separate that part of the sample from which TL primarily originates. As expected, TL measurements on separated contaminant minerals resulted in signals greater by several orders.[4-6] The normalised TL intensity was introduced to eliminate the influences caused by the amount and composition of the separated minerals.[6] Today, these normalised TL values, confirmed by international studies, are accepted in the detection of the irradiation treatment of foodstuffs.[7] Using the separated minerals, the elaboration of European standards for the TL method is now in progress being based on interlaboratory investigations.

For the investigation of irradiated foodstuffs, application of the TL method is derived from measurements on inorganic minerals, which means a solid-state physical and semi-conductor problem. Moreover, if isolated minerals are used for TL investigations, the assumption that the whole product itself is the dosimeter should be abandoned. Thus, after identification of the minerals, the results of solid-state physics can be utilised for an interpretation of the phenomenon of TL in the case of industrial food products.

It was proved in one of the experiments carried out at the College of Food Industry in Hungary that there was a group of industrial food materials where mineral separation could not be carried out in practice. Milk and egg powders, milk protein concentrate, gluten, *etc.* belong in this group. Only whole sample investigations can be performed on these products and in such cases, the application of other methods is advisable to confirm samples identified as irradiated by TL.

Kispéter and co-workers have investigated protein-containing products previously. It should be noted that the properties of such samples are characteristic of organic semi-conductors,[8] therefore, they prove to be good model materials for the detection and interpretation of radiation-induced changes.

The aim of this work was to determine the basic parameters of the shallow and deep electron traps responsible for the TL phenomenon in a chosen model material.

2 MATERIALS AND METHODS

Milk protein concentrate powder (MPCP) was used as model material in these experiments and was produced by a patented method[9] at the Hungarian Dairy Research Institute (Mosonmagyaróvár, Hungary). Pressed disc samples were prepared for the TL measurements, the samples were irradiated with the γ-rays using a ^{60}Co radiation source up to an absorbed dose of 20 kGy. TL readings commenced immediately after radiation treatment.[10,11]

TL measurements were carried out in two temperature ranges:
- by means of an NHZ-203 TL dosimeter reader (Central Research Institute for Physics, Budapest, Hungary) between 300 and 600 K, with linear heating rates of 3-15 K s^{-1}, to investigate deep electron traps.[12]
- by means of a home-built vacuum-cryostat between 80 and 300 K, with linear heating rates of 0.3-1.5 K s^{-1}, to investigate shallow electron traps.[13]

In order to determine the characteristic parameters of electron traps, the different heating rate method[1] was applied, based on the registration of TL curves at several linear heating rates. In the case of second-order kinetics (when the retrapping of thermally released electrons predominates), the rate dependence of the TL intensity is given by the expression:

$$I_m \left(\frac{T_m^2}{\beta}\right)^2 = \left(\frac{E}{2ks}\right)^2 \exp\left(\frac{E}{kT_m}\right)$$

where I_m and T_m are the intensity and temperature of the TL peak, β is the applied linear heating rate and k is the Boltzmann constant ($k = 8.616 \, 10^{-5}$ eV/K). Two fundamental parameters characterising electron traps, the activation energy (E) and the frequency factor (s), can be determined from the parameters of the corresponding line if the natural logarithm of the left-hand side of the above equation is plotted versus $1/T_m$.

3 RESULTS

Figure 1 shows TL curves measured on MPCP samples at different heating rates at a dose level of 10 kGy. With increasing β the TL peaks shift towards higher temperatures, while their intensities and areas decrease. Values of $E = 0.544$ eV and $s = 43.23$ s^{-1} were calculated for the peaks at higher temperatures[12] with the different heating rate method.

The electron traps in the samples were saturated by means of Xe-excitation at 80 K before low-temperature measurements were carried out. TL curves of non-irradiated samples (D = 0 kGy) as a function of heating rate can be seen in Figure 2. It was found that the peak temperature shifted similarly but the TL intensity and area increased with increasing heating rate. In this case,[13] values of $E = 0.093$ eV and $s = 0.35$ s^{-1} were calculated with the applied method.

This contradictory behaviour of the TL intensities and areas indicates the opposite characters of shallow ($E \approx 0.09$ eV) and deep ($E \approx 0.55$ eV) traps.

In order to verify the validity of the presumed second-order kinetics for the samples, curve fitting was performed with SigmaPlot 4.1 software.[14] These results are shown in Figure 3 for measurements above and below room temperature.

The error of fitting is less than 10%. This means that second-order kinetics can be considered acceptable for a description of the TL phenomenon occurring in such a complicated system as a industrial food product.

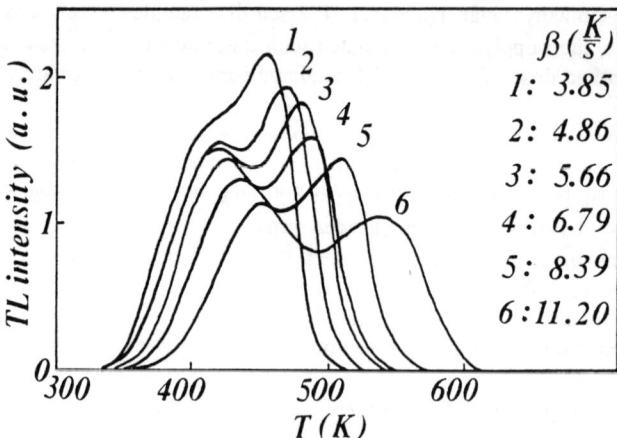

Figure 1 *High-temperature TL curves of MPCP samples (D = 10 kGy) at different heating rates*[12]

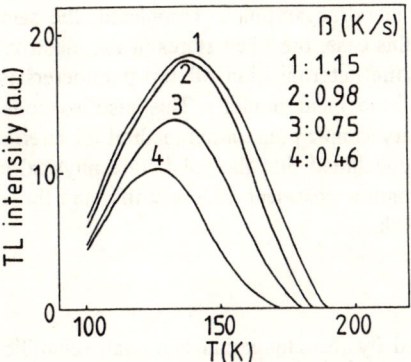

Figure 2 *Low-temperature TL curves of MPCP samples (D = 0 kGy) at different heating rates*[13]

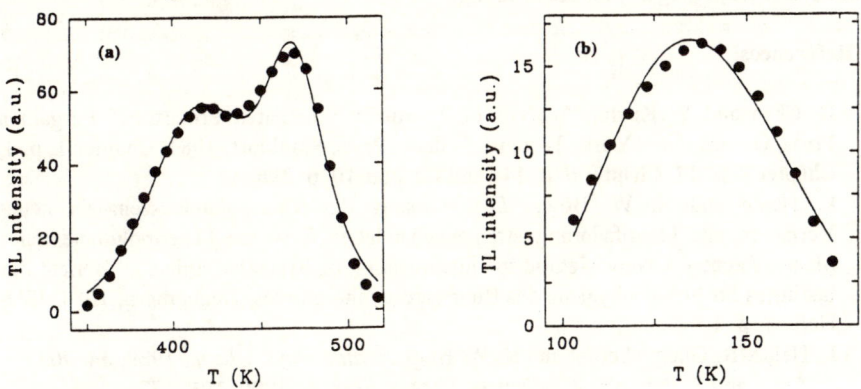

Figure 3 *Curve fitting for TL measurement of (a) an irradiated (D = 10 kGy) and (b) a non-irradiated (D = 0 kGy) MPCP sample with β = 4.86 and 0.98 K s^{-1}, respectively*
(The circles represent measured values; the continuous line is the theoretical curve obtained by the fitting method)

4 CONCLUSIONS

The presented studies relate to whole samples from which mineral separation is not possible. Irradiation treatment can be identified unambiguously with the applied TL method, as functions of absorbed dose and storage time. The influence on the TL response could also be followed by the change in content of certain microelements (Fe, Se),[10,11] thereby indicating that the compositions of products under investigation should be considered. With these industrial food samples, which have semi-conductor properties, the presented method can be successfully applied for determination of the parameters of electron traps, as was proven by the results obtained.

Supposing that the mineral separation is completed, the samples consist mainly of quartz and feldspar. In this case, the filled states in the electron traps naturally remain for a long time, so that the determination of trap parameters is obvious with the TL method, on the basis of the band model. This justifies considering the separation procedure to be the primary sample preparation method for international standards.

It can, therefore, be concluded that the solid-state physical band model permits an interpretation of the irradiation treatment of foodstuffs, and these investigations may be important in practice as well.

Acknowledgments

The work was supported by the Hungarian National Scientific Research Foundation (Contract No. I3/153 to J. K.).

The authors thank Dr. Éva Hideg (Institute of Plant Biology, Biological Research Centre, Szeged, Hungary) for curve fitting.

References

1. R. Chen and Y. Kirsh, "Analysis of Thermally Stimulated Processes," Pergamon Press, Oxford, New York, Toronto, Sydney, Paris, Frankfurt, 1981, Chapter 1, p. 1, Chapter 2, p. 17, Chapter 6, p. 144 and Chapter 10, p. 296.
2. L. Heide and K. W. Bögl, "Die Messung der Thermolumineszenz-ein neues Verfahren zur Identifizlerung strahlenbehandelter Gewurze (Thermoluminescence Measurement - A New Method for Identification of Irradiated Spices)," Bericht des Institutes fur Strahlenhygiene des Bundesgesundheitsamtes, Neuherberg, 1984, ISH-Heft 58, p. 1.
3. L. Heide, R. Guggenberger and K. W. Bögl, Radiat. Phys. Chem., 1989, **34**, 903.
4. T. Autio and S. Pinnioja, Z. Lebensm. Unters. Forsch., 1990, **191**, 177.
5. H. Y. Göksu, D. F. Regulla, B. Hietel and G. Popp, Radiat. Prot. Dos., 1990, **34**, 319.
6. D. C. W. Sanderson, "Food Irradiation and the Chemist," D. E. Johnston and M. H. Stevenson, Eds., Royal Society of Chemistry, Cambridge, 1990, p. 25.
7. G. A. Schreiber, U. Wagner, A. Leffke, N. Helle, J. Ammon, H.-V. Buchholtz, H. Delincéc, S. Estendorfer, K. Fuchs, H.-U. von Grabowski, W. Kruspe, K. Mainczyk, H. Münz, H. Nootenboom, C. Schleich, N. Vreden, C. Wiezorek and K. W. Bögl, "Thermoluminescence Analysis to Detect Irradiated Spices, Herbs and Spice-and-Herb Mixtures - An Intercomparison Study," Bericht des Institutes für Sozialmedizin und Epidemiologie des Bundesgesundheitsamtes (BGA), Berlin, 1993, SozEp Hefte 2, p. 46.
8. J. Kispéter and L. Horváth, J. Non-Cryst. Solids, 1987, **90**, 661.
9. Gy. Babella, Á. Novák, J. Jancsó, F. Mike and Gy. Molnár, "Eljárás teljes tejfehérje-koncentrátumok elõállítására (Procedure for Preparing Whole Milk Protein Concentrates)," Hungarian Patent No. 183 726, National Patent Office, Hungary, 1987.
10. J. Kispéter, L. I. Horváth and L. I. Kiss, Food Struct., 1992, **11**, 165.
11. J. Kispéter, L. I. Horváth, K. Bajúsz-Kabók, L. I. Kiss and P. Záhonyi-Racs, Food Struct., 1993, **12**, 379.

12. J. Kispéter, I. Dékány, L. I. Kiss and T. Marosi, *Food Struct.*, 1994, in press.
13. J. Kispéter and L. I. Kiss, In: "New Developments in Food, Feed and Waste Irradiation," G. A. Schreiber, N. Helle and K. W. Bögl, Eds., Bericht des Institutes für Sozialmedizin und Epidemiologie des Bundesgesundheitsamtes, (BGA), Berlin, 1993, SozEp Hefte 16, p. 71.
14. L. I. Kiss and J. Kispéter, *Acta Alimentaria,* 1994, **24**, 251.

THERMOLUMINESCENCE DETECTION OF IRRADIATED HERBS AND SPICES: AN AUSTRALASIAN TRIAL

P. B. Roberts and K. M. Hammerton*

Institute of Geological and Nuclear Sciences (GNS), Lower Hutt, New Zealand and
*Australian Nuclear Science and Technology Organisation (ANSTO), Lucas Heights Australia

1 INTRODUCTION

Thermoluminescence (TL) is generally regarded as the detection method offering most promise for irradiated herbs and spices. The method has been developed in several laboratories, especially in the United Kingdom[1] and Germany.[2] This paper describes a double blind trial of the method carried out by two Australasian laboratories (GNS and ANSTO).

2 METHODS AND RESULTS

The detection method comprises isolation of adherent minerals and measurement of their TL glow curve (glow 1). A second glow (glow 2) is carried out after re-irradiation (1 kGy). An assignment of whether the sample was or was not irradiated is then made via the glow 1/glow 2 ratio. Ratio values above 0.5 were deemed to indicate irradiated samples.

The method of mineral isolation and TL measurement are essentially as described elsewhere.[3] A Harshaw 2000 TL reader was used to integrate the glow curves over the temperature range 240-260°C at GNS. The glow curve was integrated over the whole glow curve at ANSTO. The herbs and spices chosen were taken from a test collection held at GNS. They were cinnamon, garlic powder, ground ginger, marjoram, onion powder, oreganum, paprika, sage, thyme, turmeric and white pepper.

The trial was carried out in three stages. In the first stage, samples of paprika and sage (both controls and irradiated) were exchanged. Both laboratories knew which samples were irradiated and which were controls. Similar and successful results were obtained in both laboratories providing an approximate check on the comparability of the methods and data obtained in the two laboratories.

The next stage involved the exchange of 5 samples each of sage and paprika. This time each laboratory knew only that at least one sample that it received was a control and at least one had been irradiated. Both laboratories correctly identified all samples as irradiated or control, and no difficulties were encountered.

Table 1 *Australasian Collaborative Spice Trial (GNS Results)*

Sample		COUNTS				Ratio	I/N
		Glow 1		Glow 2			
Number	Type	Mean (n=5)	SD	Mean (n=5)	SD		
100	Thyme	406	20	14927	1334	0.027	N
101	Onion	7283	1285	7696	2106	0.94	I
102	Ginger	32232	12518	9891	4078	3.2	I
103	Cinnamon	301324	44220	111675	15962	2.7	I
104	Turmeric	8055	920	62606	14052	0.13	N
105	Marjoram	585	49	52458	13934	0.011	N
106	Thyme	30562	6252	9381	1993	3.2	I
107	Marjoram	769	259	31962	6864	0.024	N
108	Paprika	284244	35072	166402	19009	1.7	I
109	Garlic	8560	2408	143851	27070	0.059	N
110	Ginger	18049	4115	8926	2565	2.0	I
111	Pepper	368	163	2391	950	0.15	N
112	Oreganum	7751	1024	98115	18459	0.078	N
113	Onion	22116	12729	9308	4625	2.3	I
114	Pepper	2998	664	1902	245	1.6	I
115	Oreganum	184467	28872	175703	23655	1.1	I
116	Paprika	5445	1621	88629	8695	0.061	N
117	Garlic	4887	1076	100019	12995	0.049	N
118	Cinnamon	2615	605	18044	2390	0.14	N
119	Turmeric	94858	16009	45812	9846	2.1	I

SD = Standard Deviation; I = Irradiated; N = Non-irradiated

Table 1 (continued) *Australasian Collaborative Spice Trial (ANSTO Results)*

Sample		COUNTS				Ratio	I/N
		Glow 1		Glow 2			
Number	Type	Mean (n=4)	SD	Mean (n=4)	SD		
200	Garlic	144	17	16800	3360	0.0086	N
201	Oregano	67100	21200	44700	9860	1.5	I
202	Paprika	73	15	8720	2040	0.0084	N
203	Ginger	159	98	1750	334	0.091	N
204	Cinnamon	45800	12200	16800	5240	2.7	I
205	Ginger	3060	515	1230	115	2.5	I
206	Marjoram	117	24	13000	1750	0.0090	N
207	Paprika	10200	3360	6300	1850	1.6	I
208	Pepper White	114	16	793	63	0.14	N
209	Onion powder	14	1	398	101	0.035	N
210	Marjoram	166	34	15400	2310	0.011	N
211	Garlic	55200	15300	48000	26400	1.2	I
212	Onion Powder	996	249	688	176	1.4	I
213	Turmeric	30600	9100	12100	7510	2.5	I
214	Oregano	1190	334	68900	23600	0.017	N
215	Cinnamon	424	109	8510	420	0.050	N
216	Thyme	107	10	9980	715	0.011	N
217	Thyme	18000	2580	9500	1530	1.9	I
218	Turmeric	26100	6950	12800	2680	2.0	I
219	Pepper White	88	37	1150	250	0.077	N

SD = Standard Deviation; I = Irradiated; N = Non-irradiated

The final stage of the trial was set up as follows. Each laboratory prepared 2 samples of each of 10 herbs or spices (20 samples in total at each laboratory). Non-participants assigned numbers to the samples at random. Other non-participants then assigned between 8-12 samples to be irradiated. At ANSTO the irradiation dose was 5 kGy; at GNS a dose of 8 kGy was applied.

The samples were then exchanged and each laboratory's standard TL protocol applied. The results are provided in Table 1. All samples were correctly identified in both laboratories. Both laboratories measured 8 irradiated samples of the same spice. The glow ratios found in the two laboratories were within a factor of 2. Eight non-irradiated samples were also measured at both laboratories. Their glow ratios, as might be expected, showed a greater difference in the values measured in the two laboratories, the difference being up to about 6 fold.

3 CONCLUSIONS

Although the total number of samples tested is relatively small, the 100% success rate for this double blind trial, and the results reported from other laboratories around the world give the confidence that TL is a detection technique which should be internationally recognised as valid when carried out by protocols developed by the ADMIT programme and work carried out by the Commission Bureau of Reference (BCR). The method is also probably usable for other foods from which it is possible to isolate minerals.

References

1. D. C. W. Sanderson, C. Slater and K. J. Cairns, *Radiat. Phys. Chem.*, 1989, **34,** 915.
2. L. Heide and K. W. Bögl, *Int. J. Radiat. Biol.*, 1990, **57,** 201.
3. G. A. Schreiber, A. Hoffman, N. Helle and K. W. Bögl, *Radiat. Phys. Chem.*, 1994, **43,** 533.

Other Physical Methods

ATTEMPTS TO ELABORATE DETECTION METHODS FOR SOME IRRADIATED FOOD AND DRY INGREDIENTS

S. Barabássy, M. Sharif, J. Farkas, J. Felföldi, Á. Koncz, Z. Formanek and K. Kaffka

University of Horticulture and Food Industry
Ménesi 45
Budapest 1118
Hungary

1 INTRODUCTION

In many countries ionising radiation is increasingly used for microbial decontamination of dry food ingredients, such as spices and herbs, because this treatment causes minimal chemical alteration and few, if any, detectable changes in the flavour of spices.[1,2] However, many health authorities and consumer organisations demand unequivocal tests for identification of irradiated foods.

Due to the diversity and delicate chemical composition of spices, there is very little chance of developing routine chemical methods to detect a specific radiolytic product in dry spices and herbs, physical methods appear to have greater potential.

Significant reduction of the gel-forming capability after irradiation could be observed in several spices.[3] Polysaccharides and their damage degradation greatly influence the texture and rheological properties of untreated and heated plant tissues. Starches, pectins and other hydrocolloids are rather vulnerable to radiation damage as small changes in their structure and chemistry can drastically alter their functionality. Radio-depolymerization of various starches is rather independent of the foodstuffs from which they are derived.[4] Radio-depolymerization effectively reduces the molecular size of starch components by random scission of internal bonds, with a lowering of the apparent viscosity as radiation dose is increased.[5]

Degradation caused by ionising radiation in the volatile oils, lipids, carotenoids and starch is indicated in the near infrared (NIR) wavelength region by changes in the reflectance spectrum. In the NIR reflectance spectra of several spices (*e.g.* black and white pepper, paprika, cinnamon, allspice) after irradiation, as a result of the excited molecules in the absorption peaks, relatively permanent changes can be noticed *i.e.* changes in amplitude and spectral shifts.

The low doses of ionising radiation (for inhibition of the sprouting of tubers and bulbs) as a consequence of some hystological characteristics, induce changes in the electrical impedance[8] and derived quantities. The potato-electrode system is basically capacitive, but its value depends strongly and non-monotonically on the frequency and its phase angle.

On the basis of this background, the scope of the present research within the framework of ADMIT programme co-ordinated jointly by the FAO/IAEA[10] was as follows:

(1) To elaborate a detection method for some irradiated dry foods and ingredients on the basis of the radiation damage of their starch content using rheological properties such as the apparent viscosity. Viscosity measurements were performed on heat-treated suspensions of several spices as a function of gamma-radiation dose. Investigations were carried out to establish the effect of moisture conditions during irradiation and post-irradiation storage on the rheological properties of heat-gelatinised suspensions of ground black pepper, the role of temperature and alkalinity of the measurement of apparent viscosity.[11,12] The variation between different batches of the same untreated ingredient was also studied. Unknown samples irradiated and coded by Agroster Enterprise, and tested by viscometry[13] and a number of analytical techniques, were applied to estimate starch damage in samples.[14]

Contributing to the protocol elaboration of the viscosity method, the laboratory participated in an inter-laboratories collaborative study.[15]

(2) To study the applicability of other physical methods such as NIR spectrophotometry and impedometry to detect whether or not a particular foodstuff has been irradiated. Investigations were carried out to establish correlations between irradiation doses, as well as, storage time, and changes in the NIR spectra.[16] NIR spectra of irradiated spices were evaluated by the polar qualification system (PQS) which is a new qualitative method.[17]

The impedometric experiments aimed to test the relationship between impedance and the radiation dose, along with the applicability of the proposed identifying parameter to Hungarian potato varieties and to choose a new dielectric parameter *i.e.* the frequency dependent phase angle (ϕvalue).[18]

The present paper describes and discusses the results obtained during the period of ADMIT activity (1989-94) which are original and may be useful for the development of the detection of irradiated foods.

2 MATERIALS AND METHODS

Non-irradiated commercial samples of various spices and potatoes were received from trade or spice processing companies (Compack Commercial Packaging Company and Hungarian Paprika Enterprise). They were prepared according to the investigation methods.

For all investigations irradiation was carried out in a self-shielded ^{60}Co irradiator, type RH-gamma-30 at the Central Food Research Institute, Budapest under aerobic conditions, at the ambient temperature. The dose rate of the treatment varied between 7.5 to 7.0 kGy h^{-1} during the period of the experiments. Samples irradiated and coded by AGROSTER were also investigated.

2.1 Rheological Measurements

2.1.1 Preparation of the Samples. Spices were roughly ground using a Moulinex kitchen grinder and then pulverised in a coffee-mill. The ground spices were sieved and certain fractions used in the investigation. The sieve fractions investigated were: black and white pepper < 0.63 mm, cinnamon < 0.16 mm, allspice < 0.5 mm, and caraway < 0.16 mm.

The samples were filled into plastic bags and irradiated at the 0, (2), 4, 8, 16 and 32 kGy doses. Samples irradiated commercially by AGROSTER at 0, 4, 8 and 16 kGy were ground following irradiation.

2.1.2 Moisture Conditioning and Storage. Ground black pepper samples were equilibrated to various relative humidities over sulphuric acid solutions at ambient temperature, then irradiated and further stored under equilibrium relative humidities (ERH) such as 25, 50 and 75%.

Within 10 days after irradiation, and at approximately one month intervals until 100 days storage time, the viscometric measurements were performed as below.

2.1.3 Heat Gelatinisation and Viscometry. Fifteen percent (W/V) suspensions were prepared from the ground spice samples in an MSE homogeniser, then the pH of the suspension was adjusted to pH 13 by 33% NaOH solution. The gelatinisation of the suspension were performed by keeping them in a water bath at 95°C for 30 min with occasional stirring. After the heat-gelatinisation, the suspensions were cooled down to room temperature and, after approximately 1 h, the apparent viscosity of replicate samples was determined in a Rheotest type 2-RV 2 rotational viscometer using the S2 marked cylinder at shear rate $Dr = 437.4$ s^{-1}. For each measurement, 30 g of heat gelatinised suspension was used.

2.1.4 Effect of Temperature. The viscosity measurement described above was carried out under constant temperatures of 20, 30, 40, 49 and 58°C.

2.1.5 Falling Number. For white pepper samples, 7 g of powder were mixed with 25 ml distilled water and the falling number was determined after 1 min gelatinisation in a standard falling number apparatus type RA-750. For black pepper, 8 g of sample powder were mixed with 25 ml distilled water and after 2 min gelatinisation the falling number was determined.

2.1.6 Other Indices of the Damaged Starch. The damaged starch was determined colorimetrically,[19] the reducing sugar content was estimated by the AOAC method,[20] the alcohol induced turbidity of aqueous extract using the method of Deschreider,[21] the saccharides content with Boehringer enzymatic method,[22] the starch content according to Hungarian Standard[23] and the gelatinisation thermograms by differential scanning calorimetry[17,24] (Du-Pont 910 DSC, SETARAM DSC).

2.2 Near Infrared Reflectance (NIR) Spectrophotometric Measurements

Paprika powder (Origin: Southern Hungary) and ground and sieve fractionated cinnamon and allspice were used for this investigation. The paprika samples were exposed to radiation doses between 0 and 17 kGy in 1 kGy steps, while the cinnamon and allspice were irradiated at 0, 2, 4, 8 and 16 kGy, respectively.

For reflectance measurements, 10 cm^3 of spice powder was placed in the sample holder. The diffuse reflectance spectra of the samples were measured immediately after irradiation and then from time to time, during storage at room temperature. NIR spectra of the samples were recorded and stored in 2 nm steps from 1000 to 2500 nm using a Spectralyzer Model 1025 PMC research composition analyser. Reference halon and ceramic were used, and the data processing was performed on a computer program using the software supplied with the NIR instrument. The applied mathematical transformations of NIR spectra were: logarithmic, Savitsky-Golay smoothing, first and second derivatives. The most characteristic wavelengths and the performance data of the calibration were determined by multivariate linear regression (MLR). Using the polar qualification system (PQS) for qualitative evaluation, the information obtained from NIR spectra of the irradiated and control samples were compared. The samples during the storage period of the experiment were stored in closed plastic bags. The moisture content of the samples was not determined.

2.3 Impedometric Measurements

A Hungarian potato variety "Kisvárdai Rózsa" was used as test material. Radiation doses of 0, 50, 100, 150 and 200 Gy were applied and the potatoes were stored at 5°C.

A Hewlett-Packard 4284 A twin-electrode system, with 10 mm long gold plated needles of 0.7 mm thickness and 10 mm distance between them was used at 20°C to measure the impedance of the material at days 20 and 25 of the post-irradiation storage. The measurements were performed continuously collecting data from the whole frequency range (30 Hz-1 MHz) and reading the complex impedance parameters: the magnitude of impedance (Z) and the phase angle (ϕ). The proposed value to be used as an identifying parameter was the Z_{k50}/Z_{k5} ratio.[8,9] The impedance ratio was measured at the apical region of the potato tuber

3 RESULTS AND DISCUSSIONS

3.1 Rheological Investigations

Drastic reduction of the apparent viscosity of heat treated suspensions was observed after an irradiation dose of 8 kGy in white and black pepper, nutmeg, ginger, marjoram, cinnamon and allspice. No viscosity changes were found in samples of garlic and onion powder and in the case of caraway the viscosity increased even at a dose of 16 kGy (Table 1).

The explanation for these observations is the damage to the starch content of the irradiated and heat treated samples calculated from the reducing sugar content and determined according to the Hungarian Standard. The radiation damaged starch changed the rheological behaviour of the samples. The radiation-induced water-absorbing capacity of onion powder and the increase in the swelling capacity of caraway could explain the increasing apparent viscosity. The starch content was calculated from the reducing sugar content by acidic hydrolysis and according to the Hungarian Standard.

Eight different white pepper shipments were investigated in order to check the variability of apparent viscosity among imported batches. The results are summarised in Table 2.

Table 1 *Apparent Viscosity (Average ± Standard Deviation) of Heat-Treated 15% Suspensions of Various Ground Spices at 25 °C*

Spice Sample	Moisture Content %	Starch Content in Dry Matter %	Apparent Viscosity (mPa.s) 0 kGy	Apparent Viscosity (mPa.s) 8 kGy
White pepper (India)	13.4	57.2	917 ± 60	167 ± 18
Black pepper (India)	13.0	42.5	387 ± 18	133 ± 34
Nutmeg	12.2	29.6	548 ± 24	232 ± 18
Ginger	12.3	42.5	316 ± 18	143 ± 0
Marjoram	9.0	11.2	25 ± 0.5	14 ± 3
Allspice	9.9	16.1	23 ± 0.5	14 ± 0.1
Cinnamon	11.5	18.1	35 ± 0.1	15 ± 0.5
Garlic powder	-	-	6 ± 0.1	6 ± 0.1
Onion powder	10.8	-	5 ± 1	5 ± 1
Caraway	-	-	25 ± 3.6	36 ± 3.8 (16 kGy)

Table 2 *Variability of Different Imported White Pepper Shipments According to the Apparent Viscosity (Average ± Standard Deviation) of the Heat-Gelatinised 15% Suspensions of Non-irradiated Samples*

White Pepper Batches	Apparent Viscosity (mPa.s) of Triplicate Samples
White Muntok Pepper	829 ± 62
White China Pepper	815 ± 103
White Muntok Pepper	718 ± 62
White Sarawak Pepper	930 ± 65
White China Pepper	361 ± 64
Quality Muntok Asta White Pepper	700 ± 44
Brazilian White Pepper	927 ± 106
Brazilian White Pepper	917 ± 60

The investigated sample batches showed comparable apparent viscosities. One batch, however, gave a much lower viscosity, although it was still a considerably higher value than that shown for the 8 kGy sample in Table 1.

The influence of humidity conditions and storage time on the apparent viscosity of untreated and irradiated black pepper are given in Table 3.

Table 3 *Apparent Viscosity (Average ± Standard Deviation) of Heat Gelatinised 15% Suspensions of Ground Black Pepper Influenced by the Radiation Dose, Equilibrium Relative Humidity (ERH) and Storage Time at Ambient Temperature*

Days	ERH	Apparent Viscosity (mPa.s)				
		0 kGy	4 kGy	8 kGy	16 kGy	32 kGy
<10	75 %	539 ± 39	233 ± 60	154 ± 26	60 ± 4	36 ± 19
	50 %	520 ± 2	534 ± 8	133 ± 8	44 ± 2	70 ± 5
	25 %	450 ± 39	249 ± 5	133 ± 0.5	50 ± 3	23 ± 4
41	75 %	430 ± 156	229 ± 3	116 ± 10	31 ± 7	19 ± 4
	50 %	483 ± 120	162 ± 30	97 ± 9	49 ± 2	20 ± 0
	25 %	596 ± 0	240 ± 2	150 ± 5	51 ± 6	24 ± 2
72	75 %	415 ± 86	173 ± 13	108 ± 9	22 ± 0	18 ± 4
	50 %	473 ± 18	198 ± 10	91 ± 2	38 ± 1	16 ± 3
	25 %	419 ± 0.5	176 ± 27	109 ± 9	41 ± 2	10 ± 0
100	75 %	250 ± 36	137 ± 19	82 ± 2	25 ± 7	14 ± 1
	50 %	458 ± 18	194 ± 33	105 ± 8	36 ± 7	17 ± 3
	25 %	571 ± 24	122 ± 16	122 ± 16	47 ± 4	26 ± 2

The water activity of samples did not significantly influence the direct effect of radiation treatment. A decrease in viscosity was observed in samples stored from 75% equilibrium relative humidity level. It is supposed that at 75% ERH some amylolytic process might occur during long-term storage. The range of individual relative viscosities of untreated samples at 75% ERH was still much higher even after 100 days of humid storage than the range of individual apparent residual viscosities of 8 kGy or higher dose irradiated samples kept under dry (25% ERH) conditions (Figure 1).

The influence of storage time on the apparent viscosity of untreated and irradiated cinnamon and allspice are shown in Table 4.

No changes during the 60 days storage period were noted and the results obtained were similar to those for heat gelatinised black pepper suspensions. With both cinnamon and allspice, during the total storage period of 60 d even at the 2 kGy dose, samples showed significantly reduced apparent viscosity in comparison with the control.

The difference between the apparent viscosities of suspensions of untreated and irradiated samples can be enlarged at elevated temperatures. Rotational viscometry was performed with ground black pepper suspensions at 25 and 50°C, and with ground white pepper suspensions from 20°C to 60°C by 10°C steps. The black pepper suspensions at 50°C showed considerably lower apparent viscosity values while the character of the viscosity versus shear-rate plots remained similar (Figure 2).

In the case of white pepper the temperature-dependency of untreated samples and irradiated ones with a dose of 8 kGy is given in Table 5.

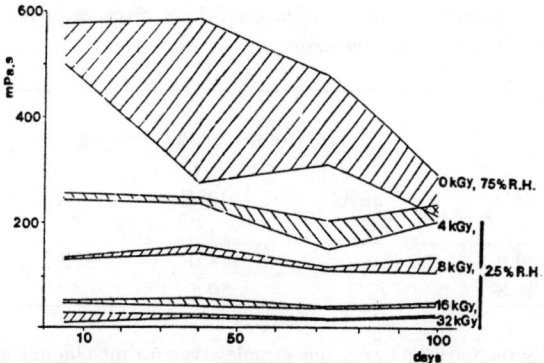

Figure 1 *Ranges of apparent viscosities of heat-gelatinised suspensions of ground black pepper samples affected by radiation doses and storage time*

Table 4 *Apparent Viscosity ± Standard Error (mPas) Values of Gelatinised Cinnamon and Allspice Samples as a Function of Storage Time*

Days	Apparent Viscosity (mPa.s)				
	0 kGy	2 kGy	4 kGy	8 kGy	16 kGy
Cinnamon					
0	35 ± 0.0	25 ± 0.0	24 ± 0.7	15 ± 0.5	8 ± 0.1
14	36 ± 0.5	22 ± 0.4	22 ± 0.4	14 ± 0.7	8 ± 0.1
60	36 ± 1.5	22 ± 0.1	23 ± 0.4	14 ± 0.0	7 ± 0.2
Allspice					
0	23 ± 0.5	19 ± 0.9	18 ± 0.0	14 ± 0.1	13 ± 0.2
14	21 ± 0.6	18 ± 0.6	17 ± 0.4	13 ± 0.0	13 ± 0.6
60	21 ± 0.4	18 ± 0.0	17 ± 0.0	14 ± 0.0	13 ± 0.4

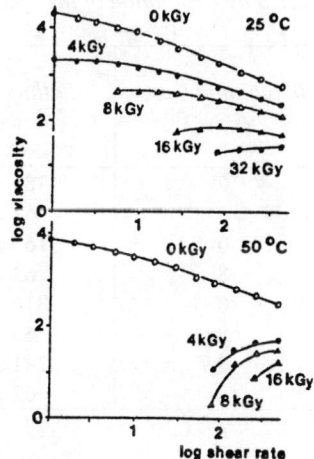

Figure 2 *Comparative rheograms of heat-treated 5% suspensions of ground black, pepper at 25 °C and 50 °C measuring temperature, respectively*

Table 5 *Apparent Viscosity of Heat Gelatinised 15% Suspensions of Ground White Pepper as a Function of Temperature*

Irradiation Dose (kGy)	Apparent Viscosity (mPa.s)				
	20°C	30°C	40°C	49°C	58°C
0	520.9	464.6	414.0	372.3	345.4
8	86.8	72.8	59.1	49.0	42.7

The apparent viscosity of the untreated samples was 6 times higher at 20°C, 8-times higher at 58°C than that of the 8 kGy irradiated sample.

In principle the falling number determination represents viscometry in a boiling-water bath. In comparison with rotational viscometry the measuring temperature is higher. Measurements were performed with white and black pepper samples untreated and irradiated at 8 kGy dose, respectively. Results are shown in Table 6.

Although falling number is being used for rapid determination of alpha-amylase activity of cereal products, in this case it was more indicative of the stability of gel formed in the spice suspension, and of the radio-depolymerization of the starch content. The great differences observed between the falling numbers of untreated and irradiated samples are encouraging to the development of this technique.

Based on these experimental results, the potential method *i.e.* viscometry, was tested with commercially irradiated samples. The coded samples were treated at AGROSTER. After decoding it could be determined that the results of the viscosity measurements followed well the order of the irradiation dose (Table 7).

Knowing that one of the samples was untreated, the investigator correctly chose the non-irradiated samples on the basis of their smallest viscosity and established the correct order of samples according to the irradiation dose.

Table 6 *Falling Number (± Standard Error Values) of Ground Black Pepper and White Pepper Samples as Affected by Irradiation of the Spice*

Spice sample	Irradiation Dose (kGy)	Falling Number (sec)
Brazilian white pepper	0	1389 ± 61
	8	62 ± 1
Indonesian black pepper	0	165 ± 17
	8	63 ± 1
Indian black pepper	0	816 ± 18
	8	63 ± 2
Indian black pepper	0	241 ± 21
	8	62 ± 0.5

Table 7 *Viscometric Results of Coded Samples Obtained from AGROSTER Inc.*

Spices	Code	Irradiation dose (kGy)	Apparent Viscosity (mPa.s) Mean	95% Confidence Intervals
Black pepper	742	0	174.7	169.4 - 180.0
	625	3	94.0	88.7 - 99.3
	593	6	52.0	46.7 - 57.3
	272	9	35.0	29.7 - 40.3
White pepper	697	0	444.7	439.4 - 450.9
	283	3	166.8	160.5 - 173.0
	121	6	106.8	100.5 - 113.1
	379	9	61.3	55.1 - 67.6
Ginger	791	0	441.0	-
	346	3	381.0	-
	951	6	334.0	-
	275	9	298.0	-

Starch gelatinisation shows an endothermic peak at calorimetry (DSC) measurements. The thermograms of untreated and irradiated black and white pepper, cinnamon and allspice were taken. The gelatinisation temperature and the heat (enthalphy) did not change significantly as an effect of the irradiation.

The apparent viscosity of the suspensions of cinnamon and allspice was measured after pH adjustment and without heat gelatinisation treatment. The results of these measurements are shown in Table 8.

For both spices irradiation caused significant changes in the apparent viscosity even at the 2 kGy dose. Making a comparison between the heat gelatinised and untreated samples, in the latter case the values of the apparent viscosity were slightly higher.

3.2 NIR Investigations

Diffuse reflectance NIR spectra of paprika, cinnamon and allspice were taken immediately after irradiation, and then from time to time: for paprika after 7, 14, 21, 30, 50 and 114 days, for cinnamon and allspice after 1, 4, 18, 30 days.

Table 8 *Apparent Viscosity ± Standard Error Values (mPas) of Non-Gelatinised Suspensions of Spices*

| Spice | Dose (kGy) | | | | |
	0	2	4	8	16
Cinnamon	46.1 ± 2.6	23.3 ± 0.0	27.6 ± 1.1	16.8 ± 0.0	9 ± 0.1
Allspice	28 ± 3.4	22.4 ± 0.7	20.7 ± 0.0	18.9 ± 0.3	18.1 ± 0.0

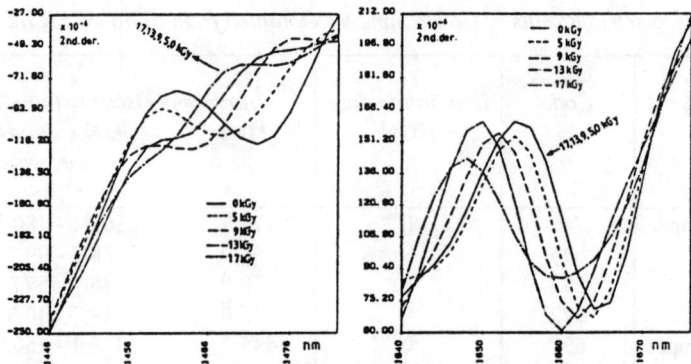

Figure 3 *Effect of gamma-irradiation on second derivative log (1/R) spectrum of paprika from 1446 to 1482 nm and 1640 to 1676 nm, respectively (Spectra were measured 7 days after irradiation)*

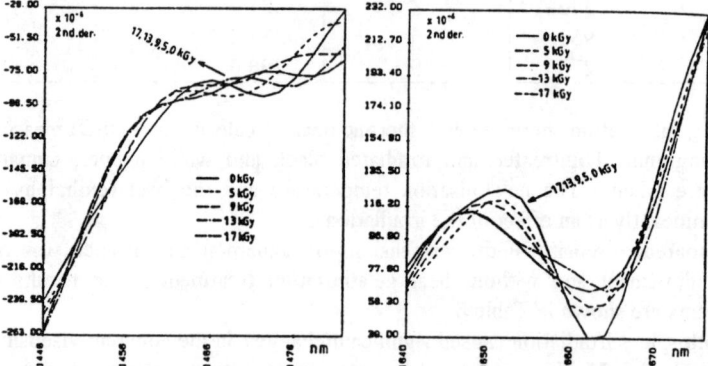

Figure 4 *Effect of gamma irradiation on second derivative log (1/R) spectrum of paprika from 1446 to 1482 nm and 1640 to 1676 nm, respectively (Spectra were measured 114 days after irradiation)*

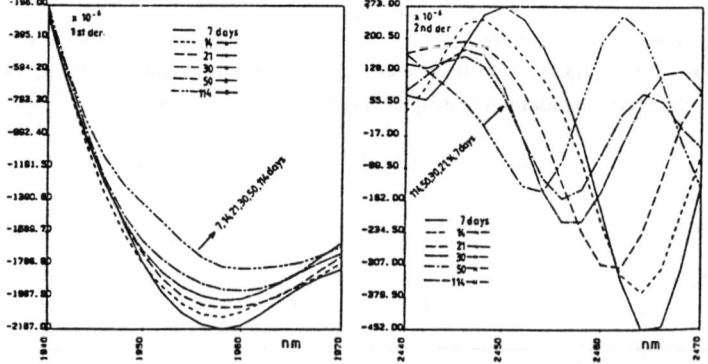

Figure 5 *Changes, at 6 intervals from 7 to 114 days after irradiation, in first and second derivative log (1/R) spectra of irradiated paprika powder from 1940 to 1970 nm and 2440 to 2470 nm, respectively (Irradiation dose, 17 kGy)*

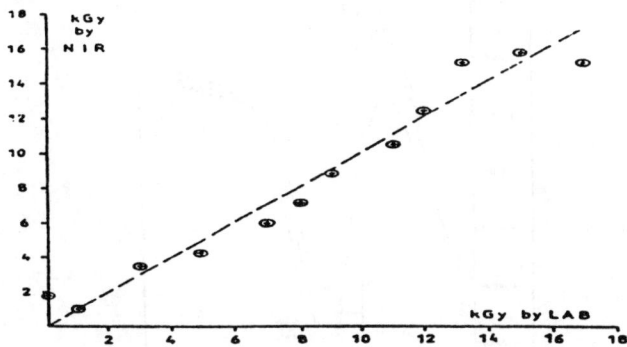

Figure 6 *Relationship between actual and NIR fitted irradiation dose 7 days after irradiation (Second derivative at 1656 nm, r = 0.975, Standard Error = 4.22 days)*

Changes caused by irradiation of the spectra were found in the wavelength regions characteristic for the oil, water and starch components. Figures 3 and 4 illustrate the effects of irradiation after 7 and 114 days, respectively, on the second derivative of the log (1/R) spectra of paprika powder in the 1452-1492 nm (water absorption) and 1644-1668 nm (oil absorption). Although the samples were exposed to radiation doses, between 0 and 17 kGy in 1 kGy steps, only the spectra of irradiated paprika with 0, 5, 9, 13, and 17 kGy are shown to demonstrate the changes.

A relatively permanent change can be noticed in the spectra. Changes in amplitude and spectral shifts are dependent on the dose applied. Figure 5 shows the changes in the paprika powder spectrum of a sample of paprika powder given a 17 kGy dose and measured at different storage times after irradiation.

The relationship between the values of the second derivative spectra at different wavelengths and irradiation doses was determined using a single term linear regression equation. The relationship 7 days after irradiation was determined by NIR and is shown in Figure 6.

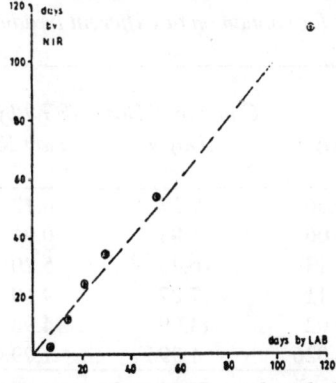

Figure 7 *Relationship between actual and NIR predicted elapsed time after irradiation (First derivative at 1952 nm, r = 0.995, Standard Error = 4.22 days)*

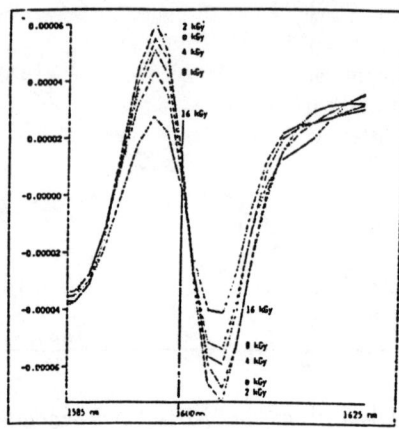

Figure 8 Log (1/R) NIR spectra of cinnamon on day 18 as affected by different irradiation doses

The correlation is good (Correlation Coefficient = 0.95, Standard Error = 1.627 kGy) but the parameters of the regression equation differed with storage time.

Results for the single term linear regression used to examine the relationship between first derivative spectral values at different wavelengths and elapsed time after irradiation are shown in Figure 7.

The question to estimate when a paprika powder was irradiated and the dose applied is not simple because the model is an equation with two unknowns.

The cinnamon and allspice samples were analysed by NIR spectrometry on different days and the most characteristic wavelength, which was determined from the second derivative of log (1/R) NIR spectra, changed from sample to sample. e.g. cinnamon on the first day was at 2360 nm, the region characteristic for the oil components, and on day 18 it was at 1600 nm which is characteristic for the starch component (Figure 8).

Table 9 shows cinnamon data evaluated by multivariate linear regression (MLR).

Table 9 Calculated Doses Obtained from the Spectral Data Measured by the NIR Spectrophotometer for Cinnamon at Different Irradiation Doses

Irradiation Doses (kGy)	Calculated Doses (kGy) by NIR			
	Day 1	Day 4	Day 18	Day 30
0	0.66	0.57	0.12	0.12
2	3.06	1.83	0.96	1.13
4	4.13	6.46	5.20	3.97
8	9.11	7.87	9.10	9.65
16	15.02	14.99	14.98	15.01
Standard Error	1.026	1.395	1.098	1.061

Other Physical Methods 197

The log (1/R) spectra of the cinnamon and allspice were also evaluated by polar qualification system (PQS). The quality points of the samples obtained as the gravity points of log (1/R) spectra represented in polar diagram, concentrated all information in one point. The arrangement of the quality points in two dimensions is shown in Figure 9 for cinnamon. Samples irradiated with higher doses are localised in the other part of the plain than the control samples.

3.3 Impedometric Investigations

The frequency-dependency of the magnitude of the complex impedance (Z) and its phase angle (ϕ) were tested followed by the dose dependence of the impedance of the potato tissue. Figure 10 illustrates the average measurements of samples irradiated with different doses.

The proposed[8,9] identifying parameter was the Z_{50k}/Z_{5k} ratio. These values were compared between two laboratories (Japanese-Hungarian)[9,18] and two varieties of potato, respectively. Table 10 gives the results for the dose range tested in the two different laboratories (countries) and there is particularly good agreement between the irradiated samples.

Although the impedance parameter significantly depends on the irradiation dose, applying this measurement to individual tubers or small lots, gives a high probability of failure due to the great variance of the parameter value. The sensitivity of the method is relatively poor as a 100 kGy dose results in only a 6-7% change of the measured parameter.

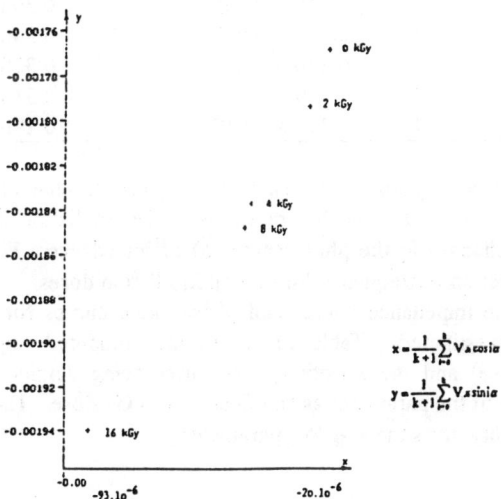

Figure 9 *The "quality points" of irradiated and control cinnamon samples determined by PQS method between 1560-1615 nm*

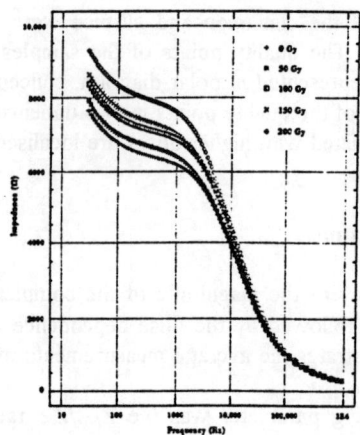

Figure 10 *Impedance curves as affected by the irradiation dose absorbed in potatoes (Curves represent average of 12-15 measurements)*

Table 10 Z_{50k}/Z_{5k} *Identifying Parameter Versus Irradiating Dose*

Irradiation Dose (Gy)	Z_{50k}/Z_{5k} Hungarian Test	Japanese Test
0	0.348 ± 0.038	0.392 ± 0.022
50	0.335 ± 0.012	0.346 ± 0.018
100	0.330 ± 0.011	0.328 ± 0.017
150	0.319 ± 0.0013	0.310 ± 0.019
200	0.298 ± 0.017	0.290 ± 0.020

The potato-electrode system has basically a capacitive character and it depends strongly and non-monotonically on the frequency. The irradiation dose influences this system and causes changes in the phase angle (ϕ) of impedance. Figure 11 shows the ratio of ϕ/ϕ_o as a function of frequency for several irradiation doses.

On this basis, the impedance curves and phase angle curves for a different pair of frequencies were investigated. Table 11 shows the computed regression equations (where d is the dose) and the sensitivity, the latter being defined as the percentage increase of the value of the parameter as an effect of 100 Gy dose. The highest sensitivity and statistical reliability gives the ϕ_{15k}/ϕ_{80} parameter.

4 CONCLUSIONS

Radiation treatment at dose levels for microbial decontamination brought about sufficient degradation in some spices, such as white pepper, black pepper, nutmeg, ginger,

Table 11 *Dose Dependence of the Identifying Parameters*

Pair of Frequencies	Parameter-Regression Equation	Sensitivity
50 kHz/5 kHz	Z_{50k}/Z_{5k} = -0.00023d + 0.349	6-7%
80 kHz/400 Hz	Z_{80k}/Z_{400} = -0.00018d + 0.188	9-10%
100 kHz/1 kHz	Z_{100k}/Z_{1k} = -0.00017d + 0.173	10%
75 kHz/80 Hz	ϕ_{75k}/ϕ_{80} = 0.00586d + 3.90	15%
15 kHz/80 Hz	ϕ_{15k}/ϕ_{80} = 0.00530d + 2.92	18%

marjoram, cinnamon and allspice, which appears in a significantly decreased viscosity of suspensions of their finely ground samples after heat treatment usually required for starch gelatinisation.

No useful results were obtained for garlic powder, onion powder and caraway. More detailed experiments with pepper samples confirmed[3,25] partial radio-depolymerization of their starch content. The viscometry was performed on particulate suspensions, therefore, both particles and the suspending medium make a contribution to rheological properties. The viscosity measurements expressed the changes by the irradiation in the starch heat gelatinization process (at 95°C, and pH 13) and also changes in the swelling capacity of the polysaccharides.

The identification limit by the viscometric method depends on the spices (*e.g.* 4 kGy for peppers and 2 kGy for cinnamon) but at 8 kGy or higher irradiation doses the method is applicable for detection of irradiated spices with considerable starch content.

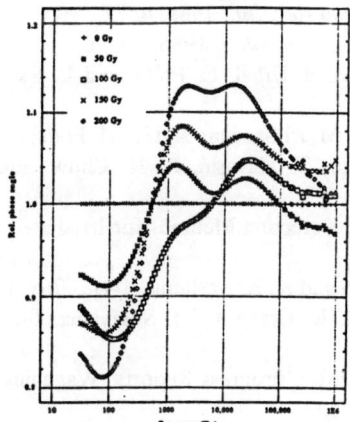

Figure 11 *Phase angle values of irradiated potatoes at various frequencies as compared to that of the non-irradiated potato (Curves represent averages of 12-15 measurements)*

The NIR spectrophotometric investigations on paprika powder, ground cinnamon and allspice showed that the changes in NIR spectra caused by irradiation were small, but detectable, and relatively permanent. The changes in the amplitude and spectral shifts were dependent on the applied dose and the elapsed time following irradiation. During storage the characteristic wavelengths of the correlation changed. The best correlation at the beginning of the storage experiments was at the wavelengths characteristic of the oil components and later at the wavelengths characteristic for water and starch components. The identification limit by the NIR method was around 3-4 kGy.

Based on some hystological characteristics of irradiated potatoes, dielectric parameters for identification were investigated *i.e.* the frequency dependent complex impedance (Z) and its phase angle (ϕ). The proposed value of Z_{50k}/Z_{5k} has shown a clear dose dependence for the Hungarian potato varieties tested, but the sensitivity of the method was limited. As new dielectric parameter of ϕ_{15k}/ϕ_{80} was suggested, which was more suitable as approximately a 3 times greater sensitivity was noted than for Z_{50k}/Z_{5k}.

The three physical methods studied in these investigations showed significant differences between irradiated and control samples. With these methods it was possible to detect the irradiated samples and place them in the correct order of dose.

The results described in this paper prove the applicability of the studied methods for spices.

The Hungarian group also participated in viscometry intercomparison studies on irradiated spices organised by ADMIT.[15]

References

1. J. Farkas, "Irradiation of Dry Food Ingredients," CRC Press, Inc., Boca Raton, Flo., USA, 1988
2. J. Kiss and J. Farkas, *Food Rev. Int.*, 1988, **4**, 77.
3. E. Mohr and G. Wichmann, *Gordian*, 1985, **85**, 96.
4. J. Raffi, B. Dauberte, M. d' Urbal, C. Pollin and L. Saint-Lèbe, *Stärke*, 1981, **33**, 301.
5. R. M. A. El Saadany, F. M. El Saadany and Y. H. Foda, *Stärke*, 1976, **28**, 208.
6. T. Suzuki, K. Yasymoto, T. Hayashi, R. K. Chow, and M Tajima, *Food Irrad. Japan*, 1988, **23**, 77.
7. H. Delincee, "Analytical Detection Methods for Irradiated Foods," IAEA-TECDOC-587, 1991, Vienna.
8. T. Hayashi, M. Iwamoto and K. Kawashima, *Agric. Biol. Chem.*, 1982, **46**, 905.
9. T. Hayashi, S. Todoriki, K. Otabe and J. Sugiyama, *Biosci. Biotechnol. Biochem.*, 1994, **56**, 1929.
10. S. Barabássy, IAEA-ADMIT Progress Reports, Wageningen 1989, Jachranka 1990, Budapest, 1992, Belfast 1994.
11. J. Farkas, M. Sharif and A. Koncz, *Radiat. Phys. Chem.*, 1990, **35**, 324.
12. J. Farkas, M. Sharif and A. Koncz, *Radiat. Phys. Chem.*, 1990, **36**, 621.
13. M. Sharif, PhD Thesis, University of Horticulture and Food. Industry, Budapest, 1994.
14. J. Farkas, M Sharif and S. Barabássy, *Acta Aliment.*, 1990, **19**, 273.
15. T. Hayashi, IAEA-ADMIT Progress Report, Belfast, 1994.

16. S. Barabássy, J. Farkas and K. Kaffka, "Making Light Work: Advances in Near Infrared Spectroscopy," VHC Ed., Weinheim, Basel, Cambridge, New York, 1992.
17. S. Barabássy, Z. Formanek and A. Koncz , *Acta Aliment.* (submitted), 1994
18. J. Felföldi, P. László, S. Barabássy and J. Farkas, *Radiat. Phys. Chem.*, 1993, **41**, 471.
19. P. C. Williams and K. S. W. Fegol, *Cereal Chem.*, 1969, **46**, 56.
20. Anon., Association of Official Agricultural Chemists, Washington DC, 1960, p. 162.
21. A. R. Deschreider, *Lebensm. Wiss. u. Technol.*, 1969, **2**, 90.
22. Boehringer, Biochemische Analytik Lebensmittelanalytik, 1989, 126.
23. Hungarian Standard MSZ 6830-66.
24. S. Barabássy and J. Farkas, *J. Food Phys.*, 1988, **52**, 23.
25. L. Heide, E. Mohr, G. Wichman and K. W. Bögl, ISH-Heft 125, 1988, p. 176.
26. L. Heide and K. W. Bögl, *Int. J. Radiat. Biol.*, 1990, **57**, 201.

DETECTION OF IRRADIATED POTATOES BY IMPEDANCE MEASUREMENT

T. Hayashi, S. Todoriki, K. Otobe and J. Sugiyama

National Food Research Institute
Ministry of Agriculture, Forestry and Fisheries
Kannondai, Tsukuba, Ibaraki 305
Japan

1 INTRODUCTION

Potato is one of the major food items to be treated with ionising radiation and potatoes are irradiated on a large scale in several countries. Every year around 15,000 t of potatoes are irradiated at doses of 60 to 150 Gy (average dose is about 100 Gy) in Japan. Although various methods to detect irradiated potatoes have been investigated,[1,2] no established method has been reported.

Measuring electrical conductivity or impedance of potatoes has been reported as a promising method for the detection of irradiated potatoes.[3-9] Scherz[4] reported that the electrical conductivity (impedance) of irradiated potatoes at a low frequency of alternating current increased for about 6 h after irradiation and then remained at a level higher than that of non-irradiated potatoes. In previous studies[7-9] it has been found that the ratio of impedance magnitude at 50 kHz to that at 5 kHz, measured immediately after puncturing a potato tuber, is dependent upon the dose applied to the tuber, independent of storage temperature and stable during storage after irradiation.

However, the impedance ratio scattered to some extent within one lot of potatoes. The conditions for impedance measurement were not investigated in detail, and it was expected that improving the measuring conditions would result in a better detection of irradiated potatoes. The aim of this study was to establish the optimum conditions for impedance measurement and to examine the applicability of the impedance measuring method to various cultivars (*cv.*) of potatoes.

2 MATERIALS AND METHODS

2.1 Potatoes

The following cultivars of potatoes were donated by Hokkaido National Agricultural Experiment Station in Eniwa, Hokkaido; May Queen, Kitaakari, Ezoakari, Waseshiro, Toyoshiro, Hokkaikogane, Norin-No.1 and Benimaru. Potatoes of *cv.* Danshaku (harvested in Shihoro, Tokoro, Kunnenpu, Makkari and Aikoku) and *cv.* Dejima (harvested in Nagasaki, Unzen, Goto and Kazusa) and the potatoes "Danshaku" irradiated at Shihoro Potato Irradiation Centre were purchased at a local market in Tsukuba. Ten to 20 potatoes (*cv*. Danshaku harvested in Shihoro, Hokkaido. Japan) were subjected to impedance measurements, unless otherwise stated.

2.2 Irradiation

Potatoes were irradiated with a Gamma Cell 220 (^{60}Co, 5.4 kGy h^{-1}, AECL, Canada). The accuracy of the dose rate was within 15%, as determined by Fricke dosimetry. Both the irradiated and non-irradiated potatoes were stored at 5°C in a dark room, unless otherwise stated.

2.3 Impedance Measurement of Potatoes

Unless otherwise stated, impedance of potatoes was measured under the following conditions. Potatoes were incubated at 22°C for 3 days and electrical measurements were carried out at that temperature with a stainless-steel two-electrode system; 1 mm diameter, 10 mm long, and 10 mm apart (Figure 1). A potato tuber was punctured with the steel electrodes connected to a Digital Spectral Analyzer TR9403 (Advantest Ltd., Japan). Alternating current of 250 Hz to 100 kHz at 1 mA was passed through the tuber and its voltage was measured. Impedance parameters were calculated from the current and the voltage and expressed as follows:-

Z_{5k}: magnitude of impedance at 5 kHz
Z_{50k}: magnitude of impedance at 50 kHz
R_{5k}: resistance at 5 kHz
R_{50k}: resistance at 50 kHz
X_{5k}: reactance at 5 kHz
X_{50k}: reactance at 50 kHz

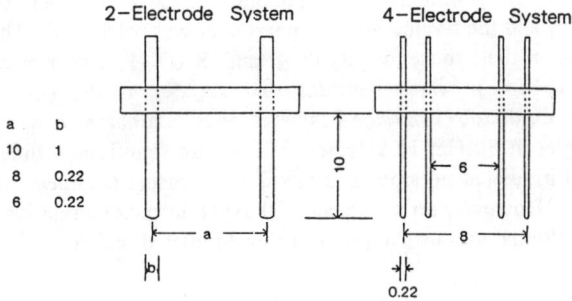

Figure 1 *Electrodes used for impedence measurements (mm)*

3 RESULTS AND DISCUSSION

3.1 Effect of Electric Current on Impedance

A Cole-Cole arc, circular locus of impedance, which was obtained by plotting resistance and reactance,[10] was determined by passing 0.1, 0.5, or 1 mA of alternating current through a potato tuber. Cole-Cole arcs from one measurement and those from an average of 8 measurements are shown in Figure 2. A higher electrical current and averaging the measurements resulted in smoother Cole-Cole arcs. The 3 averaged arcs measured at the 3 different currents coincided with each other when they were drawn on the same chart (Figure 3). These results indicate that electrical current does not essentially influence impedance values and that the fluctuation of impedance values is caused by noise attributable to the equipment. A higher current has been found to reduce the noise.

3.2 Effect of Type of Electrode on Impedance Parameters

Three different sizes of two-electrode system and one four-electrode system (Figure 1) were used for impedance measurements of potatoes. One mA of alternating current was applied to a potato tuber with the outer electrodes and the voltage was measured with the inner electrodes in the four-electrode system, and alternating current was applied to a tuber and the voltage was measured with the same electrodes in the two-electrode systems. Most of the impedance parameters at 5 kHz and 50 kHz and the ratios of impedance parameters at 5 kHz to 50 kHz were influenced by irradiation, but the parameters at 50 kHz were less sensitive to radiation treatment (Tables 1 to 3). Although impedance parameters (e.g. Z_{5k}, Z_{50k}, R_{5k}, R_{50k}, X_{5k} and X_{50k}) of potatoes irradiated at the same dose varied with the type of electrodes, the ratios of impedance parameters at 5 kHz to 50 kHz (e.g. Z_{5k}/Z_{50k}, R_{5k}/R_{50k}, and X_{5k}/X_{50k}) were independent of the type of electrodes. These results indicate that the normalised parameters (ratios at 5 kHz to 50 kHz) are free from the influence of type of electrodes.

To establish the best parameter for identifying irradiated potatoes, the data were analysed statistically and the t-value for each parameter was calculated. The t-values for Z_{5k}/Z_{50k} were higher than those for R_{5k}/R_{50k} and X_{5k}/X_{50k}, irrespective of type of electrodes (Tables 1 to 3), which indicates that Z_{5k}/Z_{50k} is the best parameter to distinguish irradiated potatoes from non-irradiated ones. Barabássy et al.[11] has reported that the phase angles at 80 Hz, 15 kHz and 75 kHz are significantly different between non-irradiated and irradiated potatoes and this is a promising parameter for identifying irradiated samples. However, no significant difference in phase angle was observed at any frequency of alternating current between non-irradiated and irradiated potatoes in this study.

3.3 Effects of Measuring Temperature on Impedance Ratio

The dependence of the impedance ratio, Z_{5k}/Z_{50k}, on measuring temperature is shown in Figure 4. The difference in the impedance ratio between non-irradiated and irradiated potatoes was at the largest level between 22 and 25°C.

Other Physical Methods

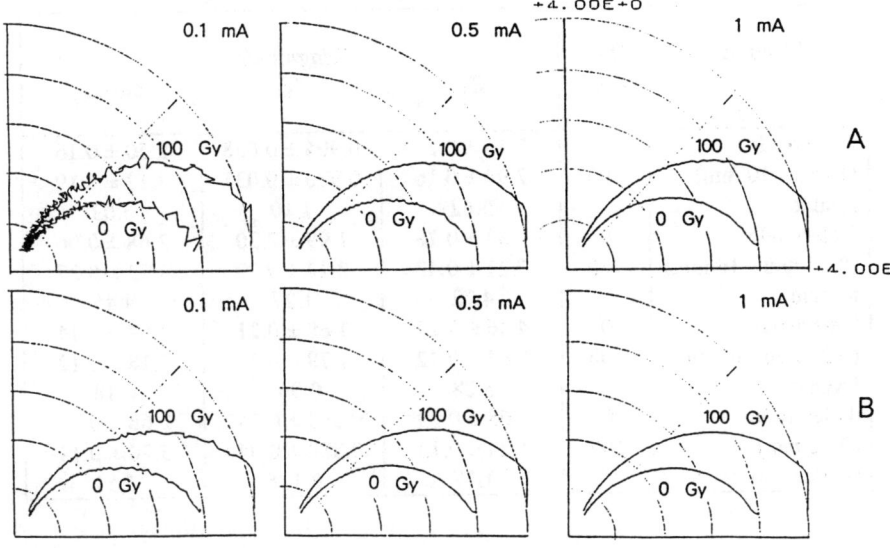

Figure 2 *Cole–Cole arcs of impedance of potatoes measured at various currents. A; One measurement. B; Average of eight measurement.*

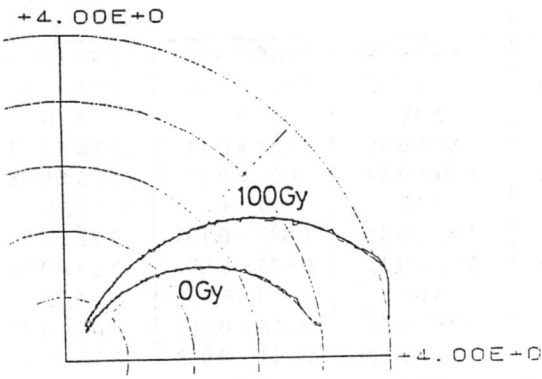

Figure 3 *Three averaged Cole–Cole arcs drawn on the same chart.*

Table 1 *Effect of Electrodes on Magnitude of Impedance*

Electrode	Dose (Gy)	Magnitude Z_{5k}	Z_{50k}	Z_{5k}/Z_{50k}
2 electrode	0	2.25 ± 0.22	0.904 ± 0.038	2.50 ± 0.16
(1 mm - 10 mm)	100	2.94 ± 0.16	0.930 ± 0.037	3.17 ± 0.19
t-value		5.62	1.10	6.03
2 electrode	0	5.33 ± 0.78	1.99 ± 0.30	2.68 ± 0.06
(0.22 mm - 10 mm)	100	7.21 ± 0.43	2.27 ± 0.39	3.23 ± 0.27
t-value		4.72	1.27	4.45
2 electrode	0	4.46 ± 0.27	1.68 ± 0.21	2.67 ± 0.14
(0.22 mm - 6 mm)	100	5.64 ± 0.42	1.78 ± 0.10	3.18 ± 0.12
t-value		5.28	0.961	6.18
4 electrode	0	1.05 ± 0.23	0.393 ± 0.080	2.68 ± 0.18
(0.22 mm)	100	1.24 ± 0.13	0.388 ± 0.040	3.20 ± 0.11
t-value		1.57	0.125	5.51

Table 2 *Effects of Electrodes on Resistance of Impedance*

Electrode	Dose (Gy)	Resistance R_{5k}	R_{50k}	R_{5k}/R_{50k}
2 electrode	0	2.09 ± 0.18	0.610 ± 0.090	3.43 ± 0.30
(1 mm - 10 mm)	100	2.62 ± 0.12	0.517 ± 0.065	5.15 ± 0.89
t-value		5.48	1.87	4.10
2 electrode	0	4.82 ± 0.78	1.24 ± 0.14	3.88 ± 0.22
(0.22 mm - 10 mm)	100	6.40 ± 0.52	1.23 ± 0.31	5.38 ± 0.88
t-value		3.77	0.066	3.70
2 electrode	0	4.03 ± 0.31	1.097 ± 0.163	3.72 ± 0.36
(0.22 mm - 6 mm)	100	5.03 ± 0.37	0.975 ± 0.091	5.20 ± 0.60
t-value		4.63	1.46	4.73
4 electrode	0	0.95 ± 0.20	0.258 ± 0.047	3.68 ± 0.42
(0.22 mm)	100	1.10 ± 0.12	0.211 ± 0.022	5.22 ± 0.58
t-value		1.39	2.03	4.81

Table 3 *Effects of Electrodes on Reactance of Impedance*

Electrode	Dose (Gy)	Reactance X_{5k}	X_{50k}	X_{5k}/X_{50k}
2 electrode	0	-0.836 ± 0.123	-0.598 ± 0.049	1.45 ± 0.10
(1 mm - 10 mm)	100	-1.271 ± 0.226	-0.756 ± 0.026	1.75 ± 0.25
t-value		3.78	3.55	4.15
2 electrode	0	-2.27 ± 0.20	-1.56 ± 0.27	1.48 ± 0.16
(0.22 mm - 10 mm)	100	-3.30 ± 0.39	-1.90 ± 0.26	1.77 ± 0.36
t-value		5.26	2.03	1.65
2 electrode	0	-1.88 ± 0.21	-1.28 ± 0.14	1.50 ± 0.03
(0.22 mm - 6 mm)	100	-2.55 ± 0.24	-1.49 ± 0.07	1.72 ± 0.17
t-value		4.70	3.00	1.43
4 electrode	0	-0.454 ± 0.112	-0.296 ± 0.069	1.54 ± 0.20
(0.22 mm)	100	-0.557 ± 0.065	-0.326 ± 0.037	1.72 ± 0.21
t-value		1.78	0.857	1.39

3.4 Effect of Pre-Incubation at 22°C on the Impedance Ratio

The difference in the impedance ratio between non-irradiated and irradiated potatoes became larger with the increase in the incubation period at 22°C prior to the impedance measurement (Figure 5). The influence of pre-incubation at 22°C on the

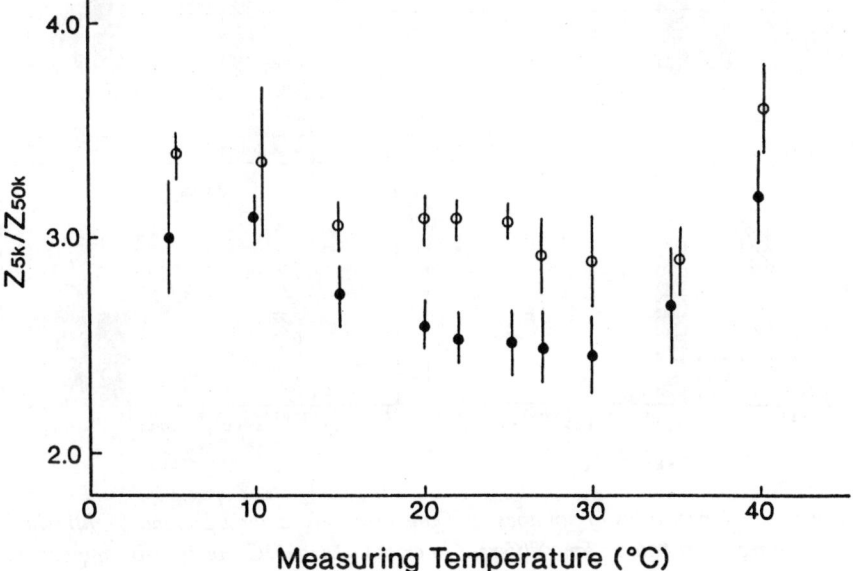

Figure 4 *Impedance ratios measured at various temperatures*
●, *0 Gy;* ○, *100 Gy*

impedance ratio of Dejima was similar to that of Danshaku; the potatoes of *cv.* Dejima incubated for 1 day at 22°C showed less distinct difference in the impedance ratio between non-irradiated and irradiated tubers, as compared with those incubated for 3 or 7 days at 22°C (Figure 5).

These results indicate that potatoes should be incubated at 22°C for 3 days or longer prior to the impedance measurement in order to clearly differentiate irradiated from non-irradiated potatoes.

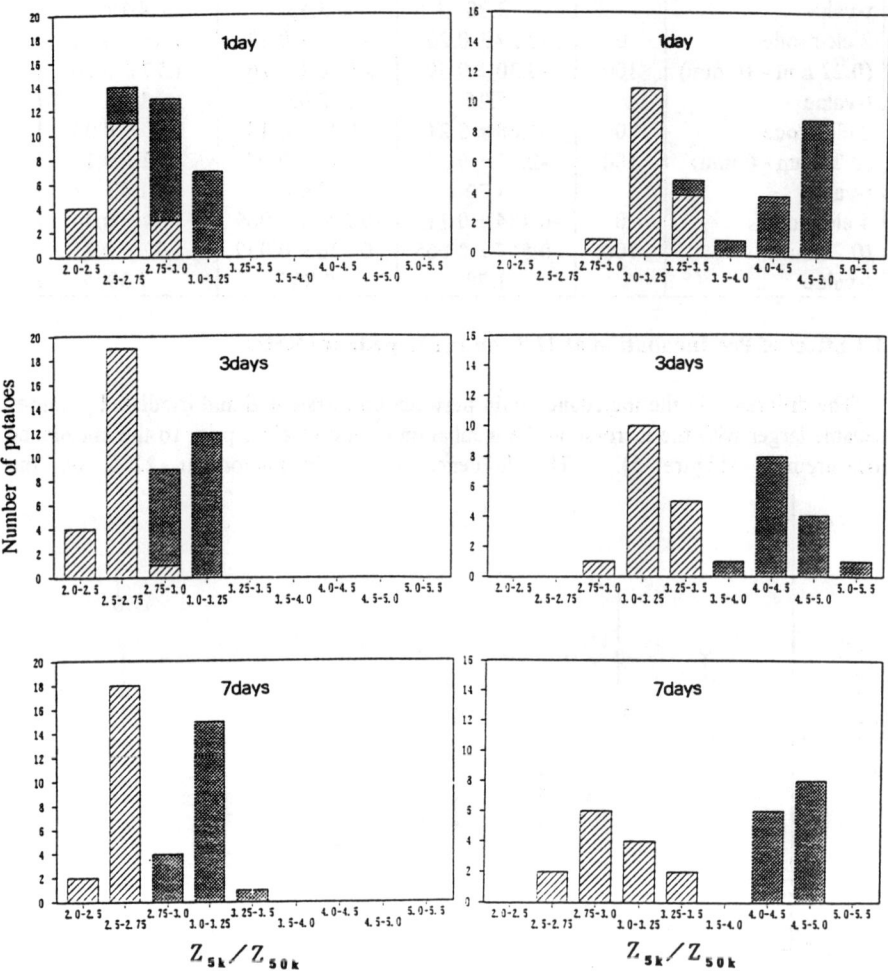

Figure 5 *Impedance ratios of potatoes cv. Danshaku (left) and cv. Dejima (right) which were incubated for different periods at 27°C prior to impedance measurement.*
▨, 0 Gy; ▬, 100 Gy

3.5 Effect of Measuring Region of Potato Tuber on Impedance Ratio

Impedance was measured at 4 different regions (A, Apical; B, Central; C, Side; D, Basal) (Figure 6) for 3 cultivars of potatoes. The impedance ratio, Z_{5k}/Z_{50k}, measured at the apical region showed the largest difference between non-irradiated and irradiated potatoes, irrespective of potato cultivar (Figure 6). The impedance ratio measured at an apical region was almost constant during storage after irradiation and allowed differentiation between non-irradiated and irradiated potatoes for up to 6 months.

3.6 Dose Dependency of the Impedance Ratio

The impedance ratios (Z_{5k}/Z_{50k}) of potatoes determined after storage for one month at 5°C increased with the dose (Figure 7).

3.7 Impedance Ratios of Potatoes from Different Planting Localities

The impedance ratios (Z_{5k}/Z_{50k}) of non-irradiated and irradiated (100 Gy) potatoes of *cv.* Danshaku and *cv.* Dejima from different planting localities are shown in Figure 8. Most of the impedance ratios of the non-irradiated potatoes of *cv.* Danshaku were lower than 2.75 and most of those of the irradiated ones were higher than 2.75, irrespective of planting locality. Most of the impedance ratios of the non-irradiated potatoes of *cv.* Dejima were lower than 3.25 and most of those of the irradiated samples were higher

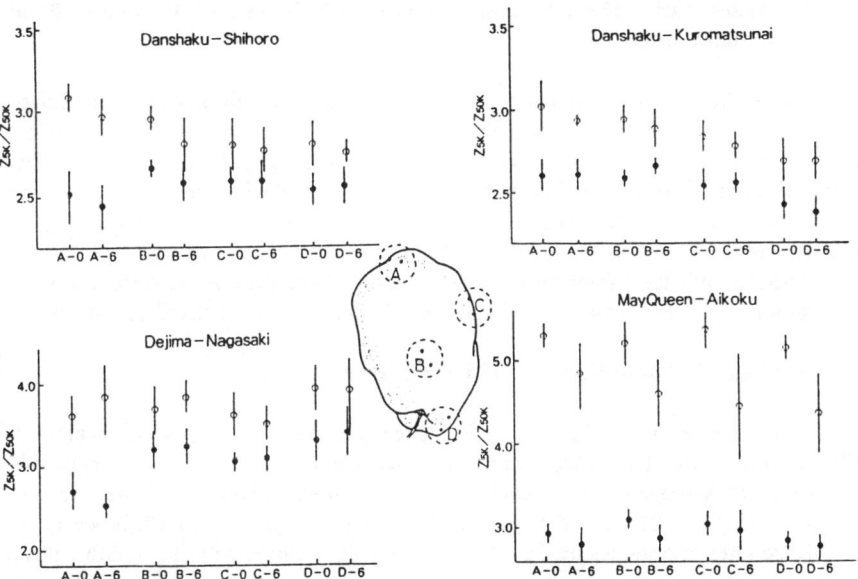

Figure 6 *Impedance ratios at different regions of potato tuber*
●, *0 Gy;* ○, *100 Gy; A, apical; B, central; C, side; D, basal;*
0, 0 month (one week) after irradiation; 6, 6 months after irradiation

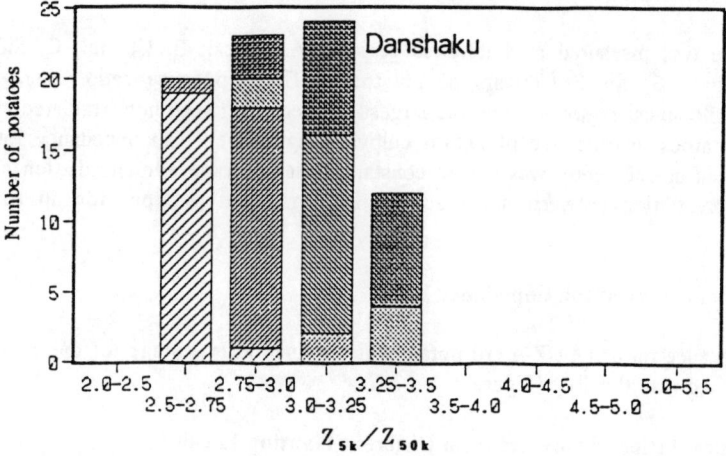

Figure 7 *Impedance ratios of potatoes which were irradiated at different doses* ▨, 0 Gy; ▦, 60 Gy; ▨, 100 Gy; ■, 150 Gy

than 3.50, irrespective of planting locality. These results suggest that the irradiated potatoes can be detected by the impedance ratio, without any information about planting locality.

3.8 Impedance Ratios of the Potatoes Commercially Irradiated at Shihoro Potato Irradiation Centre

Most of the impedance ratios of the potatoes commercially irradiated at Shihoro Potato Irradiation Centre were in a range of 2.75 and 3.25, irrespective of irradiation date (28 November 1991 and 3 February 1992) (Figure 9). There was a significant difference in the impedance ratio between the non-irradiated and commercially irradiated potatoes at a level of 5%. The results shown in Figures 7, 8 and 9 suggest that the commercially irradiated potatoes can be discriminated from non-irradiated potatoes of cv. Danshaku with the parameter of Z_{5k}/Z_{50k}; most of the parameters were lower than 2.75 for non-irradiated potatoes and higher than 2.75 for those irradiated commercially.

3.9 Impedance Ratios of Various Cultivars of Potatoes

The impedance ratios, Z_{5k}/Z_{50k}, of irradiated potatoes (100 Gy) were significantly different from those of non-irradiated potatoes at a significant level of 1% for cultivars of Benimaru, Hokkaikogane, May Queen, Toyoshiro, Ezoakari, Kitaakari, Norin-No.1 and Waseshiro (Figure 10). Based on the results shown in Figures 7 to 10, however, the impedance ratio of potatoes irradiated at the same dose varied with the potato cultivar, which suggests that the detection of irradiated potatoes is impossible without any information about the cultivar.

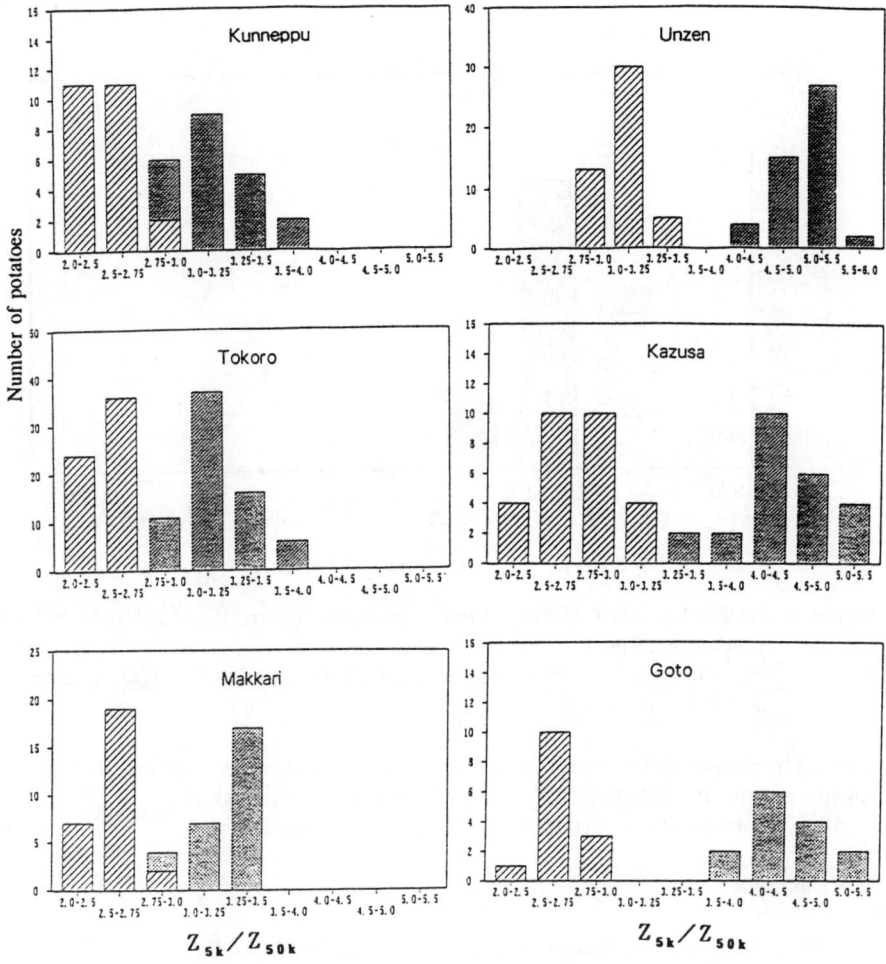

Figure 8 *Impedance ratios of potatoes cv. Danshaku (left) and cv. Dejima (right) at different planting localities*
▨, 0 Gy; ▩, 100 Gy

4 CONCLUSION

The results of this study have led to the conclusion that the ratio of impedance magnitude at 5 kHz to 50 kHz (Z_{5k}/Z_{50k}), measured at 22-25°C at the apical region of potato tuber with 1 mA of alternating current, results in the best detection of irradiated potatoes. Because the identification parameter, Z_{5k}/Z_{50k}, showed different values depending upon potato cultivar, the detection of irradiated potatoes is impossible without any information about the cultivar. However, irradiated potatoes can be

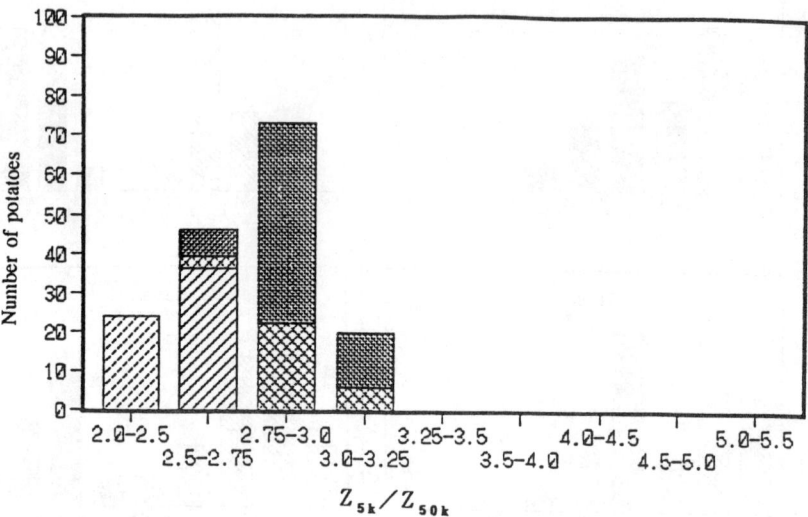

Figure 9 *Impedance ratios of the potatoes commercially irradiated at Shihoro Potato Irradiation Centre*
▨ , non-irradiated; ▧ , irradiated on 28 November, 1991; ▬ , irradiated on 3 February, 1992

detected by the impedance measuring method if the potato variety is known, because the impedance ratio at 5 kHz to 50 kHz is not influenced by planting locality.
All of the data in this report have been published in previous papers.[12,13]

References

1. IAEA, Analytical Detection Methods for Irradiated Foods, 1991.
2. R. Jona and A. Fronda, *Radiat. Phys. Chem.*, 1990, **35**, 317.
3. R. V. Dongen, D. Onderlinden and L. Strackee, "The Identification of Irradiated Foodstuffs," Office for Official Publications of the European Communities, Luxembourg, 1974, p. 203.
4. H. Scherz, "Colloquium on the Identification of Irradiated Foodstuffs," Office for Official Publications of the European Communities, Luxembourg, 1970, p. 13.
5. H. Scherz, "The Identification of Irradiated Foodstuffs," Office for Official Publications of the European Communities, Luxembourg, 1974, p. 194.
6. T. Hayashi and D. Ehlermann,. *Rept. Natl. Food Res. Instit.*, 1980, **36**, 91.
7. T. Hayashi, M. Iwamoto and K. Kawashima, *Agric. Biol. Chem.*, 1982, **46**, 905.
8. T. Hayashi and K. Kawashima, *J. Japn. Soc. Food Sci. Technol.*, 1983, **30**, 51.
9. T. Hayashi, "Health Impact, Identification, and Dosimetry of Irradiated Foods," WHO Working Group, Institut fuer Strahlenhygiene, Neuherberg, Germany, 1988, p. 432.
10. K. S. Cole and R. H. Cole, *J. Chem. Phys.*, 1941, **9**, 341.

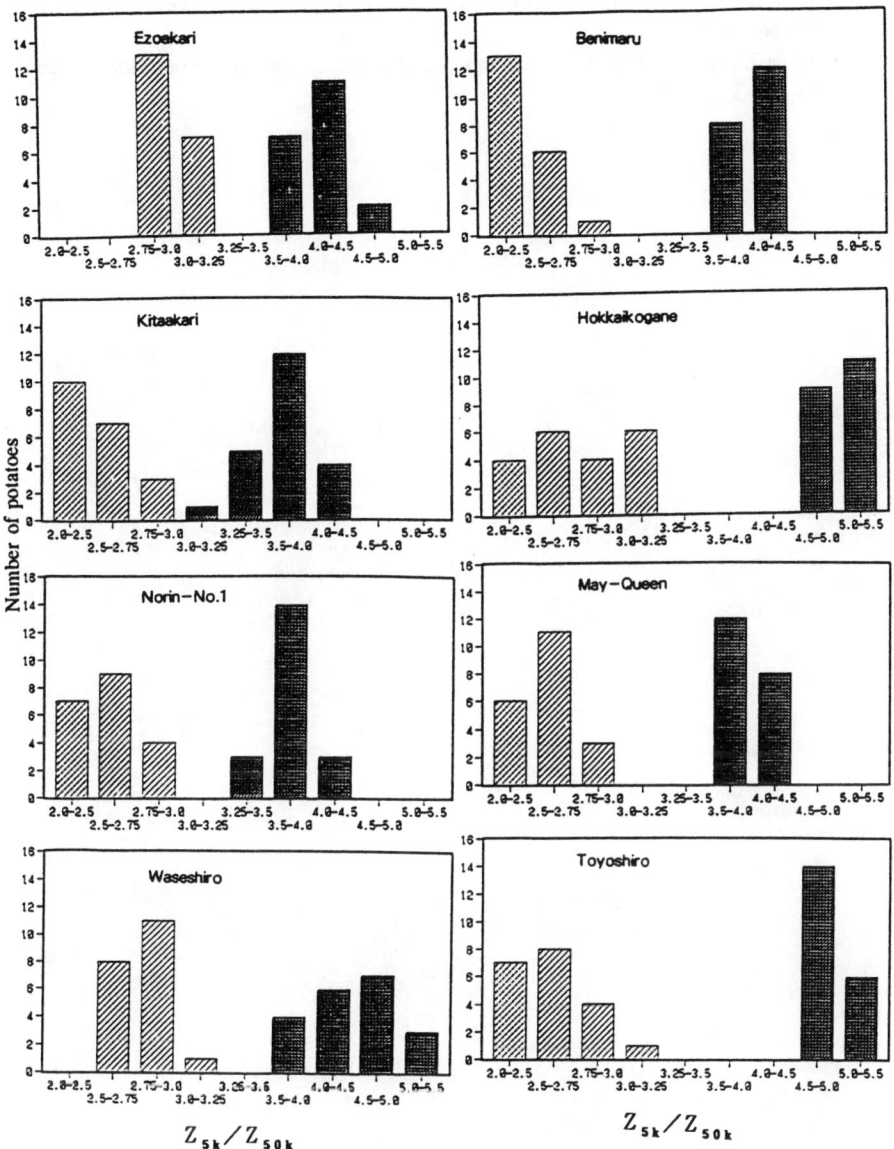

Figure 10 *Impedance ratios of potatoes of various cultivars.*

⧄ ; *0 Gy*, ■ ; *100Gy*

11. J. Felfoldi, P. Laszlo, S. Barabássy and J. Farkas, *Radiat. Phys. Chem.*, 1993, **41**, 471.
12. T. Hayashi, S. Todoriki, K. Otobe and J. Sugiyama, *Biosci. Biotechnol. Biochem.*, 1994, **56**, 1929.
13. T. Hayashi, S. Todoriki, K. Otobe and J. Sugiyama, *J. Jpn. Soc. Food Sci. Technol.*, 1994, **40**, 378.

APPLICABILITY OF VISCOSITY MEASUREMENT TO THE DETECTION OF IRRADIATED PEPPERS

T. Hayashi, S. Todoriki and K. Kohyama

National Food Research Institute, Ministry of Agriculture, Forestry and Fisheries
Kannondai
Tsukuba
Ibaraki 305
Japan

1 INTRODUCTION

Starch is degraded by ionising radiation, resulting in a decrease in viscosity.[1-3] The viscosities of black and white peppers which contain large amounts of starch are reduced by irradiation so, therefore, viscosity measurement has been proposed as a method to detect the irradiation treatment of these food products.[4-8] Although detection of irradiated spices by thermoluminescence measurement has been established,[7-13] it is useful to establish the viscosity measuring technique for detecting irradiated peppers, as this method is carried out widely in the laboratories of food controlling authorities and food processing companies.

In the studies by Farkas et al.[5,6] and Heide et al.,[8] however, some non-irradiated pepper samples showed viscosity values as low as those of irradiated samples. Viscosity values are influenced by the conditions for viscosity measurement such as shear rate, temperature and type of viscometer and it is not easy for laboratories who do not possess an irradiator to establish such criteria for judging whether or not peppers have been irradiated. It is important to establish a method which distinguishes irradiated peppers from those which are non-irradiated with a parameter value which is little influenced by the planting locality, batch of peppers and the conditions for viscosity measurement. The aim of this study was to clarify the effects of viscometric conditions on viscosity value and to establish parameters for detecting irradiated peppers which are independent of viscosity measuring conditions.

2 MATERIALS AND METHODS

2.1 Spices

Spice samples harvested in various countries were donated by three Japanese spice companies (A, B and C) and labelled as follows:

Black-1: Black pepper powder from Malaysia donated by A
Black-2: Black pepper powder from India donated by A

Black-3: Black pepper powder from Malaysia donated by B
Black-4: Black pepper powder from Malaysia donated by B
Black-5: Black pepper powder from Indonesia donated by C
Black-6: Black pepper powder from India donated by B
Black-7: Black pepper powder from India donated by B
Black-8: Black pepper powder from Brazil donated by A
Black-W1: Whole black pepper from Malaysia donated by B
Black-W2: Whole black pepper from Brazil donated by A
White-1: White pepper powder from Malaysia donated by A
White-2: White pepper powder from Malaysia donated by C
White-3: White pepper powder from Malaysia donated by B
White-4: White pepper powder from Malaysia donated by B
White-5: White pepper powder from Indonesia donated by B
White-6: White pepper powder from Brazil donated by A
White-W1: Whole white pepper from Malaysia donated by B
White-W2: Whole white pepper from Indonesia donated by B
White-W3: Whole white pepper from Brazil donated by A

Prior to decontamination treatments, the pepper powder samples were sieved to remove contaminants and collect samples smaller than 0.5 mm.

Unless otherwise stated, Black-4 and White-4 were used for the experiments.

2.2 Starch

Corn starch produced by Wako Pure Chemical Industries, Ltd., Japan was used as a standard.

2.3 Irradiation of Peppers

Peppers packed in polyethylene bags were irradiated under aerobic conditions in a Gammacell 220 (2.1×10^2 TBq of ^{60}Co, 4.6 kGy h^{-1}, AECL, Canada). The dose was evaluated with radiochromic film dosimeters (FWT-60-00, Far West Technology, Inc., USA) and cellulose triacetate film dosimeters (FTR-195, Fuji Photo Film Co., Japan). Both irradiated and non-irradiated samples were stored at 30°C and relative humidity of (R.H.) 60%.

2.4 Fumigation of Peppers

Pepper samples of Black-4 and White-4 packed in paper bags were exposed to ethylene oxide gas (EOG) at 1 kg cm^{-3} for 3 h at 50°C in an Automatic EOG Sterilizer Es-15 (Hirayama Manufacturing Corporation, Japan). The EOG fumigated samples were stored at 30°C and R.H. 60%.

2.5 Super-Heated Steam Treatment

Pepper samples of Black-4 and White-4 were treated with super-heated steam for 4 sec at 150°C and 2.5 kg cm^{-3} in a Super-Heated Steam Sterilize System KP-30

(Rikkoman Co., Japan). The samples treated with super-heated steam were stored at 30°C and R.H. 60%.

2.6 Viscosity Measurement

Prior to viscosity measurement, the whole pepper samples were ground to sizes smaller than 0.5 mm. The viscosity was measured according to Farkas et al.[5] and Heide et al.[8] with a slight modification, as follows:

Unless otherwise stated, the viscosity was measured with a HAAKE Rotovisco RV 1' equipped with a PG 149 (HAAKE Mess-Technik GmbH u.Co., Germany). Suspensions of pepper (10%, w/v) were homogenized with a Polytron PCD-2 (KINETICA GmbH, Switzerland) for 30 sec at the maximum speed and the pH was adjusted to 13.0 with 33% NaOH. The suspensions were heated in boiling water (99°C) with an occasional agitation for 30 min, followed by incubation for 3 h at 25°C. The viscosity of the suspension was measured at 25°C with the HAAKE Rotovisco equipped with a coaxial cylinder type Rotar NV. The gelatinised suspension (9 ml) was put in the rotor cup and the viscometer was operated at 100 rpm (541 s^{-1}) and 25°C for 30 sec, and then the apparent viscosity (mPa.s) was measured under that condition.

A Newport Rapid Visco Analyzer (Newport Scientific P/L, Australia) and a Brabender Rapid Amylogram (Brabender OHG Duisburg, Germany) were used to study the viscosities of pepper suspensions determined with different types of viscometers. The gelatinised suspension (27 ml) was put in a cup with a paddle and the viscosity was measured with the Newport Rapid Visco Analyzer at 25°C. A 100 ml aliquot of the suspension was put in a measuring bowl and the viscosity was measured at 75 rpm and 25°C with the Brabender Rapid Amylogram equipped with a Sensitivity Cartridge 125 cmg.

2.7 Determination of Starch Content

A 100 mg sample of pepper was heated with 10 ml of 1N HCl for 2.5 h in a boiling waterbath and then the glucose formed was determined with the aid of enzyme reactions of glucose oxidase and peroxidase according to Hayashi and Kawashima.[14] The starch content was calculated from the glucose content thus determined.

2.8 Determination of Microorganisms

Pepper samples (1 g) were suspended in 5 ml of 0.85% NaCl with 0.001% Triton X-100 followed by dilution with 0.85% NaCl. The diluted suspensions (0.2 ml) were incubated for 2 days at 30°C on nutrient agar, and the colonies formed were counted.

3 RESULTS AND DISCUSSION

3.1 Starch Contents of Peppers

The starch contents of black and white peppers varied with the batch and planting locality (Table 1). The starch contents of black peppers were in the range of 36 to 46%

Table 1 *Starch Contents of Peppers One Week After Irradiation (%)*[15]

Dose	0 kGy	5 kGy	10 kGy	15 kGy
Black-1	40.6 ± 2.0	40.6 ± 2.0	41.2 ± 2.3	41.0 ± 1.1
Black-2	36.8 ± 1.6	37.8 ± 1.2	37.4 ± 2.0	36.8 ± 1.8
Black-3	39.6 ± 1.3	40.2 ± 1.2	39.4 ± 0.9	40.1 ± 1.2
Black-4	41.7 ± 0.5	41.2 ± 1.3	40.8 ± 1.3	41.1 ± 1.7
Black-5	38.0 ± 1.6	38.4 ± 1.4	39.3 ± 0.6	39.1 ± 2.0
Black-6	43.8 ± 1.0	43.7 ± 0.5	43.7 ± 1.0	43.7 ± 0.8
Black-7	39.2 ± 2.1	40.0 ± 0.1	39.3 ± 0.6	38.8 ± 0.8
Black-8	43.5 ± 1.1	43.8 ± 0.8	43.9 ± 0.5	43.5 ± 0.5
Black-W1	45.2 ± 2.2	45.7 ± 1.5	45.4 ± 1.3	44.8 ± 2.0
Black-W2	46.0 ± 1.2	46.5 ± 0.9	45.7 ± 2.1	45.9 ± 1.1
White-1	58.6 ± 1.4	59.3 ± 1.8	59.0 ± 2.3	57.7 ± 1.0
White-2	57.6 ± 0.3	57.0 ± 0.9	57.0 ± 0.9	57.1 ± 1.0
White-3	59.1 ± 2.9	57.8 ± 1.3	58.8 ± 3.0	59.7 ± 2.1
White-4	54.1 ± 1.3	53.8 ± 1.2	53.4 ± 1.4	54.7 ± 0.5
White-5	55.6 ± 1.9	56.6 ± 0.9	55.0 ± 0.7	56.5 ± 1.7
White-6	55.2 ± 0.5	55.3 ± 0.5	55.8 ± 0.9	55.5 ± 0.5
White-W1	60.8 ± 1.0	59.5 ± 2.5	60.0 ± 3.1	61.6 ± 1.6
White-W2	64.9 ± 1.9	64.4 ± 2.9	63.4 ± 0.8	64.2 ± 1.6
White-W3	56.0 ± 1.5	55.8 ± 0.6	55.9 ± 0.5	56.3 ± 1.0

Mean value ± Standard Deviation for 3 measurements

and those of white peppers were in the range of 53 to 65%. The starch contents of the peppers were not influenced by irradiation. The starch contents of samples fumigated with EOG and those treated with super-heated steam showed almost the same values as the untreated pepper samples.

3.2 Dependence of Viscosity on Dose

The viscosity of pepper suspensions decreased with dose for all pepper samples. Most of the viscosities of the non-irradiated black peppers were higher than those of the irradiated samples (Figure 1). However, the viscosity of irradiated Black-8 (5 kGy) was higher than those of non-irradiated Black-2 and Black-5 and the viscosity of irradiated Black-W2 (5 kGy) was higher than those of non-irradiated Black-2, Black-3 and Black-5. The viscosities of the non-irradiated Black-2 and Black-5 were not significantly different from that of the Black-W1 irradiated at 5 kGy ($p > 5\%$). These results indicate that the viscosity value cannot completely differentiate black peppers irradiated at 5 kGy from non-irradiated ones. All of the irradiated white peppers (5 kGy) showed viscosity values significantly lower than the non-irradiated ones (Figure 2).

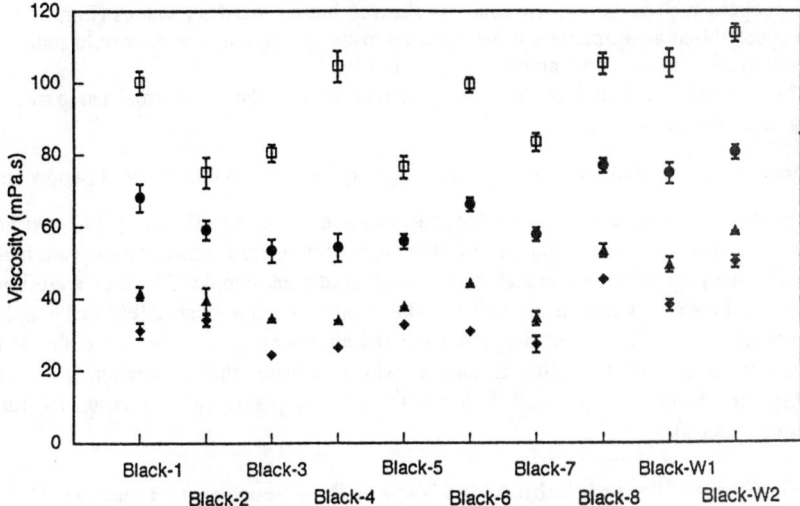

Figure 1 *Viscosities of black peppers one week after irradiation*[15]
□, *0 Gy*; ●, *5 Gy*; ▲, *10 Gy*; ◆, *15 Gy*

3.3 Dependence of Parameter, Viscosity/Starch Amount, on Irradiation Dose

The data shown in Table 1 indicated that the starch content of pepper varied remarkably with sample. Degradation of starch has been considered to be responsible for the viscosity decrease of peppers caused by irradiation.[5,6,8,16] Generally, non-irradiated

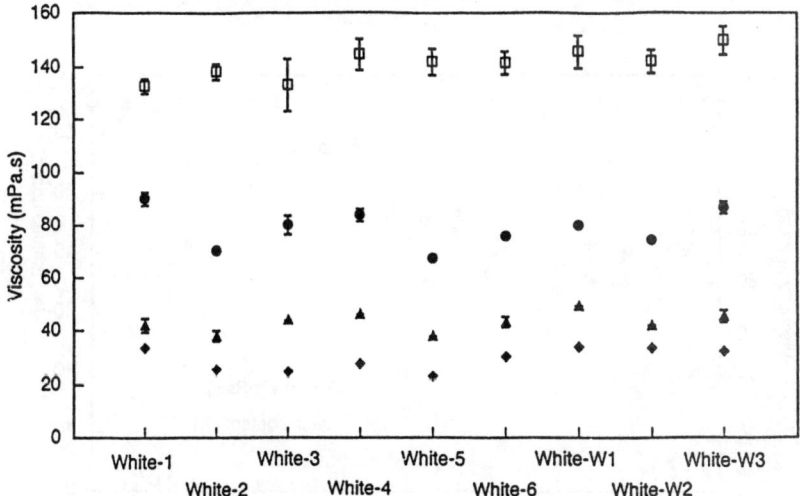

Figure 2 *Viscosities of white peppers one week after irradiation*[15]
□, *0 Gy*; ●, *5 Gy*; ▲, *10 Gy*; ◆, *15 Gy*

black peppers with higher starch contents showed higher viscosity values (Figure 3). It was expected that a parameter consisting of viscosity and starch content would reduce the fluctuation of viscosity value among pepper samples.

The viscosity was divided by starch content to calculate a normalised parameter (Parameter A), as follows:

Parameter A = Viscosity of 10% pepper (mPa.s) / Starch amount in 1 g of pepper (g)

Parameter A showed a dose dependent relation and a clearer difference between non-irradiated and irradiated samples; all of the values of non-irradiated samples were higher than 190 mPa.s/g of starch and all of those of irradiated samples (5 kGy) were lower than 180 mPa.s/g of starch, irrespective of the lot and planting locality (Figures 4 and 5). The values of Parameter A of peppers 6 months after storage at 30°C were almost the same as those immediately after irradiation, which indicates that Parameter A does not notably vary during storage at 30°C for 6 months, irrespective of irradiation treatment (Figures 6 and 7).

3.4 Viscosities of Peppers Subjected to Various Decontamination Treatments

The viscosities of pepper samples which had been subjected to EOG fumigation and super-heated steam treatment were at the same level as those of non-irradiated ones. Peppers subjected to these decontamination treatments of which the sterilisation efficiencies were higher than 5 kGy irradiation and lower than 10 kGy irradiation, gave values higher than 200 mPa.s g^{-1} of starch (Table 2).

These results indicate that the decontamination treatments other than irradiation do not lower viscosity and that irradiated peppers can be distinguished from those decontaminated with a treatment other than irradiation by determining Parameter A.

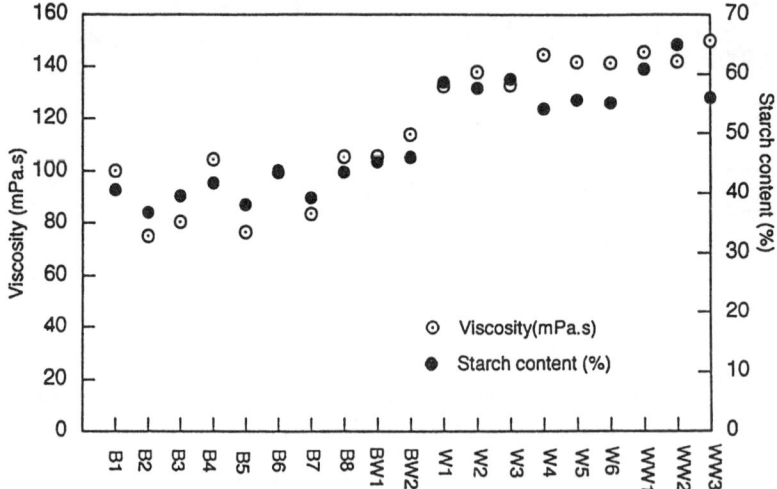

Figure 3 *Viscosities and starch contents of irradiated peppers*[15]

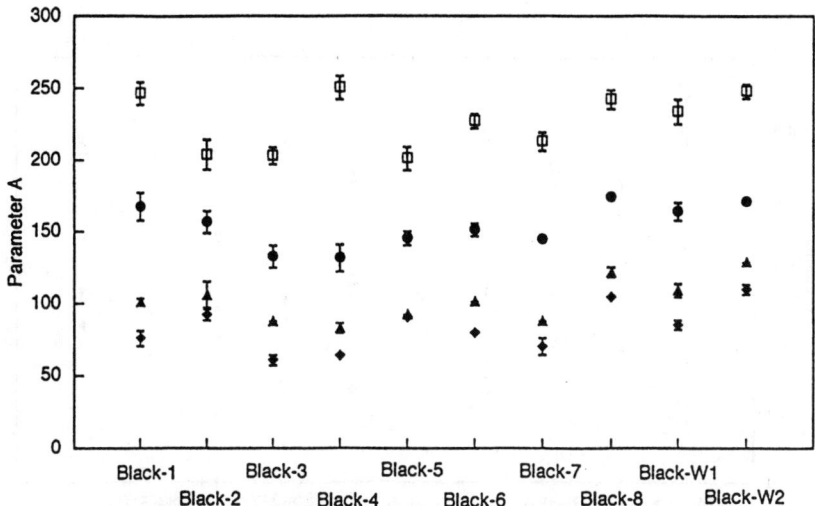

Figure 4 *Parameter A of black peppers one week after irradiation*[15]
□, *0 Gy;* ●, *5 Gy;* ▲, *10 Gy;* ◆, *15 Gy*

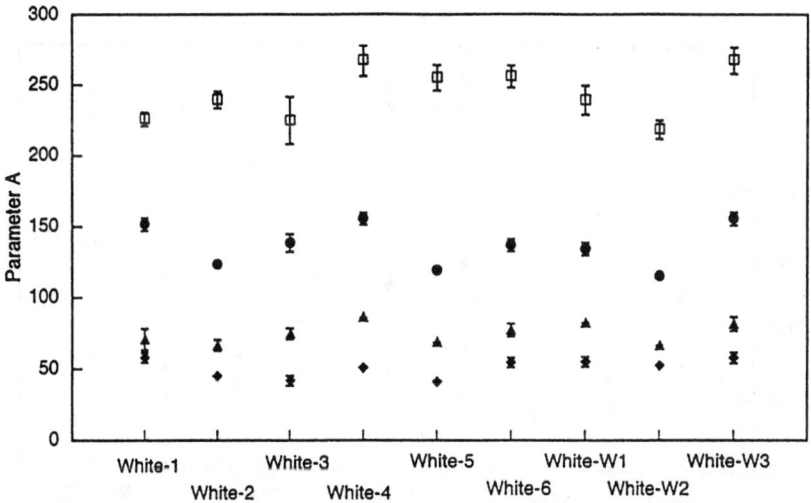

Figure 5 *Parameter A of white peppers 1 week after irradiation*[15]
□, *0 Gy;* ●, *5 Gy;* ▲, *10 Gy;* ◆, *15 Gy*

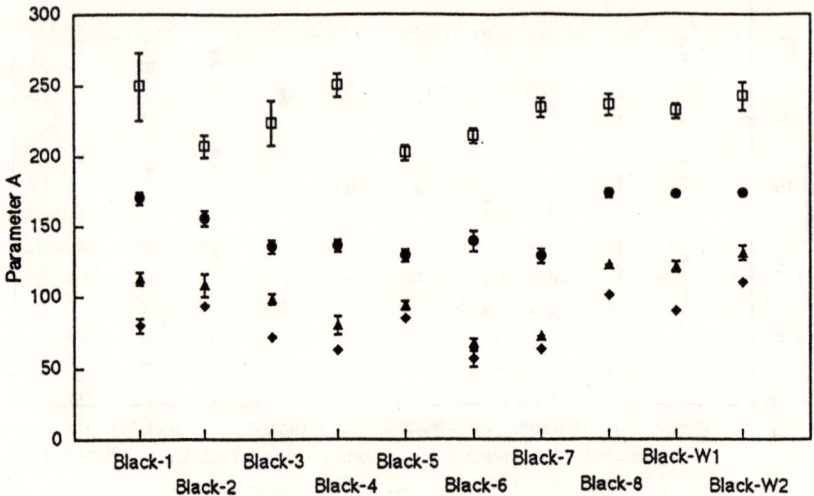

Figure 6 *Parameter A of black peppers 6 months after irradiation*[15]
□, *0 Gy;* ●, *5 Gy;* ▲, *10 Gy;* ◆, *15 Gy*

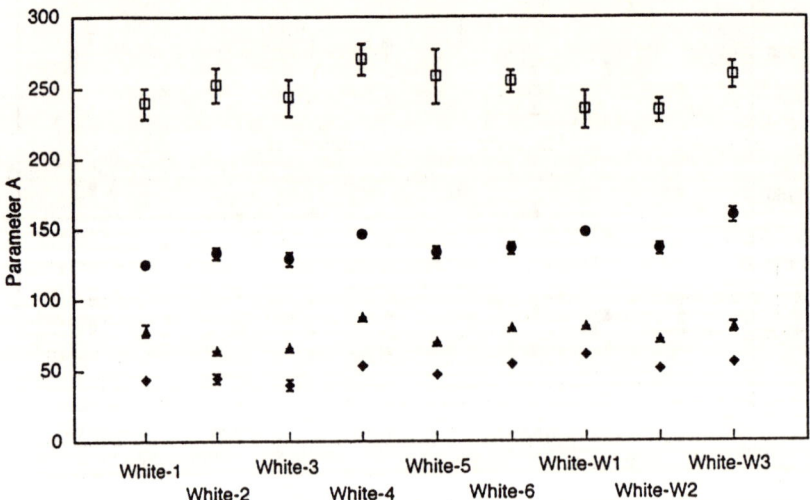

Figure 7 *Parameter A of white peppers 6 months after irradiation*[15]
□, *0 Gy;* ●, *5 Gy;* ▲, *10 Gy;* ◆, *15 Gy*

3.5 Effects of Gelatinising Conditions on Viscosity

The pH and temperature for gelatinisation greatly influenced the viscosity of black pepper (Figure 8). Higher pH resulted in lowered viscosity values of pepper suspensions and higher temperatures resulted in larger differences between non-irradiated and irradiated peppers.

3.6 Effects of Measuring Conditions on Viscosity

The temperature for viscosity measurement did not significantly influence the viscosity value in a temperature range of 18°C to 33°C, whereas the revolutionary rate (shear rate) of the viscometer had a great influence on the value (Figure 9).

The results shown in Figures 8 and 9 suggest that the viscosity values obtained under different conditions for gelatinisation and viscosity measurement are not comparable with each other. Each viscosity measuring system should establish its own criteria on which the judgment of irradiation treatment is based, but laboratories which do not own any irradiation facility will not be able to easily establish such criteria. Establishing a parameter which is not influenced by the conditions for viscosity measurement is helpful for any laboratory to utilise the viscosity method for detecting irradiated peppers.

3.7 Parameter for Detecting Irradiated Peppers without Any Influence of Viscosity Measuring Condition

A 5% starch suspension was gelatinised and the viscosity was measured under the

Table 2 *Viscosity of Peppers Subjected to Decontamination Treatments*[1,5]

Sample	Viscosity (mPa.s)	Parameter A (mPa. s g^{-1} of Starch)	Number of Microorganisms (Counts g^{-1})
Black pepper			
untreated	104.3 ± 3.4	250.1 ± 8.2	9.0×10^7
5 kGy irradiation	54.1 ± 3.9	131.3 ± 9.5	2.4×10^4
10 kGy irradiation	33.7 ± 1.3	82.6 ± 3.2	>100
super-heated steam	110.9 ± 3.7	252.6 ± 8.4	1.3×10^3
fumigation with EOG	111.1 ± 2.8	245.3 ± 6.2	9.6×10^2
White pepper			
untreated	144.3 ± 5.9	266.7 ± 10.9	1.4×10^6
5 kGy irradiation	83.6 ± 2.1	155.4 ± 3.9	3.8×10^2
10 kGy irradiation	46.3 ± 0.3	86.1 ± 0.6	>10
super-heated steam	143.6 ± 5.3	266.1 ± 14.2	>100
fumigation with EOG	147.6 ± 5.2	266.9 ± 9.4	>100

Mean value ± Standard Deviation for 3 measurements

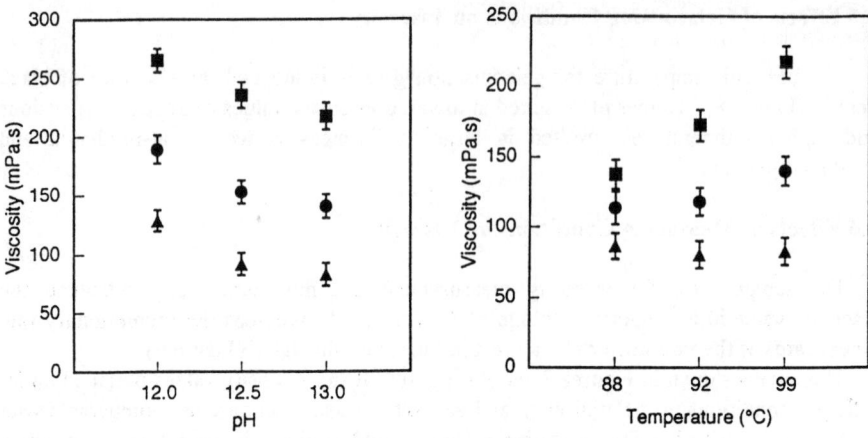

Figure 8 *Effects of gelatinising conditions on the viscosity of black pepper*[17]
■, 0 kGy; ●, 5 kGy; ▲, 10 kGy

same conditions as the pepper suspension, and Parameter B was determined by dividing Parameter A by viscosity of a 5% suspension of corn starch (mPa.s), as follows:

Parameter B = Parameter A / Viscosity of 5% corn starch

Parameter B was dependent upon the dose but was independent of the measuring temperature and the revolutionary rate (Figure 10). However, the parameter was still influenced by the conditions for gelatinisation (Figure 11).

The parameter values for 10% suspensions of black pepper determined at 25°C with different viscometers after gelatinisation at pH 13.0 in boiling water (99°C) are shown in

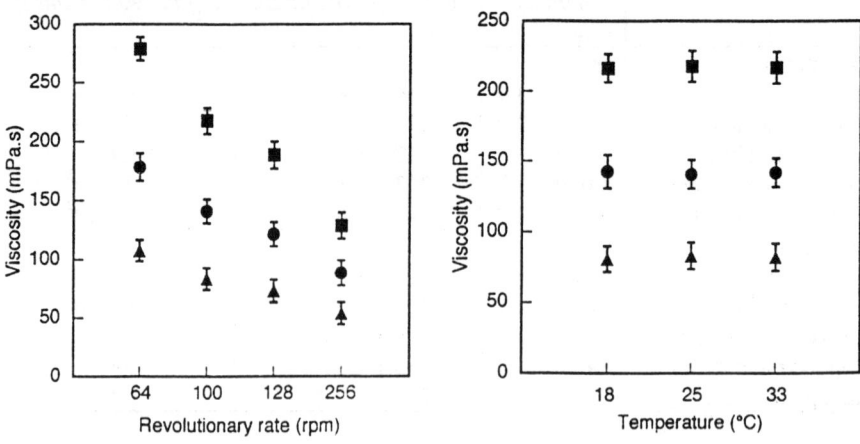

Figure 9 *Effects of viscosity measuring conditions on the viscosity of black pepper*[17]
■, 0 kGy; ●, 5 kGy; ▲, 10 kGy

Figure 12. The parameter was dependent upon dose and was not significantly influenced by the type of viscometer.

These results suggest that the parameter determined with a 10% suspension of pepper and a 5% suspension of corn starch can eliminate the effect of the condition for viscosity measurement, if the suspensions are gelatinised for 30 min under the same condition; *i.e.* pH 13.0 in boiling water (99°C).

3.8 Parameter Values for Peppers from Different Planting Localities

The values of Parameter B of non-irradiated peppers decreased during storage for one year (Table 3). However, the parameter values of non-irradiated peppers were always higher than those of peppers irradiated at 5 kGy or higher. The parameter values of black peppers were in ranges of 2.2 to 3.2, 1.7 to 2.2, and 0.7 to 1.7 for non-irradiated, 5 kGy irradiated and 10 kGy irradiated samples, respectively. Those of white peppers were in ranges of 2.5 to 3.4, 1.2 to 2.0, and 0.6 to 1.1 for non-irradiated, 5 kGy irradiated and 10 kGy for irradiated samples, respectively.

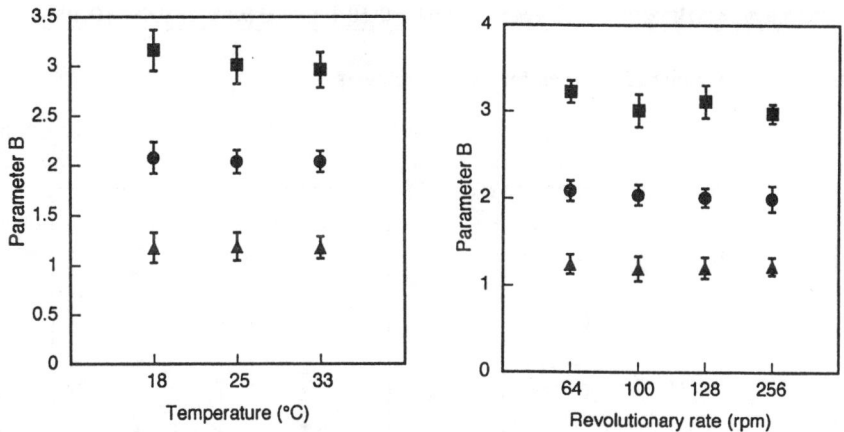

Figure 10 *Effects of viscosity measuring conditions on Parameter B of black pepper*[17]
■, *0 kGy;* ●, *5 kGy;* ▲, *10 kGy*

Table 3 *Values of Parameter B of Black and White Peppers*[17]

Sample	Storage Period	0 kGy	5 kGy	10 kGy
Black pepper (Sarawak)	1 week	3.29 ± 0.20	2.14 ± 0.12	1.26 ± 0.08
Black pepper (Sarawak)	1 year	3.10 ± 0.25	1.95 ± 0.09	1.48 ± 0.09
Black pepper (India)	1 week	3.15 ± 0.21	2.09 ± 0.09	1.69 ± 0.03
Black pepper (India)	1 year	2.73 ± 0.19	2.11 ± 0.12	1.41 ± 0.06
Black pepper (Indonesia)	1 week	2.76 ± 0.15	1.58 ± 0.10	0.93 ± 0.07
Black pepper Indonesia)	1 year	2.43 ± 0.11	1.55 ± 0.09	0.87 ± 0.08
Black pepper (Brazil)	1 week	2.78 ± 0.12	2.09 ± 0.10	1.61 ± 0.08
Black pepper (Brazil)	1 year	2.53 ± 0.14	1.99 ± 0.11	1.55 ± 0.10
White pepper (Sarawak)	1 week	2.85 ± 0.21	1.36 ± 0.15	0.75 ± 0.09
White pepper (Sarawak)	1 year	2.61 ± 0.22	1.30 ± 0.10	0.72 ± 0.08
White pepper (Indonesia)	1 week	3.25 ± 0.32	1.33 ± 0.12	0.61 ± 0.08
White pepper (Indonesia)	1 year	2.85 ± 0.21	1.28 ± 0.11	0.65 ± 0.09
White pepper (Brazil)	1 week	2.87 ± 0.15	1.68 ± 0.12	1.01 ± 0.12
White pepper (Brazil)	1 year	2.62 ± 0.10	1.71 ± 0.15	0.99 ± 0.10

Mean value ± Standard Deviation for 3 measurements

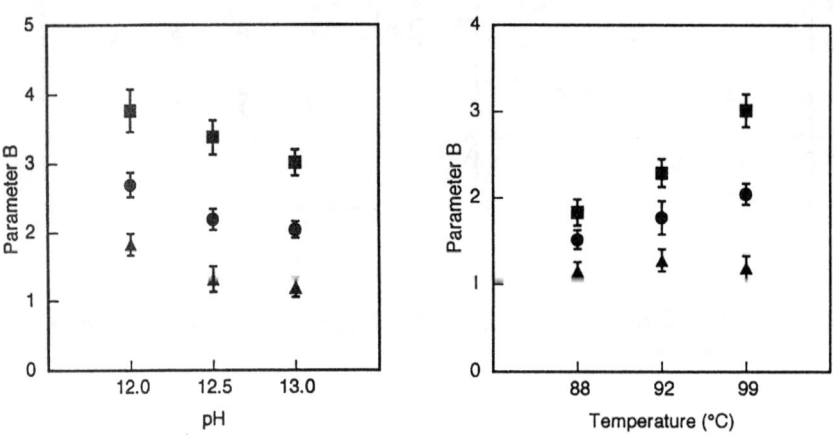

Figure 11 *Effects of gelatinising conditions on Parameter B of black pepper*[17]
■, 0 kGy; ●, 5 kGy; ▲, 10 kGy

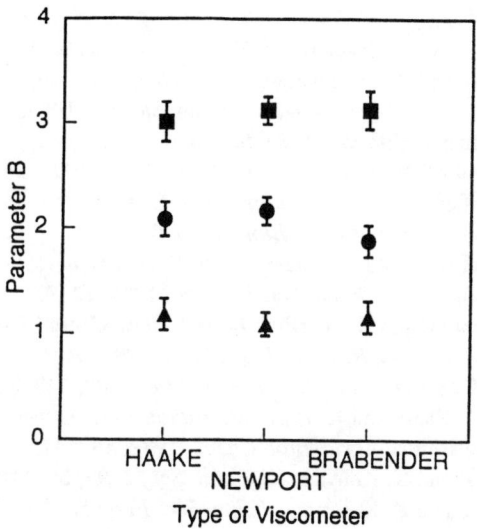

Figure 12 *Parameter B of black pepper determined by different viscosmeters*[17]
■, *0 kGy;* ●, *5 kGy;* ▲, *10 kGy*

4 CONCLUSION

A normalised parameter (Parameter A x viscosity/starch amount) of black and white peppers was a better parameter for detecting irradiation treatment than a viscosity value itself. The normalised parameter was lowered by gamma-irradiation but was not influenced by other decontamination treatments such as fumigation with EOG treatment with super-heated steam. The parameter was dependent upon dose but was little influenced by the planting locality and remained almost constant during storage. Generally, peppers are irradiated at doses of 5 to 10 kGy on a commercial basis. Parameter B, which is determined by dividing Parameter A by viscosity of a 5% suspension of corn starch, can detect irradiated peppers without any information about the peppers or standard pepper samples, only if the suspensions are gelatinised for 30 min at pH 13.0 in boiling water.

Peppers with values of Parameter A higher than 2.0 or values of Parameter B higher than 2.5 can be judged to be non-irradiated samples and those with values of Parameter A lower than 1.7 or values of Parameter B lower than 2.0 can be judged to have been irradiated. Peppers with values in ranges of 1.7 to 2.0 and 2.0 to 2.5 for Parameter A and B respectively, should be subjected to other detection techniques, such as thermoluminescence analysis, to confirm the irradiation treatment.

References

1. R. M. A. El-Saadany, F. M. El-Saadany and Y. H. Foda, *Staerke*, 1976, **28**, 208.
2. S. P. Nene, U. K. Vakil and A. Sreenivasan, *J. Food Sci*, 1975, **40**, 943.
3. M. Roushdi, A. Harras, A. El-Meligi and M. Bassim, *Staerke*, 1983, **35**, 15.
4. E. Mohr and G. Wichmann, *Gordian*, 1985, **85**, 96.
5. J. Farkas, A. Koncz and M. M. Sharif, *Radiat. Phys. Chem.*, 1990, **35**, 324.
6. J. Farkas, M. M. Sharif and A. Koncz, *Radiat. Phys. Chem.*, 1990, **36**, 621.
7. L. Heide and K. W. Bögl, *Int. J. Radiat. Biol.*, 1990. **57**, 201.
8. L. Heide, E. Nuernberger and K. W. Bögl, *Radiat. Phys. Chem.*, 1990, **36**, 613.
9. L. Heide and K. W. Bögl, *Int. J. Food Sci. Technol.*, 1987, **22**, 93.
10. L. Heide, R. Guggenberger and K. W. Bögl, *Radiat. Phys. Chem.*, 1989, **34**, 903.
11. J. M. Oduko and N. M. Spyrou, *Radiat. Phys. Chem.*, 1990, **36**, 603.
12. D. C. W. Sanderson, C. Slater and K. J. Cairns, *Nature*, 1989, **340**, 23.
13. D. C. W. Sanderson, C. Slater and K. J. Cairns, *Radiat. Phys. Chem.*, 1989, **34**, 915.
14. T. Hayashi and K. Kawashima, *Agric. Biol. Chem.*, 1982, **46**, 1475.
15. T. Hayashi, S. Todoriki and K. Kohyama, *J. Food Sci.*, 1994, **59**, 118.
16. T. Hayashi, S. Todoriki and K. Kohyama, *J. Jpn. Soc. Food Sci. Technol.*, 1993, **40**, 456.
17. T. Hayashi, S. Todoriki and K. Kohyama, *Radiat. Phys. Chem.*, 1995, **45**, 665.

COLLABORATIVE STUDY OF VISCOSITY MEASUREMENT OF BLACK AND WHITE PEPPERS

T. Hayashi

National Food Research Institute
Ministry of Agriculture, Forestry and Fisheries
Kannondai, Tsukuba, Ibaraki 305
Japan

1 INTRODUCTION

Viscosity measurement has been reported to be a promising method for detecting irradiated peppers.[1-8] Hayashi and co-workers have proposed detection parameters which are little influenced by the planting locality, batch of peppers and the conditions for viscosity measurement.[7,8] Based on the results of previous studies,[7,8] a protocol for determining the parameter values for detecting irradiated peppers by viscosity measurement was prepared and a collaborative study conducted to examine the usefulness of the viscometric method.

2 MATERIALS AND METHODS

2.1 Protocol

A protocol for determining parameters for differentiating irradiated from non-irradiated peppers was prepared (see Appendix).

2.2 Pepper Samples

Eight pepper samples labelled as follows were subjected to the collaborative test. Both black and white peppers harvested in Malaysia were donated by Lion Co., Japan.

Sample 1: Non-irradiated black pepper
Sample 2: Black pepper irradiated at 5 kGy with gamma-rays
Sample 3: Black pepper irradiated at 10 kGy with gamma-rays
Sample 4: Black pepper fumigated with ethylene oxide gas
Sample 5: Non-irradiated white pepper
Sample 6: White pepper irradiated at 5 kGy with gamma-rays
Sample 7: White pepper irradiated at 10 kGy with gamma-rays
Sample 8: White pepper fumigated with ethylene oxide gas

2.3 Irradiation of Peppers

Peppers were irradiated in July 1993. The samples were packed in polyethylene bags and irradiated in a Gammacell 290 (2.1×10^2 TBq of ^{60}Co, 4.6 kGy h^{-1}, AECL, Canada) at the National Food Research Institute, Tsukuba, Japan. The dose was evaluated with radiochromic film dosimeters (FWT-60-00, Far West Technology, Inc., USA) and cellulose triacetate film dosimeters (FTR-125, Fuji Photo Film Co., Japan). The variation of the dose rate in the irradiation chamber was less than 15%.

2.4 Fumigation of Peppers

Peppers were fumigated in July 1993. Peppers were exposed to ethylene oxide gas (EOG) at 1 kg cm^3 for 3 h at 50°C.

2.5 Collaborators and Viscosity Measuring Conditions at Each Laboratory

The following 10 laboratories were involved in the collaborative test. Each laboratory determined the viscosities and starch contents according to the protocol (Appendix) with their own viscometer.

Laboratory 1
    Collaborator: Z. Formanek, R. Jori and S. Barabássy
    Institution: University of Horticulture & Food Industry, Hungary
    Instrument: Rheotest Rotational Viscometer Type 2 RV 2
    Shear rate: 437.4 s^{-1}
    Measuring temperature: 25°C
    Date: February-March 1994

Laboratory 2
    Collaborator: H. Delincée
    Institution: Federal Research Centre for Nutrition, Germany
    Instrument: HAAKE Rotovisco RV 12
    Shear rate: 69.248 s^{-1}
    Measuring temperature: 30°C
    Date: October 1993

Laboratory 3
    Collaborator: E. Marchioni
    Institution: Louis Pasteur University and CRITT AERIAL, France
    Instrument: Contraves Rheomat 115A
    Shear rate: 450 s^{-1}
    Measuring temperature: 25°C
    Date: March-April 1994

Laboratory 4
    Collaborator: G. A. Schreiber
    Institution: Institute for Social Medicine & Epidemiology, Germany

Instrument: Brookfield RVT, Spindle 6
Revolutionary rate: 50 rpm
Measuring temperature: 30°C
Date: October 1993

Laboratory 5
Collaborator: U. Matsukura
Institution: National Agriculture Research Centre, Japan
Instrument: Newport Rapid Visco Analyzer
Shear rate: not determined
Measuring temperature: 30°C
Date: September 1993

Laboratory 6
Collaborator: Y. Hirata
Institution: Japan Food Research Laboratories, Japan
Instrument: B Type Revolutionary Viscometer
Revolutionary rate: 12 rpm
Measuring temperature: 25°C
Date: September-October 1993

Laboratory 7
Collaborator: M. Otaka
Institution: Japan Grain Inspection Association, Japan
Instrument: Brabender Viscograph
Sensitivity Cartridge: 700 cmg
Measuring temperature: 25°C
Date: September 1993

Laboratory 8
Collaborator: K. Hirasa
Institution: Lion Co., Japan
Instrument: HAAKE Rotovisco RV 19
Shear rate: 541 s^{-1}
Measuring temperature: 25°C
Date: October 1993

Laboratory 9
Collaborator: Y. Higashimura
Institution: San-Eigen FFI Co., Japan
Instrument: BL Type Revolutionary Viscometer
Revolutionary rate: 12 rpm
Measuring temperature: 25°C
Date: November 1993

Laboratory 10
   Collaborator: T. Hayashi
   Institution: National Food Research Institute, Japan
   Instrument: HAAKE Rotovisco RV 12
   Shear rate: 541 s^{-1}
   Measuring temperature: 25°C
   Date: August 1993

3 RESULTS AND DISCUSSION

The results of the collaborative study are shown in Tables 1 and 2. The values of Parameter A (Table 1) and Parameter B (Table 2) for non-irradiated and fumigated peppers were always higher than those for irradiated samples for each laboratory. However, the values for the same pepper sample were different amongst the laboratories, although a single value of Parameter B for the same sample irrespective of laboratory had been expected. The inconsistency of the value was attributed to the difference in the method for adjusting pH of the pepper and starch suspensions with 33% NaOH - slower addition of the NaOH solution to the suspensions consumed a larger amount of the NaOH solution for adjusting pH. Larger amounts of the NaOH solution added to the suspensions resulted in lower viscosity values for irradiated peppers and corn starch, while the viscosity values for non-irradiated peppers were almost constant (Table 3).

These results indicate that the description of the method for pH adjustment in the protocol was not adequate. The results shown in Table 3 and those reported previously[5,6] indicate that the difference in the viscosity between non-irradiated and irradiated peppers is larger at a higher pH. The starch in peppers which have been degraded by irradiation will be further degraded by gelatinisation under alkaline conditions to a greater degree than the starch in non-irradiated peppers.

Table 1 *Values of Parameter A Obtained at Each Laboratory*[c]

Laboratory	Sample							
	1	2	3	4	5	6	7	8
1	1.74	1.09	0.65	1.96	3.01	1.04	0.88	3.03
2	1.43	0.80	0.75	1.74	1.96	0.85	0.48	2.73
3	3.56	0.85	0.46	3.67	5.16	1.16	0.61	5.67
4	2.60^b	0.70^b	0.35^b	6.30^b	5.25^b	1.00^b	0.05^b	4.90^b
5	1.13	0.57	0.27	1.19	1.37	0.62	0.27	1.44
6	7.65	1.90	0.36	a	23.75	1.69	0.45	a
7	1.26	0.43	0.17	a	2.70	0.77	0.24	a
8	3.19	1.21	0.47	2.66	3.16	1.26	0.48	3.81
9	1.39	0.20	0.05	1.30	2.96	0.40	0.08	2.48
10	1.06	0.64	0.48	1.25	1.87	0.80	0.44	1.79

a: Sample was not subjected to the viscosity measurement
b: Viscosity value, Parameter could not be determined
c: Mean value of two measurements

Table 2 *Values of Parameter B Obtained at Each Laboratory*

Laboratory	Sample							
	1	2	3	4	5	6	7	8
1	3.04	1.88	1.05	2.64	6.53	1.79	1.40	5.29
2	3.81	2.25	2.12	4.53	4.09	1.81	1.04	6.06
3	7.82	1.94	1.01	7.91	8.64	1.89	1.01	8.95
4	b	b	b	b	b	b	b	b
5	2.83	1.42	0.68	2.97	2.54	1.15	0.51	2.66
6	19.40	4.81	0.91	a	41.80	2.97	0.79	a
7	3.21	1.08	0.43	a	4.76	1.36	0.43	a
8	8.07	3.06	1.19	6.73	5.56	2.37	0.84	6.70
9	4.40	0.62	0.16	4.12	6.41	0.86	0.18	5.38
10	2.69	1.62	1.21	3.12	3.30	1.40	0.77	3.09

a: Sample was not subjected to the viscosity measurement
b: Parameter could not be determined
c: Mean value of two measurements

Table 3 *Viscosities of 10% Black Pepper Suspension and 5% Starch Suspension at Different pH (mPa.s)[b]*

Amount of NaOH[a]	pH	0 kGy	5 kGy	Starch
0.5 ml	13.0-13.1	70.3	36.6	48.2
1.0 ml	13.4-13.5	69.3	28.5	25.2
2.0 ml	13.8-13.9	68.1	25.8	21.2

a: Amount of 33% NaOH added to 40 ml of pepper suspension or 80 ml of starch suspension
b: Mean value of two measurements

One collaborator (Laboratory 4) could not determine the values of Parameters A and B because the viscosity of the starch suspension could not be quantified.[9] However, the viscosity values determined according to the Protocol for the non-irradiated and EOG-fumigated peppers were remarkably higher than those for the irradiated ones. The viscosity values determined with this collaborators own method showed a clearer difference between non-irradiated and irradiated peppers.[9,10] The difficulty in the measurement of viscosity of the corn starch suspension was also reported by a Japanese collaborator (Laboratory 6), who used an auto-titrater for adjusting the pH of the suspension.[11]

4 CONCLUSION

Taking the results in the present collaborative study and those reported so to date[1-8] into consideration, it may be concluded that the viscosity measurement can be used, at least as a screening method, for detecting irradiated black and white peppers. It is more practical that the procedures for gelatinising pepper suspensions and viscosity measurement as well as the detection parameters, are established for each viscometric system. This is because the optimum conditions for viscosity measurement, such as the concentration of pepper suspension and the shear rate of the viscometer, are different for each system used for viscosity measurement. Although the protocol and previous reports recommend that the pH of the pepper suspension is adjusted to 13, the results of this work suggest that the pH of the suspension should be adjusted to a value as high as possible for better differentiation of irradiated from non-irradiated peppers.

References

1. E. Mohr and G. Wichmann, *Gordian*, 1985, **85**, 96.
2. J. Farkas, A. Koncz and M. M. Sharif, *Radiat. Phys. Chem.*, 1990, **35**, 324.
3. J. Farkas, M. M. Sharif and A. Koncz, *Radiat. Phys. Chem.*, 1991, **36**, 621.
4. L. Heide and W. Bögl, *Int. J. Radiat. Biol.*, 1990, **57**, 201.
5. L. Heide, E. Nuernberger and K. W. Bögl, *Radiat. Phys. Chem.*, 1990, **36**, 613.
6. T. Hayashi, S. Todoriki and K. Kohyama, *J. Jpn. Soc. Food Sci. Technol.*, 1993, **40**, 456.
7. T. Hayashi, S. Todoriki and K. Kohyama, *J. Food Sci.*, 1994, **59**, 118.
8. T. Hayashi, S. Todoriki and K. Kohyama, *Radiat. Phys. Chem.*, 1995, **45**, 665.
9. G. A. Schreiber, Personal communication, 1993.
10. G. A. Schreiber, A. Leffke, M. Mager, N. Helle and K. W. Bögl, *Radiat. Phys. Chem.*, 1994, **44**, 467.
11. Y. Hirata, Personal communication, 1993.

APPENDIX

PROTOCOL FOR THE DETECTION OF IRRADIATED BLACK AND WHITE PEPPERS BY VISCOSITY MEASUREMENT

1. Equipment

   Use a revolutionary viscometer for viscosity measurement.

   Example 1: Brookfield Viscometer Model RTV and Spindle No.6
   Example 2: Rheotest Rotational Viscometer Type 2-RV 2 and S2 Cylinder
   Example 3: Haake Rotovisco RV 12 and Coaxial Cylinder Type Rotar NV
   Example 4: Newport Scientific Rapid Visco Analyzer
   Example 5: Brabender Rapid Amylogram

2. Sample Preparation

2-1 Grind whole pepper sample (granules) with a grinder.
   Example 1: Sibata Slicer: 20,000 rpm for 1 min.
   Example 2: Moulinex kitchen grinder: 3,000 rpm for 5 min.
   Pepper powder sample is directly subjected to sieving.

2-2 Sieve the pepper samples to collect sample with sizes smaller than 500 μm.

3. Viscosity measurement

3-1 Prepare a 10% (v/w) aqueous suspension in a glass vessel.

3-2 Homogenise the suspension with a homogeniser.
   Example 1: Kinetika Polytron PCD-2: 20,000 rpm for 30 sec.
   Example 2: MSE laboratory homogeniser: 15,000 rpm for 30 sec.

3-3 Adjust the pH to 12.8-13.0 with 33% NaOH.

3-4 Heat the suspension for 30 min in boiling water. The suspension should be occasionally agitated in order to prepare a uniformly gelatinised sample. Evaporation during heating should be prevented by covering the glass vessel with a screw cap or a glass plate. The suspension should not be excessively agitated, because an excess agitation reduces viscosity.

3-5 Incubate the gelatinised sample for 3 h at a constant temperature between 20 and 30°C.

3-6 Operate a viscometer at a constant shear rate between 350 and 600 s^{-1}, if shear rate can be controlled, and at the constant temperature between 20 and 30°C for 30 sec, and then measure the apparent viscosity (mPa.s) of the suspension at the shear rate and the temperature.

Example 1: For the Brookfield viscometer: 250 ml of suspension in a low-form beaker (250 ml), 50 rpm.

Example 2: For the Rheotest viscometer: 30 ml of suspension in a S2 cylinder, 81 rpm (437 s^{-1}).

Example 3: For the Haake Viscometer: 9 ml of suspension in a Rotar NV, 100 rpm (541 s^{-1}).

Example 4: For the Newport viscometer: 27 ml of suspension in a cup with a paddle.

Example 5: For the Brabender viscometer: 100 ml of suspension in a measuring bowl, 75 rpm, Sensitivity Cartridge 125 cmg.

4. Determination of viscosity of standard starch suspension

<u>4-1</u> Use corn starch (Wako Pure Chemical Industries, Ltd.) as a standard.

<u>4-2</u> Prepare 5% (w/v) suspension of the standard corn starch and conduct the procedures of 3-3, 3-4, 3-5 and 3-6 to determine the apparent viscosity (mPa.s) of the standard suspension under the same conditions as pepper samples.

<u>4-3</u> Calculate PARAMETER A (Normalised Viscosity), as follows:

$$\text{PARAMETER A} = \frac{\text{Viscosity of 10\% pepper suspension (mPa.s)}}{\text{Viscosity of 5\% standard starch suspension (mPa.s)}}$$

5. Determination of starch content of spices

<u>5-1</u> Reflux 100 mg of pepper with 10 ml of 1N HCl for 2.5 h in boiling water

<u>5-2</u> Add water to the hydrolysed pepper suspension to make up to 100 ml.

<u>5-3</u> Incubate 10 µl of the diluted suspension and 3.5 ml of reaction solution for 1 h at 37°C. The reaction solution is composed of 0.1 M phosphate buffer (pH 6.0), 100 µg ml^{-1} glucose oxidase (GOD grade 2, 100U/mg, Boehringer Mannheim GmbH), 50 µg ml^{-1} peroxidase (POD grade 2, 100 U/mg, Boehringer Mannheim GmbH) and 1 mg ml^{-1} 2,2'-Azinodi(3-ethylbenzthiazoline-6-sulfonate) (ABTS, Boehringer Mannheim GmbH).

<u>5-4</u> Measure the absorbance of the incubated solution at 420 nm (OD_{sample}).

<u>5-5</u> Conduct the procedures of 5-1 to 5-4 except refluxing to determine the absorbance of blank (OD_{blank}).

<u>5-6</u> Conduct the procedures of 5-3 to 5-4 with 10 µl of 0.1% glucose instead of the diluted suspension to determine the absorbance of the standard ($OD_{standard}$).

5-7 Calculate the starch content by multiplying glucose content by 0.9, as follows:-

Starch content (g of starch/g of pepper) = 0.9 x $(Od_{sample} - OD_{blank})/OD_{standard}$

6. Calculation of identification parameter

Calculate PARAMETER B (Identification Parameter), as follows:

$$\text{PARAMETER B} = \frac{\text{PARAMETER A (Normalised Viscosity)}}{\text{Starch amount in 1 g of pepper (g)}}$$

6. Miscellaneous

6-1 The measurements should be carried out in duplicate.

6-2 Starch content can be determined by other methods:"GOD and oxygen electrodes" etc.

6-3 Standard glucose solution (0.1%) should be prepared at each measurement, or stored in a freezer but not in a refrigerator, to avoid reduction of glucose due to microbiological degradation.

6-4 Viscosity of gelatinised starch is dependent upon the source and the producer. Corn starch produced by Wako Pure Chemical Industries, Ltd., Osaka, Japan is recommended as a standard.

6-5 Any type of revolutionary viscometer which enables the viscosity measurement of 5% starch and 10% peppers can be used.

6-6 The conditions for gelatinisation (pH of suspension, and heating temperature and period) greatly affect the viscosity value.

6-7 The conditions for the viscosity measurement described in this protocol have been established for obtaining PARAMETERS A and B which are independent of the condition for viscosity measurement. The optimum condition for viscosity measurement for each system can be found in section 7.

7. References

7-1 Brookfield viscometer
L. Heide, E. Nuernberger and K. W. Bögl, *Radiat. Phys. Chem.*, 1990, **36**, 613.

7-2 Rheotest viscometer
J. Farkas, M. M. Sharif and A. Koncz, *Radiat. Phys. Chem.* 1990, **36**, 621.

7-3 Haake viscometer
T. Hayashi, S. Todoriki and K. Kohyama, *J. Jpn. Soc. Food Sci. Technol.*, 1993, **40**, 456.

Chemical Methods
Lipids

PROGRESS IN THE DETECTION OF IRRADIATED FOODS BY MEASUREMENT OF LIPID-DERIVED VOLATILES

W.W. Nawar, Z. Zhu, H. Wan, E. DeGroote, Y. Chen and T. Aciukewicz

Department of Food Science
University of Massachusetts
Amherst, MA 01003
U.S.A.

1 INTRODUCTION

When exposed to high-energy radiation, fatty acids undergo preferential cleavage in the ester-carbonyl region giving rise to certain radiolytic compounds, typical for each fatty acid. Of these, two hydrocarbons are produced in relatively large quantities. One has a carbon atom less than the parent fatty acid and results from cleavage at the carbon-carbon bond alpha to the carbonyl group; the other has two carbon atoms less and one extra double bond, and results from cleavage beta to the carbonyl. This phenomenon, originally observed in experiments with model systems of fatty acids, esters and triacylglycerols,[1,2] was clearly detectable in the radiolytic products of beef and pork fats,[3] fish oil,[4] and vegetable oils.[5] The mechanism responsible for the formation of these compounds has been established and documented.

In 1970 Nawar and Balboni proposed a method based on the above for the detection of irradiation of food.[6] The technique,[7] which involves solvent extraction of the lipids, collection of the volatiles and measurement of the hydrocarbons by gas chromatography, was first examined for reliability, sensitivity and practicality when applied to meats and poultry.[7]

Since oleic, palmitic and stearic acids are the major fatty acids present in meats the six expected "key hydrocarbons", heptadecene, hexadecadiene, pentadecane, tetradecene, heptadecane and hexadecene, were easily detected and measured.

Although a linear relationship between radiolytic products and dose was observed for all the hydrocarbons tested, hexadecadiene, heptadecene and tetradecene proved to be the most suitable indicators of irradiation. These compounds satisfy the two conditions required for an effective marker, i.e. absence, or low level, in the non-irradiated control, and a high sensitivity to irradiation. The relative error based on 95% confidence level was approximately 10%. Two storage studies were conducted to investigate possible changes in the concentration of the radiolytic hydrocarbons with storage time after the irradiation treatment. The samples were stored at -20°C for 16 weeks. Although there appeared to be a declining trend in concentration of some volatiles, such a decrease was small compared to the amounts formed during irradiation and thus did not compromise the reliability of the method for the identification of irradiation in meats.[8]

Blind tests were conducted to assess the ability of the technique to determine (a) whether a meat sample of unknown history had been irradiated, and (b) the approximate dose received. Samples of chicken, beef and pork were irradiated at 0.0, 0.25, 0.5, 1 and 2 kGy at ambient temperature and stored at -20°C. For a total of 81 meat samples, the technique gave 100% correct determinations of whether irradiation had been applied. For the chicken, 2 samples irradiated at 0.25 kGy were misjudged as having received 0.5 kGy, and 2 others irradiated at 0.5 kGy misjudged as having received 1 kGy. For beef, one sample was misjudged as having received 2 kGy instead of 1 kGy. For pork, 4 samples were misjudged as having received 0.5 kGy instead of 1 kGy, and 3 samples as having received 1 kGy instead of 2 kGy.

2 APPLICATION TO DIFFERENT FOODS

In addition to the work by Nawar and co-workers with chicken, beef and pork, new applications were also investigated. Foods under study included spices (nutmeg, basil, oregano, cassia, black pepper, fennel, chilies), fish (mackerel and blue fish), eggs and shrimp. Each of these items has its own features, *e.g.* composition and native volatiles, requiring special considerations in technique and/or interpretation. Obviously the choice of the marker or markers must be based on the fatty acid composition of the food item, resolution of the radiolytic hydrocarbons relative to the volatiles naturally present in the food, and response, *i.e.* sensitivity and linearity, of the marker.

2.1 Spices

The detection method worked well with nutmeg, fennel, red pepper and cumin. Typical gas chromatographs for fennel are given in Figure 1. Success of the method was documented with in-house blind testing. Samples of cumin were irradiated at 0, 5, 10 and 30 kGy. Five analysts were each provided with 5 coded samples. One hundred percent correct determinations were obtained for whether a given sample was or was not irradiated. Dose estimation was accurate in 90% of the samples analysed (Table 1).

For basil, black pepper and oregano, analysis by our present method was complicated. The control samples gave extremely complicated gas chromatographic patterns due to the presence of a large number of native flavor components. This made the detection of radiolytic products very difficult. Attempts to overcome this difficulty using pre-fractionation by such methods as thin layer chromatography (TLC), urea-adduction, supercritical fluid chromatography and column chromatography simplified the gas chromatographic (GC) pattern somewhat, but not enough to allow detection and quantitative measurement of the radiolytic products. In addition to the problem of peak overlap, the very low yields of the radiolytic hydrocarbons produced from the polyunsaturated fatty acids which are present in these spices make detection of such compounds even more difficult. Therefore, for basil, black pepper and oregano other methods of irradiation detection may be more suitable. Thermoluminescence (TL) or electron spin resonance (ESR) spectroscopy are recommended.

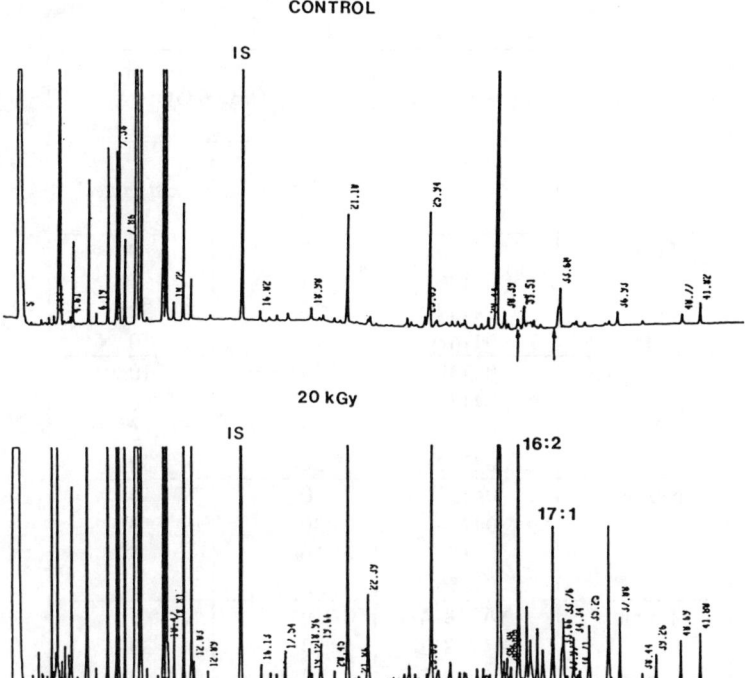

Figure 1 *Volatile analysis of non-irradiated and irradiated (20 kGy) fennel Supelcowex 10 capillary column, 30 m x 0.32 mm i.d., 0.5 μm film thickness, temperature programme: 40-245°C, 2.5°C min^{-1}*
IS: C13:1 as internal standard
16:2, 17:1: radiolytic hydrocarbons

2.2 Fish

Preliminary work with mackerel indicates that while the technique has great potential, several difficulties, specific to fish, must be resolved. GC analysis of the volatiles from non-irradiated fillets exhibited a unique pattern where pristane was present in a much greater quantity than other volatile compounds. Pentadecane is known to arise also by non-radiolytic decarboxylation and, therefore, may be encountered in the non-irradiated samples. The yield of radiolytic hydrocarbons produced from long-chain polyunsaturated fatty acids is relatively low. Research designed to overcome these difficulties is in progress.

2.3 Model Systems

Results indicate that most of the problems encountered in the application of the method to different foods can be attributed to: (a) overlap from volatiles native to the food, (b) reduction in yield of radiolytic hydrocarbons with increased unsaturation of the

Table 1 *Blind Study with Spices*

Spices	Code Number	Dose (kGy)	
		Actual	Estimated
Fennel	04145	0	0
	04146	5	5
	04147	30	30
	04148	10	10
Red Pepper	04149	10	10
	04150	0	0
	04151	5	5
	04152	30	30
Nutmeg	04153	0	0
	04154	30	10
	04155	10	5
	04156	5	5
Cumin	AA718-A	10	10.6
	AA718-B	30	>30
	AA718-C	0	0
	AA718-D	5	6
	AA718-E	0	0

fatty acid substrates, and (c) differences in yield of radiolytic hydrocarbons depending on the food in which the substrate fatty acids exist. In view of the above, a study with model systems has been initiated to investigate these factors.

Table 2 demonstrates the significant reduction in the two major radiolytic hydrocarbons (*i.e.* the n-1 and the monounsaturated n-2) with increase in number of double bonds. Table 3 shows the variation in radiolytic yield with source of substrate fatty acid.

3 INTERLABORATORY COMPARISON

A number of interlaboratory tests using the volatile hydrocarbon method were co-ordinated by various organisations with several food products. These include the following:

Table 2 *Effect of Unsaturation on the Yield of Radiolytic Hydrocarbons ($\mu g\ g^{-1}$ Substrate FA (ME)/kGy)*

	18:1	18:2	18:3	22:6
n-1 from TG	1.8	1.0	0.61	
n-1 from ME	*	1.4	0.31	Not detectable
n-2 from TG	2.2	0.56	0.31	
n-2 from ME	0.71	0.18	0.02	Not detectable

* Analysis in progress; TG = triglyceride; ME = methyl ester

Table 3 *Effect of Medium on the Yield of Radiolytic Hydrocarbons (HC) from Oleic Acid ($\mu g\ g^{-1}$ FA/kGy)*

HC	TG	Beef	Chicken	Egg Yolk	Pork	Fennel*
16:2	2.2	2.1	2.0	1.4	2.5	2.1
17:1	1.8	1.0	1.3	0.85	1.4	1.0

* Major fatty acid in fennel is a mixture of 18:1(9) and 18:1(6); HC = hydrocarbon

3.1 University of Massachusetts Intercomparison with Chicken and Egg Powder

Limited experiments with chicken, started in 1991, involved three laboratories, the Queen's University of Belfast (QUB), the Federal Drug Administration (FDA) in Washington and the University of Massachusetts (UM) in Amherst.

Table 4 shows the estimations of irradiation dose by UM in chicken samples sent by the QUB laboratory. Table 5 shows dose estimation by the FDA in Washington in chicken samples supplied by the UM.

Table 4 *Blind Test: Irradiated Chicken from the Queen's University of Belfast (QUB) Analysed in the University of Massachusetts (UM)*

Number	Estimated in UM (kGy)	Actual from QUB (kGy)
1	0	0
2	2	1.9
3	>3	3.8
4	2	1.9
5	>3	3.7
6	0	0
7	0	0
8	2	1.9

Table 5 *Blind Test: Irradiated Chicken from the University of Massachusetts (UM) Analysed by the FDA in Washington*

Number	Estimated in FDA (kGy)	Actual from UM (kGy)
1	0.25	0.6
2	2	3
3	1	2
4	0	0
5	2	3
6	0.5	1
7	0.5	1
8	2	1.9

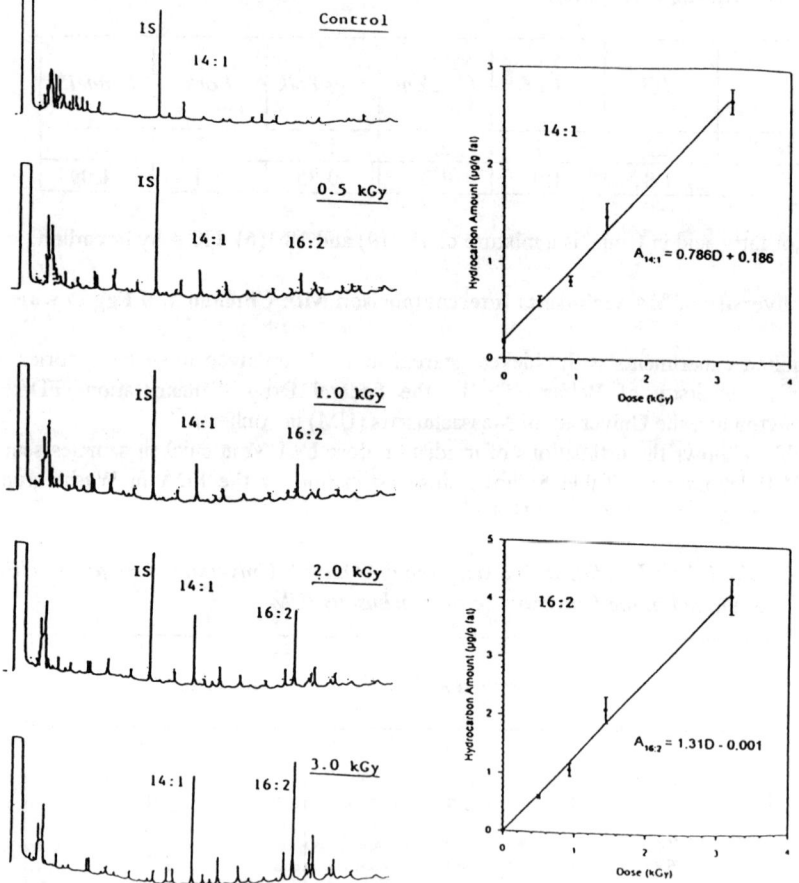

Figure 2 *Gas chromatograms and calibration curves for the radiolytic hydrocarbons (HC) in egg powder (GC conditions same as in Figure 1)*

For the experiments with egg powder, 13 laboratories (from Germany, Sweden, Finland, France and the USA) participated. A set of guidelines, 5 coded samples (actual doses were 0, 0.5, 1, and two 2 kGy samples); one non-irradiated control sample (for procedure check), were shipped by airmail to each laboratory. Eight laboratories provided data. Figure 2 shows calibration curves for the 16:2 and 14:1 hydrocarbons and GC analysis conducted in the UM laboratory on the coded samples. The bottom GC trace is that of a sample irradiated at 3 kGy for which no internal standard was added. This trace confirms the absence of peak overlap with the internal standard peak.

Based on the criteria suggested by the UM for positive determinations, no false positives were reported for the non-irradiated controls and no false negatives were reported for samples irradiated at 2 kGy. Although some laboratories reported negative results for the 0.5 kGy samples according to the recommended criteria, the differences between the controls and the 0.5 kGy samples were significant. The criteria suggested (0.65 $\mu g\ g^{-1}$ lipid for 16:2, 0.58 $\mu g\ g^{-1}$ lipid for 14:1) were based on a 5 person in-house study of a single batch of irradiated samples.

Significant problems were encountered by participating laboratories. These are attributed to one or more of the following difficulties: extensive noise or contamination (from Florisil, poor blank, lipid overload); faulty addition of internal standard; variation in amount of lipid extracted; misunderstanding of the instructions; peak misidentification; or mistakes in calculation.

A new study is under way to investigate the polar volatile compounds from irradiated egg powder. Several compounds appear to increase with radiation dose. Some of these compounds reach a peak then plateau or decrease with dose. Further efforts are being made to identify and study these compounds.

3.2 BCR Interlaboratory Test on Chicken Meat

Eight laboratories participated in a European Community Bureau of Reference (BCR) study to identify irradiated chicken. Samples were either non-irradiated or treated with 0.5, 3 or 5 kGy. Samples had to be examined one and 6 months after irradiation. Out of 239 samples analyzed 93% were correctly identified. Two false positive identifications could be attributed to sample mix ups. Fifteen samples irradiated with 0.5 kGy were not identified as irradiated.

3.3 Interlaboratory Test of the German Federal Health Office (BGA) on Camembert, Avocados, Papayas and Mangos

Twenty-one laboratories participated in a BGA study to identify irradiated Camembert, avocados, papayas and mangos. Camembert samples were either non-irradiated or treated with 0.5 or 1 kGy. Avocados, papayas and mango samples were either non-irradiated or treated with 0.35, 0.55 or 1 kGy. Of the 406 samples examined, 98% were correctly identified. One false positive result could be attributed to sample mix-up. Out of 7 false negative identifications, 4 were reported from one laboratory. The other 3 samples were irradiated with 0.35 kGy.

4 RECENT ADVANCES IN THE VOLATILE HYDROCARBON TECHNIQUE

4.1 On-Line Coupled LC-GC

An improved method for sample clean up previously described by Biedermann et al.[9] has been successfully applied by Schulzki et al.[10] to the detection of volatile hydrocarbons. The method uses a normal phase HPLC (LC) to separate the hydrocarbons from the lipid. The LC is coupled directly to the GC. The on-line coupled LC-GC system can be fully automated which is a further step to make the hydrocarbon determination more amenable to routine analysis. Furthermore, a higher separation efficiency is available with this technique. Long chain hydrocarbons similar to the radiation-induced compounds may occur in association with natural fat or as ubiquitous contaminants. Various saturated, unsaturated or branch chained patterns occur as contaminants from packaging material. Most of these hydrocarbons are alkanes or alkenes. An increased efficiency of the LC separation can be achieved by inserting a second LC-column into the system (LC-LC). on this additional column the hydrocarbon fraction is separated into classes. This allows the selection of, e.g. the higher unsaturated radiation-induced hydrocarbons (dienes, trienes) for transfer to the GC. An additional advantage over Florisil is the possibility of increasing sensitivity by allowing more lipid extract to be applied to the LC column.

4.2 Direct Headspace Analysis

This technique is simple, fast and minimises contamination. It is, however, less sensitive than methods employing lipid extraction.

References

1. M. F. Dubravcic and W. W. Nawar, *J. Amer. Oil Chem. Soc.*, 1968, **45**, 656.
2. P. R. LeTellier and W. W. Nawar, *J. Agr. Food Chem.*, 1972, **20**, 129.
3. J. R. Champagne and W. W. Nawar, *J. Food Sci.*, 1969, **34**, 335.
4. M. F. Dubravcic and W. W. Nawar, *J. Amer. Oil Chem. Soc.*, 1969, **17**, 639.
5. J. P. Kavalam and W. W. Nawar, *J. Amer. Oil Chem. Soc.*, 1969, **46**, 387.
6. W. W. Nawar and J. J. Balboni, *J. Assoc. Off. Anal. Chem.*, 1970, **53**, 726.
7. ADMIT, "Co-ordinated Research Programme on Analytical Detection Methods for Irradiation Treatment of Foods," Joint FAO/IAEA Division Report, Belfast, UK, 1994, p. 58.
8. W. W. Nawar, Z. R. Zhu and Y. J. Yoo, "Food Irradiation and the Chemist," D. E. Johnston and M. H. Stevenson, Eds., Royal Society of Chemistry, Cambridge, 1990, p. 13.
9. M. Biedermann, K. Grob, D. Frohlich and W. Meier, *Z. Lebens. Unters. Forsch.*, 1992, **195**, 409.
10. G. Sculzki, A. Spiegelberg, N. Helle, K. W. Bögl and G. A. Schreiber, "New Developments in Food, Feed and Waste Irradiation," G. A. Schreiber, N. Helle and K. W. Bögl, Eds., BGA, Berlin, SozEp-Heft 16, 1993, p. 55.

IDENTIFICATION OF IRRADIATED SEAFOOD

K. M. Morehouse

U.S. Food and Drug Administration, Center for Food Safety and Applied Nutrition
Office of Premarket Approval, Chemistry Methods Branch, 200 C Street, S.W.
Washington DC 20204
U.S.A.

1 INTRODUCTION

Interest in the use of ionising radiation for the treatment and preservation of food is increasing throughout the world. Foods are treated with ionising radiation to decrease microbial and insect infestations, inhibit maturation and extend shelf-life.[1,2] Ionising radiation can be used in place of, or in conjunction with, chemical treatment and other processes currently used to preserve foods. The treatment of food by ionising radiation is accepted for specific purposes in several countries, although in other countries the sale of irradiated food for human consumption is prohibited. The United States Food and Drug Administration (FDA) has established regulations to allow the treatment of several foods with ionising radiation.[3-5] It would be advantageous if a method was available to determine whether a commercial food has been treated with ionising radiation and is within regulatory limitations for permissible food types and maximum allowable absorbed dose. Because of differences in the composition of the food commodities that potentially could be treated by irradiation, several analytical procedures will probably have to be developed.

An ideal identification method would meet several requirements: (1) the measured response would be specific for radiation and not be induced by other techniques, (2) the method would be reliable, rapid and easy to perform, (3) the method would be inexpensive and not involve the use of expensive, complicated instruments, (4) the sensitivity and range of the method would permit identification over a wide range of absorbed dose, and (5) the method would permit an estimation of the absorbed dose in the food.[6]

There have been many recent advances in the development of analytical procedures to identify foods treated with ionising radiation and estimate the absorbed dosages.[6] At present several methods are undergoing interlaboratory validation studies. The FDA is interested in the development of analytical methods to determine if a commercial food has been treated with ionising radiation and research efforts have centered on two approaches. The first involves the isolation of specific hydrocarbons generated by radiolysis of the fats present in foods, followed by the separation and measurement of these compounds by capillary gas chromatography (GC). This procedure is based on an approach originally proposed by Nawar and Balboni.[7] The results obtained from the GC procedure will be compared with results from a procedure proposed by Dodd et al.[8,9] that uses electron spin resonance (ESR) spectroscopy to measure radiation-induced free radicals trapped in the hard matrix of calcified tissues. ESR spectroscopy is currently one of the most promising techniques to identify bone-containing foods that have been treated with ionising radiation. The utility of these two techniques will be discussed, including the limitations encountered during the development of the procedures and their application to various kinds of finfish and shellfish.

2 EXPERIMENTAL PROCEDURES

2.1 Sample Collection

Finfish (cod, haddock, perch, mackerel, salmon and catfish) and shellfish (shrimp, clams and oysters) were purchased from Washington, DC area grocery stores. The fish were purchased either whole or in a dressing. The whole finfish were cleaned and filleted before treatment with ionising radiation. Some of the finfish were treated with the bones present. The shellfish were purchased, and irradiated, either with or without the shell. After processing, the fish products were kept frozen.

2.2 Chemicals

All solvents and chemicals used were of the highest purity available. Petroleum ether, high purity, distilled in glass, with a boiling range of 30-60°C, was obtained from Burdick and Jackson Laboratories, Inc. (Muskegon, MI). Granular anhydrous sodium sulfate obtained from EM Science (Gibbstown, NJ) was heated at 700°C for 12 h before use. Florisil, obtained from the FDA Minneapolis, MN District, was heated at 130°C for 24 h the day before use. The n-alkanes and 1-alkenes used as standards were obtained from Aldrich Chemical Co. (Milwaukee, WI) or Sigma Chemical Co. (St. Louis, MO). The fatty acid methyl ester standards were obtained from Sigma Chemical Co.

2.3 γ-Radiolysis

Non-irradiated fish were packed with dry ice and irradiated using a Gammacell 220 (0.1 kGy min^{-1} in water, at the National Institute of Standards and Technology (NIST), Gaithersburg, MD). The fish were irradiated at various absorbed doses (0.5 to 10 kGy). The absorbed dose was calculated with respect to water. No corrections were made for dose-depth distribution or the difference between the stopping power or absorption coefficient of the test sample and that of the dosimeter.

2.4 Extraction

The radiolytically generated hydrocarbons and the fat were extracted from the fish meat by using either a modification of the acetonitrile extraction procedure for determining organochlorine pesticides in non-fatty foods (OMA, Method 970.52, K, a and e)[10-12] or a modification of the extraction procedure for determining organochlorine pesticides in fatty foods (OMA, Method 290.52, L, e).[10,13]

Approximately 20 to 100 g of fish was used for each analysis. When appropriate, the fish was deboned and minced before the lipids and radiolytically generated hydrocarbons were extracted. A reagent blank and a non-irradiated control were analysed along with each set of radiation treated fish to determine and minimise experimental contamination. At least two test portions of fish, at each radiation dose, were extracted. Each extract was analysed in duplicate, by capillary GC, for radiolytically generated hydrocarbons.

In the modified extraction procedure for fatty foods that was used for this investigation, approximately 20 to 100 g of fish (depending on the fat content) was blended with 50 to 100 g of anhydrous sodium sulfate (depending on the amount of fish extracted and its water content). The fish were extracted with 200 ml of petroleum ether, followed by two 100 ml portions of petroleum ether. The petroleum ether was evaporated by using a Kaderna-Danish apparatus equipped with a 3-ball Snyder column and heated with a steam bath. The petroleum ether extract was brought to a final volume of 100 ml with petroleum ether in a graduated cylinder, equipped with a glass stopper.

2.5 Florisil Column Cleanup

The radiolytic hydrocarbons were separated from the extracted lipid by Florisil column chromatography (OMA, Method 970.52, O).[10-13] The volume of petroleum ether extract applied to the Florisil column depended on the concentration of fat in the fish being analysed and was adjusted so that less than 1 g of fat was used. The eluate was concentrated to 1 ml after the addition of 1 ml of isooctane by using a Kaderna-Danish concentrator equipped with a 3-ball Snyder column and heated with a steam bath.

2.6 Fatty Acid Methyl Ester Preparation

The lipid concentration and fatty acid composition of each test sample of fish were determined according to a modification of the procedure of Eining and Ackman,[14,15] as previously described.[12] The concentration of each radiolytic hydrocarbon was then reported as ng mg^{-1} of precursor fatty acid. The fatty acid methyl esters were quantitated by using an internal standards method and a DB-23 capillary column as previously described.[12]

2.7 Gas Chromatography

The radiolytic hydrocarbons were quantitated by using a GC (HP 5890A, Hewlett-Packard Co., Avondale, PA), equipped with an HP 5895A workstation, a HP 7673A autosampler using a split/splitless injector (200°C) and a flame ionisation detector (250°C). A 1 µl aliquot of the concentrated Florisil eluate was injected into the GC, which was operated in the splitless mode. Two capillary columns were used: DB-23 (50% cyanopropyl polysiloxane, 30 m, 0.25 mm i.d., 0.25 µm film thickness, J&W Scientific, Folsom, CA) and Ultra-2 (5% phenyl, 95% methyl polysiloxane, 25 m, 0.2 mm i.d., 0.3 µm film thickness, Hewlett-Packard Co.). For the radiolytic hydrocarbon determination, the capillary columns were used with the following temperature programs: DB-23: 50°C, 5°C min^{-1} to 200°C and hold 10 min; Ultra-2: 80°C for 1 min, 5°C min^{-1} to 200°C and hold 10 min. The concentrations of the radiolytic hydrocarbons in the extracts were quantitated by using an external standard containing known concentrations of C_{14}(1-ene), C_{15}, C_{15}(7-ene), C_{16}(1-ene), C_{16}(1,7-diene), C_{16}(1,7,10-triene), C_{17}, C_{17}(8-ene) and C_{17}(6,9-diene) (Table 1).

The radiolytic hydrocarbons were identified by GC/mass spectrometry (MS) or by comparing their retention times with those of authentic hydrocarbon standards on both capillary GC columns. The radiolytic hydrocarbons (C_{15}(7-ene), C_{16}(1,7-diene), C_{16}(1,7,10-triene), C_{17}(8-ene) and C_{17}(6,9-diene)) isolated from γ-irradiated triglycerides or fatty acids and commercially available hydrocarbons were used as reference standards. The identities of the radiolytic hydrocarbons isolated from the irradiated triglycerides were confirmed by GC/MS and GC/Fourier transform infrared spectroscopy.

2.8 Fortification Experiments

Several radiolytic hydrocarbons were added at various concentrations to non-irradiated fish. The hydrocarbons were extracted from these spiked controls and recoveries were determined.[12]

2.9 ESR Spectroscopy

For ESR spectroscopic analysis, the fish bones or shells were cleaned, free of meat and fat, dried under vacuum at room temperature in a bulk tray dryer (FYS Systems, Inc., Stone Ridge, NY) and broken into small pieces.

A weighed portion (approximately 100 mg, but less than 2 cm long) was analysed by ESR spectroscopy (Varian E109 X-band spectrometer equipped with a TE_{102} cavity). The intensity of the ESR signal was normalised for the weight of bone or shell used in the analysis.[11,13]

Table 1 *Radiolytic Hydrocarbons and their Precursor Fatty Acids*

Precursor Fatty Acid	Radiolytic Hydrocarbon	
	n-1	*n-2*
Palmitic acid	C_{15}	C_{14}(1-ene)
Palmitoleic acid	C_{15}(8-ene)	C_{14}(1,7-diene)
Stearic acid	C_{17}	C_{16}(1-ene)
Oleic acid	C_{17}(8-ene)	C_{16}(1,7-diene)
Linoleic acid	C_{17}(6,9-diene)	C_{16}(1,7,10-triene)

Abbreviations: 1-tetradecene, C_{14}(1-ene); 1,7-tetradecadiene, C_{14}(1,7-diene); n-pentadecane, C_{15}; 7-pentadecene, C_{15}(7-ene); 1-hexadecene, C_{16}(1-ene); 1,7-hexadecadiene, C_{16}(1,7-diene); 1,7,10-hexadecatriene, C_{16}(1,7,10-triene); n-heptadecane, C_{17}; 8-heptadecene, C_{17}(8-ene); 6,9-heptadecadiene, C_{17}(6,9-diene).

3 RESULTS AND DISCUSSION

3.1 GC Method

A series of saturated and unsaturated hydrocarbons arise from the termination of alkyl radicals formed during the radiolysis of lipids (Table 1).[7,16-22] The FDA recently reported a fairly simple analytical procedure for the isolation, identification and quantitation of the specific hydrocarbons that are formed during the radiolysis of lipids.[11-13] This procedure, which is based on an approach originally proposed by Nawar and Balboni,[7] uses extraction procedures similar to those normally employed for determining organochlorine pesticides.[10-13]

Morehouse and co-workers previously reported the utility of monitoring the radiolytically generated hydrocarbons formed during the radiolysis of lipids, by capillary GC with flame ionisation detection, to identify irradiated shrimp[12] and frog legs.[11] These initial investigations demonstrated that low-fat foods could be identified as having been irradiated and the absorbed dose could be estimated. For the shrimp, co-extractants occasionally interfered with the identification of the radiolytic products.

To extend this study, several different types of finfish and shellfish were investigated. The extractable lipid content of the fish investigated ranged from 0.1 to 5%. Because the concentration of radiolytically generated products is proportional to the concentration of precursor compound in the food being analysed, the fish containing small amounts of extractable lipid yielded very low concentrations of radiolytically generated hydrocarbons. This made analysis difficult but, as with the shrimp and frog legs, analysis of a clean extract was possible. Figure 1 displays the gas chromatograms for the cod fillets irradiated at 2 kGy and the non-irradiated controls (0.1% extractable lipid). The concentration of the radiolytically generated hydrocarbons, especially the n-2 products, was very small (>10 ppb).

As the amount of lipid present in the fish increased, the concentration of the radiolytically generated hydrocarbons also increased. However, for the fish with the highest lipid content, i.e. mackerel (5% extractable lipid), the presence of co-extractants interfered with the identification and quantitation of the radiolytic products. As the content of extractable fat increased, so did the amount of co-extractants present in the fatty tissue of the finfish. These co-extractants may have been due to the absorption and deposition of hydrocarbon material present in the water or formed enzymatically. Figure 2 displays the gas chromatograms for the coho salmon fillets irradiated at 4 kGy and the non-irradiated controls (1.5% extractable lipid). Figure 3 displays the gas chromatograms for the chinook salmon fillets irradiated at 4 kGy and the non-irradiated controls (2% extractable lipid). With extreme care, it was possible to determine several of the radiolytically generated hydrocarbons in both of these salmon species, however, the determination of the products in the chinook salmon was very difficult. For the mackerel, even larger amounts of co-extractants were found and the determination of the radiolytic products was not possible.

For clams and oysters, the determination of the radiolytically generated hydrocarbons was also difficult because of the presence of co-extracted compounds. The large amount of co-extractants present in the clams and oysters is probably indicative of the environment in which these shellfish developed and the waters from which they were

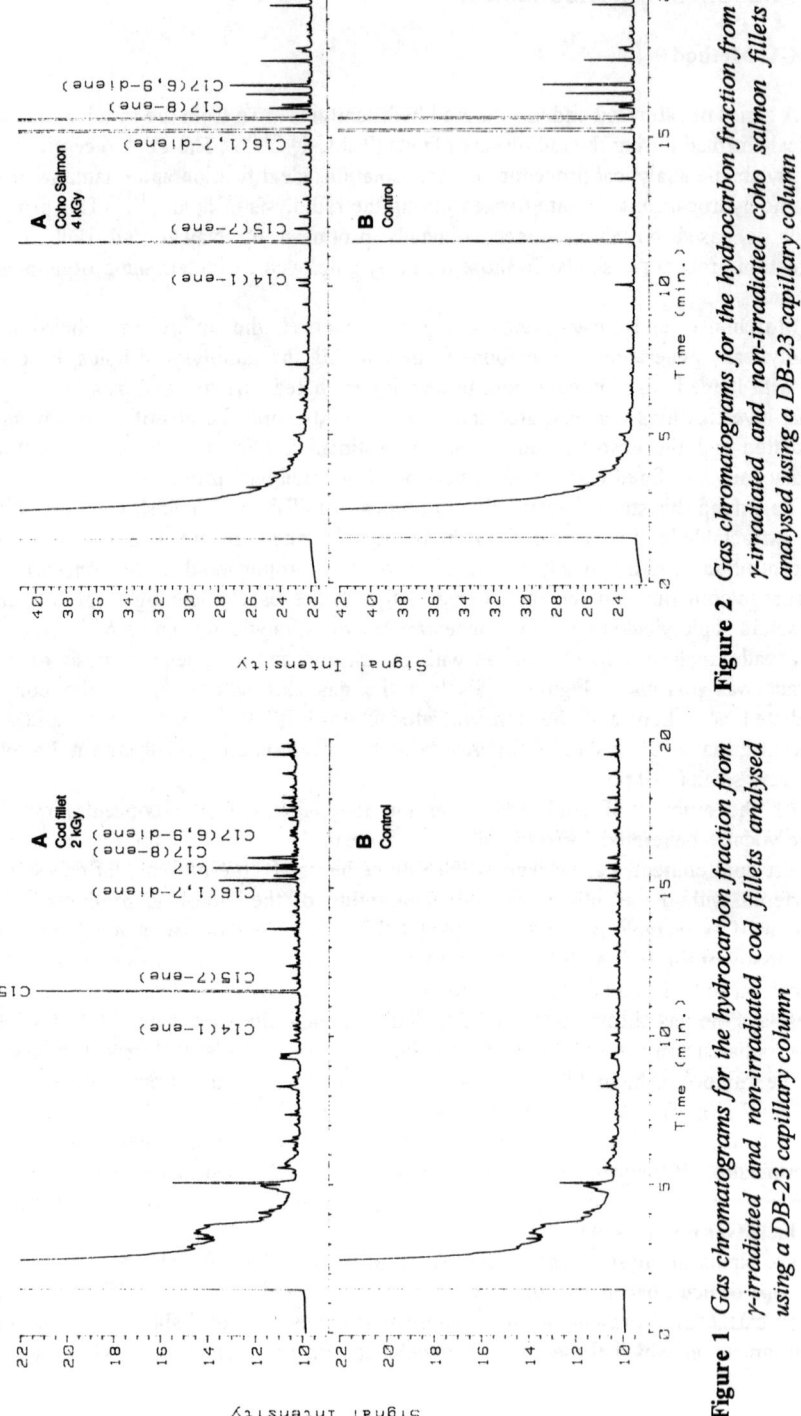

Figure 1 Gas chromatograms for the hydrocarbon fraction from γ-irradiated and non-irradiated cod fillets analysed using a DB-23 capillary column

Figure 2 Gas chromatograms for the hydrocarbon fraction from γ-irradiated and non-irradiated coho salmon fillets analysed using a DB-23 capillary column

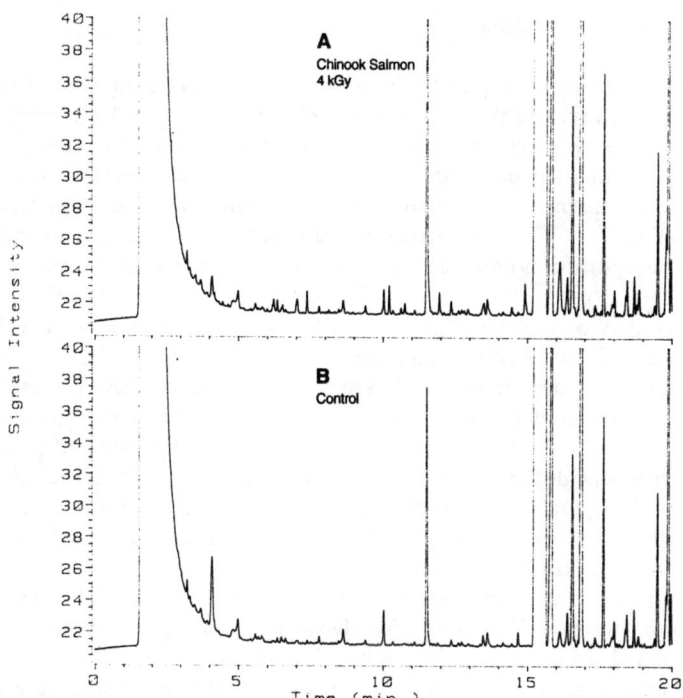

Figure 3 *Gas chromatograms for the hydrocarbon fraction from γ-irradiated and non-irradiated chinook salmon fillets analysed using a DB-23 capillary column*

harvested. As with the high-fat finfish, the radiolytically generated hydrocarbons could not always be determined (data not shown).

Capillary GC with mass selective detection, instead of flame ionisation detection, was used to analyse the seafood containing large amounts of co-extractable compounds. The use of GC/MS with electron ionisation did not help the identification of the radiolytically generated hydrocarbons because most of the co-extractants gave similar mass spectra with weak peaks for the parent ions. However, when GC/MS with positive ion chemical ionisation with methane was used the radiolytic products from their parent ions could be identified. Also, by monitoring the ions at m/z 111 and 113, it was possible to determine the unsaturated compounds (*i.e.* C_{14}(1-ene), C_{15}(7-ene), C_{16}(1-ene), C_{16}(1,7-diene), C_{16}(1,7,10-triene), C_{17}(8-ene) and C_{17}(6,9-diene)). By using mass selective detection it was possible to identify the radiolytically generated hydrocarbons in the high-fat foods that contained co-extractants (data not shown). Further research on the use of mass selective detection with chemical ionisation for the identification of radiolytic products is in progress.

Research at FDA[11-13] and that of others[7,23-27] have employed many different procedures for the extraction and isolation of radiolytically generated hydrocarbons and lipids from foods. As long as the recovery yield of the radiolytically generated hydrocarbons is acceptable, any of these procedures may be used, although recoveries will be better for some procedures than for others.

3.2 ESR Spectroscopic Method

When calcified tissue is treated with ionising radiation, a radiation-induced free radical is trapped in the matrix and this free radical can be monitored by ESR spectroscopy.[28] The application of ESR spectroscopy to the identification of foods that have been treated with ionising radiation has been extensively studied. ESR spectroscopy exhibits great promise for the identification of bone-containing foods that have been treated with ionising radiation.[8,9,11,29,30] The relative intensity of the ESR signal, corrected for the weight of the test portion, is dose dependent and displays a linear relationship to absorbed dose.[8,11,29]

It has been previously demonstrated that ESR spectroscopy is a good technique to identify irradiated foods which contain bone,[11,13] but that some problems do exist when ESR spectroscopy is applied to shrimp.[31] Several research groups have previously shown that the bones present in finfish are suitable for analysis by ESR spectroscopy and that this technique can be used to identify irradiated finfish.[8,30,32] The application of this technique to finfish is very straightforward and can be easily applied to the cartilaginous bones present in the fish (Figure 4). The ESR signal was found to be very stable for more than 6 months, whether the bones were stored frozen with the meat intact or at room temperature after the meat had been removed.

For shellfish (clams and oysters), it has been shown that a stable ESR signal is formed by the radiation treatment.[30,32] This ESR signal was also monitored and found to be easily detected and stable for more than 6 months with only small changes in the overall spectral characteristics, even when the shell was stored at room temperature (data not shown).

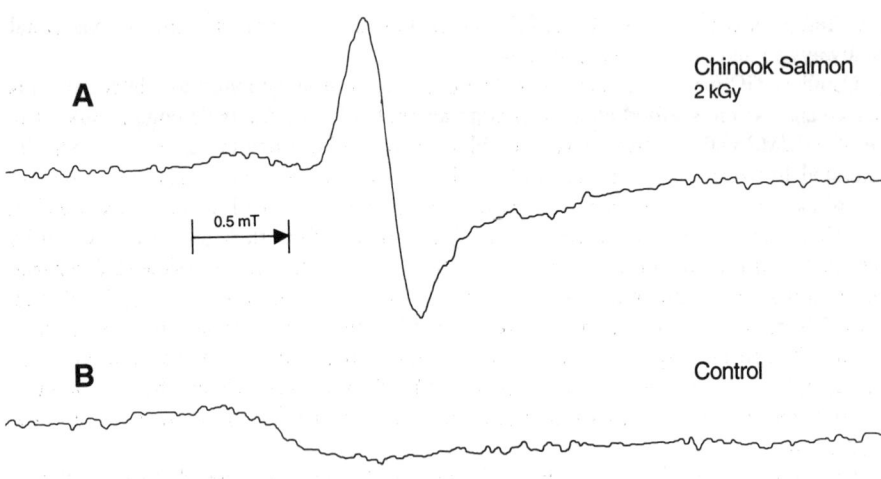

Figure 4 *ESR spectra for the cartilaginous bones from γ-irradiated and non-irradiated chinook salmon fillets*

The limitation on the application of the ESR spectroscopic technique to the identification of irradiated foods is that bone or shell must have been present during radiolysis. The ESR spectroscopic technique is not applicable to shucked clams or oysters, de-shelled shrimp, the exoskeleton of some species of shrimp[12,31] and filleted finfish from which the bones have been removed. However, GC analysis of radiolytically generated hydrocarbons may be a suitable technique to identify these irradiated commodities.

4 CONCLUSIONS

These investigations demonstrate that either the GC procedure, which isolates and then identifies radiolytically generated hydrocarbons from lipids, or the ESR spectroscopic procedure, which monitors radiation-induced free radicals trapped in calcified tissue, can be used to identify irradiated seafood. It has previously been demonstrated that in some instances it is possible to estimate the absorbed dose by both procedures and obtain agreement within the limits of precision of the two procedures.[11-13] Furthermore, these investigations demonstrate that for food commodities containing bone or hard shells, the ESR spectroscopic procedure is superior to the radiolytic hydrocarbon procedure in terms of ease of use and precision. A limitation in the use of ESR spectroscopy to identify irradiated foods, however, is that bone or shell which traps the radiation-induced free radicals must be present. Therefore, this procedure is not applicable to some shrimp, shucked shellfish or filleted finfish, but the radiolytic hydrocarbon procedure may be used. For filleted finfish and for shucked shellfish, that contain large concentrations of co-extractable compounds, a technique such as GC/MS with positive ion chemical ionisation may need to be used to avoid interfering co-extractives.

References

1. International Atomic Energy Agency (IAEA), "Food Preservation by Irradiation," IAEA, Vienna, Austria, Volumes I and II, 1978.
2. E. S. Josephson and M. S. Peterson, Eds., "Preservation of Food by Ionising Radiation," CRC Press, Inc., Boca Raton, FL, Volumes I-III, 1982.
3. *Fed. Regist.*, April 18, 1986, **51**, p. 13376.
4. *Fed. Regist.*, Dec. 30, 1988, **53**, p. 53176.
5. Code of Federal Regulations (CFR), US Government Printing Office, Washington, DC, Title 21, Part 179, 1993.
6. H. Delincée, "Monograph on Identification Methods of Irradiated Food," IAEA, Vienna, Austria, 1988.
7. W. W. Nawar and J. J. Balboni, *J. Assoc. Off. Anal. Chem.*, 1970, **53**, 726.
8. N. J. F. Dodd, A. J. Swallow and F. J. Ley, *Radiat. Phys. Chem.*, 1985, **26**, 451.
9. N. J. F. Dodd, J. S. Lea and A. J. Swallow, *Nature*, 1988, **334**, 387.
10. Official Methods of Analysis (OMA), 15th Ed., Chapter 10, Method 970.52, K and O, p. 274, A.O.A.C., Arlington, VA, 1990.
11. K. M. Morehouse, Y. Ku, H. L. Albrecht and G. C. Yang, *Radiat. Phys. Chem.*, 1991, **38**, 61.

12. K. M. Morehouse and Y. Ku, *J. Agric. Food Chem.*, 1992, **40**, 1963.
13. K. M. Morehouse, M. Kiesel and Y. Ku, *J. Agric. Food Chem.*, 1993, **41**, 758.
14. R. G. Eining and R. G. Ackman, *J. Amer. Oil Chem. Soc.*, 1987, **64**, 499.
15. D. J. Joseph and R. G. Ackman, *J. AOAC Int.*, 1992, **75**, 488.
16. M. F. Dubravcic and W. W. Nawar, *J. Amer. Oil Chem. Soc.*, 1968, **45**, 656.
17. A. Faucitano, P. Locatelli, A. Perotti and F. Faucitano-Martinotti, *J. Chem. Soc., Perkin Trans.*, 1972, **2**, 1786.
18. A. P. Handel and W. W. Nawar, *Radiat. Res.*, 1981, **86**, 428.
19. D. R. Howton and G.-S. Wu, *J. Amer. Chem. Soc.*, 1967, **89**, 516.
20. C. Merritt, Jr., P. Angelini and R. A. Graham, *J. Agric. Food Chem.*, 1978, **26**, 29.
21. C. Merritt, Jr., M. Vajdi and P. Angelini, *J. Amer. Oil Chem. Soc.*, 1985, **62**, 708.
22. W. W. Nawar, *J. Agric. Food Chem.*, 1978, **26**, 21.
23. G. Schulzki, A. Spiegelberg, N. Helle, K. W. Bögl and G. A. Scheiber, "Recent Advances on the Detection of Irradiated Food," M. Leonardi, J. J. Raffi and J.-J. Belliardo, Eds., Commission of the European Communities (BCR), Brussels, Luxembourg, EUR 14315 EN, 1992, p. 250.
24. A-M. Sjoberg, J. P. Tuominen, T. Kiutamo and S. M. Luukkonen, *J. Sci. Food Agric.*, 1992, **59**, 65.
25. M. Biedermann, K. Grob, D. Fröhlich and W. Meier, *Z. Lebensm. Unters. Forsch.*, 1992, **195**, 409.
26. G. Lesgards, J. Raffi, I. Pouliquen, A.-A. Chaouch, P. Giamarchi and M. Prost, *Amer. Oil Chem. Soc.*, 1993, **70**, 179.
27. G. A. Schreiber, N. Helle and K. W. Bögl, *Int. J. Radiat. Biol.*, 1993, **63**, 105.
28. W. Stachowicz, K. Ostrowski, A. Dziedzic-Goclawska and A. Komender, *Nucleonika*, 1970, **15**, 131.
29. M. F. Desrosiers and M. D. Simic, *J. Agric. Food Chem.*, 1988, **36**, 601.
30. N. J. F. Dodd, J. S. Lea and A. J. Swallow, *Appl. Radiat. Isot.*, 1989, **40**, 1211.
31. K. M. Morehouse and M. F. Desrosiers, *Appl. Radiat. Isot.*, 1993, **44**, 429.
32. M. F. Desrosiers, *J. Agric. Food Chem.*, 1989, **37**, 96.

IRRADIATION DETECTION IN COMPLEX LIPID MATRICES BY MEANS OF ON-LINE COUPLED (LC-)LC-GC

G. Schulzki, A. Spiegelberg, K. W. Bögl and G. A. Schreiber

BgVV - Federal Institute for Health Protection of Consumers and Veterinary Medicine
FGr21/FG 212
Postfach 33 00 13
D-14191 Berlin
Germany

1 INTRODUCTION

The analysis of radiation-induced hydrocarbons is one of the best studied methods for the detection of irradiated food containing fat. Approximately 20 years ago, basic work on reaction mechanisms and radiation products in irradiated lipids was published by Nawar and co-workers.[1-4] Besides, "cold finger distillation," separation of the hydrocarbons from the lipid by Florisil chromatography and subsequent gas chromatographic analysis has proved to be a reliable method to identify irradiated food.[5-10] It has been tested successfully, or is being tested in various interlaboratory studies e.g. to detect irradiation treatment of raw meat,[11] soft cheese, egg powder and exotic fruits.[12-13]

An improved method for sample clean-up described by Biedermann[14] uses a normal phase HPLC (LC) to separate the hydrocarbons from the lipid. The LC is coupled directly to the gas chromatograph (GC). The on-line coupled LC-GC system[15] may operate in a fully automated mode which can be considered as a further step to render hydrocarbon determination more convenient for routine analysis. Furthermore, a higher separation efficiency is available with this technique.

It is known that long-chain hydrocarbons, similar to the radiation-induced ones, may occur as natural fat attendant substances as well as ubiquitous contaminants. Various saturated, unsaturated or branch-chained hydrocarbons have been found in fish oils,[16] and terpenes and azulenes in fruit lipids. A specific pattern occurs as a consequence of contamination by mineral oil which can penetrate the food during processing, storage or from packaging material.[17] Most of these hydrocarbons are alkanes or alkenes. An increased efficiency of LC separation of hydrocarbons can be achieved by inserting a second LC column into the system (LC-LC). On this additional column the hydrocarbon fraction is separated into classes. This allows a selection of e.g. the higher unsaturated radiation-induced hydrocarbons (dienes, trienes) for transfer to the GC.

In the following paper, the application of on-line coupled LC-LC-GC for identification of irradiated fish and shrimps will be described. The relevance of these studies was shown in 1993/94 when several shrimp products on the German market were identified as irradiated by thermoluminescence (TL) measurement. Since at this time it was illegal to sell irradiated shrimp it was quite important to prove irradiation treatment by a second method. In the case of shrimps in an oily dressing, it was also important to elucidate whether only the shrimps had been irradiated or the whole product.

In addition, experiments on irradiated Camembert have been carried out. Cheese prepared with unpasteurized milk may be infected with the bacterium *Listeria monocytogenes*. The surface of cream cheese like Camembert with its fungal culture coating offers a good chance of multiplication to this bacterium which may cause the infectious disease listeriosis. Treatment by ionising radiation prevents growth of the bacteria but the cheese should be irradiated as early as possible. To reveal possible differences in hydrocarbon formation depending on the time of irradiation, the cheese was irradiated before and after maturation. These samples were also used to compare the off-line determination of the hydrocarbons by Florisil chromatography and the on-line LC-GC method.

2 MATERIALS AND METHODS

Cod *(Gadus morhua)* and Saithe *(Pollachius virens)* were purchased at a retail store and stored in frozen condition. Irradiation was carried out using a ^{60}Co source at a dose rate of 25 Gy min^{-1}. The doses applied ranged from 0.7 to 5 kGy.

Shrimps in oily dressing and raw shrimp (deep frozen) were bought at a local supermarket. For analysis of the shrimp product, the shrimps (shrimps in dressing) were separated from the dressing. Dressing was irradiated at 0.5 and 2 kGy.

Camembert cheese (45% fat in dry matter) was produced at a dairy from pasteurized milk. The freshly prepared cheese was divided into two parts. One part was irradiated before maturation, the other part after the maturation period of 7 days. The doses applied were 0.25, 0.5, 1 and 2 kGy at a dose rate of 17 Gy min^{-1}.

Lipid extraction from fish and seafood was carried out using a mixture of chloroform/methanol (2:1). The milk fat from Camembert was extracted by n-hexane after mixing the cheese with anhydrous sodium sulphate.

Fatty acid determination was carried out according to DGF-method C-VI- 11 a.[18]

LC system:
Electrically actuated rotating switching valves C6W and C10W (Valco) Phoenix 20 syringe major and slave pump (Fisons Instruments)
UV detector LC 55 (Perkin Elmer), detection at 200 nm
Column I: LiChrospher Si 60 5μm, 50 mm x 4 mm ID (Merck)
Column II: LiChrospher Si 60 5 μm, 125 mm x 4 mm ID (Merck) Eluent n-hexane AR (Merck), flow rate 200 μl min^{-1}
Backflush eluent tert-butylmethyl ether, LiChrosolv (Merok), flow rate 400 μl/min
Internal standards to mark the hydrocarbon fraction 1-hexane and 1,5-hexadiene (Fluka)
Samples: 20% solutions of the extracted fat in n-hexane
Specifications for LC-GC analysis:
10 μl of the lipid solution were injected. The hydrocarbons were separated from the lipid on column I. The entire hydrocarbon fraction (400 μl) was then transferred to the GC and the LC column was backflushed.

Specifications for LC-LC-GC analysis:

20 µl and 50 µl of the fish samples and the shrimp samples were injected, respectively. The hydrocarbons were separated from the lipid on column II and transferred to column I. While column II was backflushed, the hydrocarbons were separated on column II. The alkadiene fraction was then transferred to the GC.

On-column LC-GC-interface: The on-column interface has been described in detail by Grob.[15] The following column system was used to evaporate the eluent: an uncoated pre-column (10 m x 0.53 mm ID) was attached to a retaining pre-column (3 m x 0.32 mm ID x 0.17 µm film of 5% phenyl 95% methyl silicone). The retaining precolumn was linked through a Y-connector with the separation column as well as the early vapour exit. The conditions for solvent evaporation were as follows: column head pressure 0.9 bar, oven temperature 78°C, evaporation rate 155 µl min^{-1}.

GC-system for on-line analysis: Gas chromatograph 5890 with mass selective detector 5970B Hewlett Packard, carrier gas helium 5.0, separation column DB 5 (J&W, 30 m x 0.25 mm x 0.25 µm), temperature 78°C for 3 min; rate 5°C min^{-1} to 240°C.

Florisil Chromatography was carried out as described in reference 10 using 3% deactivated Florisil. One gramme of lipid mixed with 1 µg of the internal standard n-eicosane was used.

Off-line GC analysis: Gas chromatograph 5890 with mass selective detector 5970B and automatic sampler 7673A (Hewlett Packard), carrier gas helium 5.0, separation column HP Ultra II (25 m x 0.2 mm x 0.33 µm), injector temperature 200°C, splitless injection of 1 µl, temperature programme 55 °C for 2 min. 12°C min^{-1} to 155°C, 5°C min^{-1} to 230°C

3 RESULTS

3.1 Camembert

The main fatty acids in the Camembert analysed were myristic acid (11.8%), palmitic acid (36.7%), stearic acid (12.8%) and oleic acid (23.4%). The radiation-induced hydrocarbons considered were 1-12:1, 1-14:1, 1-16:1, 1,7-16:2 and 8-17:1, whereas the alkanes were excluded from evaluation of irradiation treatment due to their ubiquitous presence. (Figure 1). When applying both methods, Florisil chromatography as well as on-line LC-GC, none of the unsaturated radiation-induced hydrocarbons could be detected in non-irradiated samples, but all of them were detectable at and above a dose of 500 Gy (Figure 1). 1-Tetradecene, 1,7-hexadecadiene and 8-heptadecene were generated in decreasing amounts. These markers could even be clearly identified at 250 Gy. For both samples, irradiated before and after maturation, similar hydrocarbon concentrations were determined which increased with the dose in a linear mode (Figure 2). Comparison of both analytical methods, *i.e.* Florisil chromatography and on-line LC-GC, revealed a good correspondence between quantitative results (Figure 3).

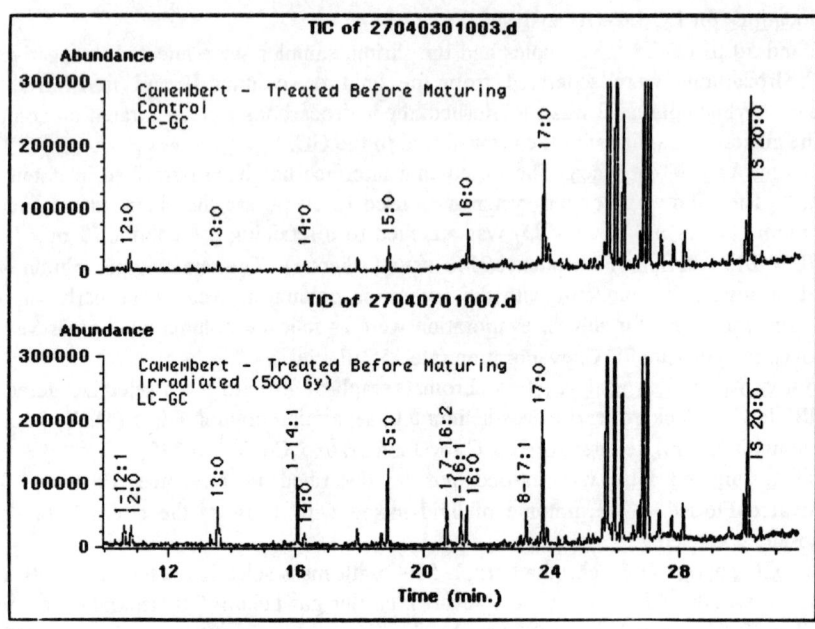

Figure 1 *Gas chromatograms of the hydrocarbon fraction from non-irradiated Camembert (top) and Camembert irradiated at 500 Gy (bottom) analysed using LC-GC*

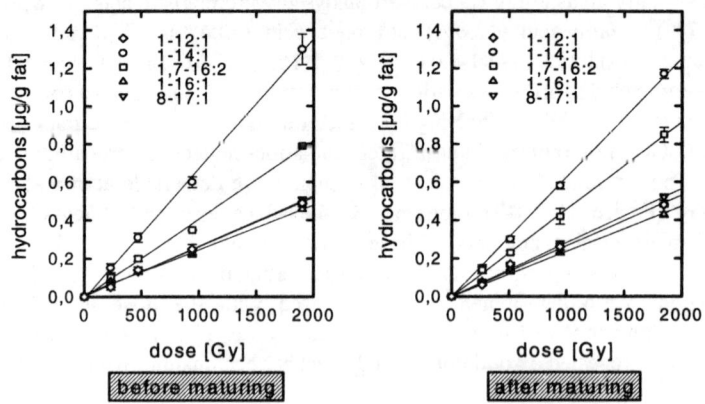

Figure 2 *Dose dependency of the radiation-induced hydrocarbons in Camembert. Comparison of the results obtained when irradiation was carried out before and after maturing.*

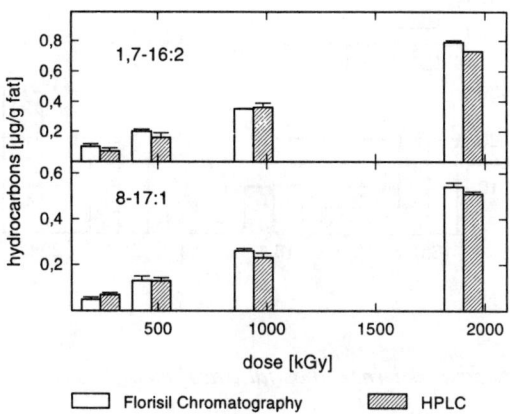

Figure 3 *Direct comparison of Florisil chromatography and on-line LC separation by means of quantification of the hydrocarbons 1,7-16:2 and 8-17:1*

3.2 Cod and Saithe

Cod and saithe have a similar fatty acid composition and a low fat content which is about 0.5%. The main fatty acids are shown in Figure 4. Despite the high proportion of polyunsaturated fatty acids, the extracted lipid was as soft as wax and not oily. Only 65% of the total lipid could be attributed to fatty acids whereas the other 35 % were of unknown origin. Oleic acid being the precursor of 1,7-hexadecadiene was present in both species accounting for 10% of the total fatty acids.

In none of the controls, 17-16:2 could be detected whereas an unequivocal identification was possible even at 0.1 kGy. The dose dependency within the range examined was almost linear (Figure 5). The yields of 1,7-16:2 were nearly the same for both fish species which is in accordance with their similar oleic acid content.

In the gas chromatograms of the irradiated samples another alkadiene was discovered in the group of the straight-chained C17 hydrocarbons (Figure 6). It was not detected in the controls and also showed a linear dose dependency. The molecular ion in the mass spectrum of this compound was 236 which corresponded to the molecular ion of heptadecadiene (17:2). The fragmentation pattern corresponded to that of 6,9-heptadecadiene obtained from irradiated linoleic acid methyl ester as well as chicken fat. However, its formation from linoleic acid is unlikely, since the content was below 1 % of the total lipid. A similar appearance was observed in former studies of other fish species[19] which also contained only small amounts of linoleic acid.

3.3 Shrimps in Oily Dressing

Analysis of the lipid of raw shrimp revealed the same effect that was observed for the fish fat mentioned previously. The lipid content of raw shrimp was between 0.5 and

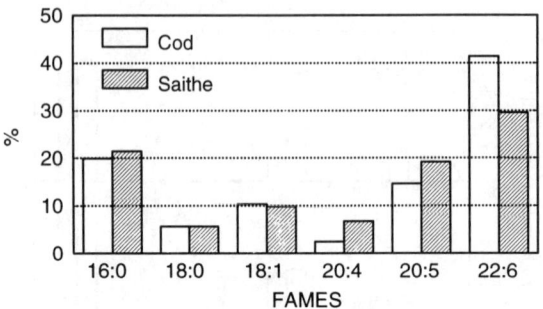

Figure 4 *Main fatty acids determined as fatty acid methyl esters (FAMES) in cod and saithe*

0.8% but only 40% consisted of fatty acids. In contrast, the total lipid extracted from the shrimp in dressing was unexpectedly high (2%). It was concluded that the shrimp had absorbed oil from the dressing and the abnormal fatty acid composition found proved this assumption. Table 1 compares the main fatty acids of 3 types of raw shrimp, the isolated shrimp with dressing and the oily dressing. Despite the naturally occurring variety, the content of oleic, linoleic and linolenic acid seemed to be increased in the shrimp with

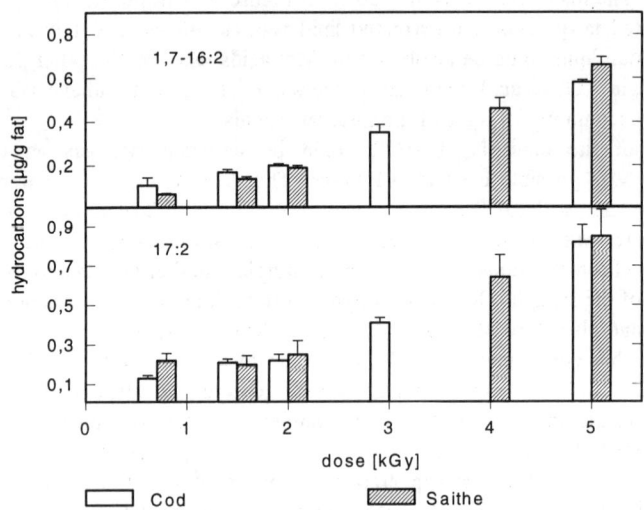

Figure 5 *Dose dependency of the alkadiene formation in irradiated cod and saithe analysed by LC-LC-GC*

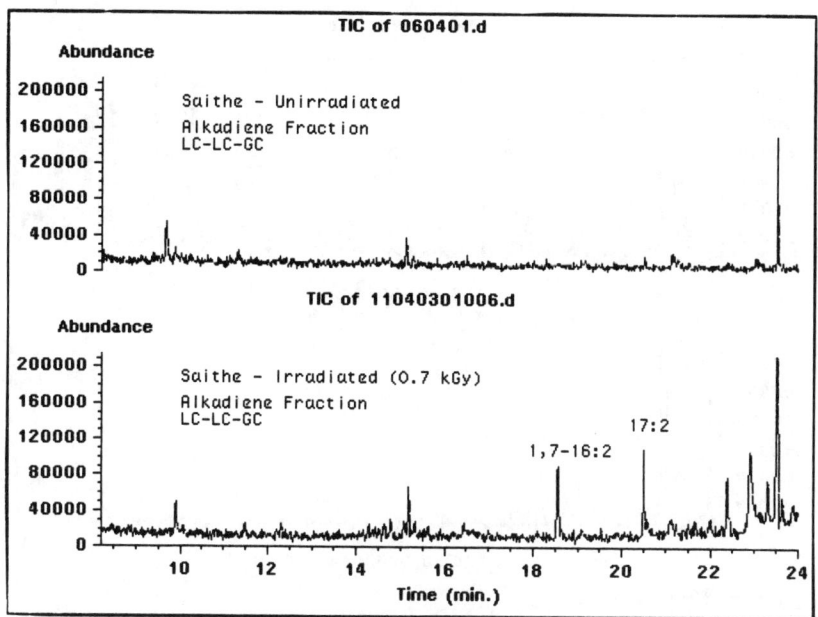

Figure 6 *Gas chromatograms of the alkadiene fraction from non-irradiated saithe (top) and saithe irradiated at 0.7 kGy (bottom) analysed by LC-LC-GC*

dressing whereas *e.g.* the content of palmitic acid was relatively low. In general, the most abundant fatty acids of the dressing were elevated in the shrimp fat whereas levels of fatty acids typical of seafood (C20:4, 20:5, 22:6) were lower.

The absorption of foreign lipid as well as the original low proportion of fatty acids in the shrimp fat resulted in a very low concentration of the radiation-induced hydrocarbons. To obtain a higher sensitivity, the injected amount of lipid was increased up to 10 mg for LC-LC-GC analysis. In the gas chromatograms of the alkadiene fraction

Table 1 *Main Fatty Acids of Raw Shrimp, Shrimp in Dressing and Oily Dressing*

Sample	% Fatty Acid Methyl Esters (FAMES)								
	16:0	16:1	18:0	18:1w-9	18:2	18:3	20:4	20:5	22:6
Raw shrimp 1	19.6	6.6	11.9	10.7	3.7	1.0	10.9	14.4	11.8
Raw shrimp 2	19.2	4.1	6.5	20.2	ND	ND	4.1	20.7	12.0
Raw shrimp 3	18.3	4.1	12.1	9.5	4.8	1.1	9.1	12.6	7.7
Shrimp in Dressing	13.0	2.8	7.5	31.3	12.0	4.2	5.5	8.4	7.4
Dressing	5.1	ND	2.2	57.0	22.6	8.7	ND	ND	ND

ND = not detected

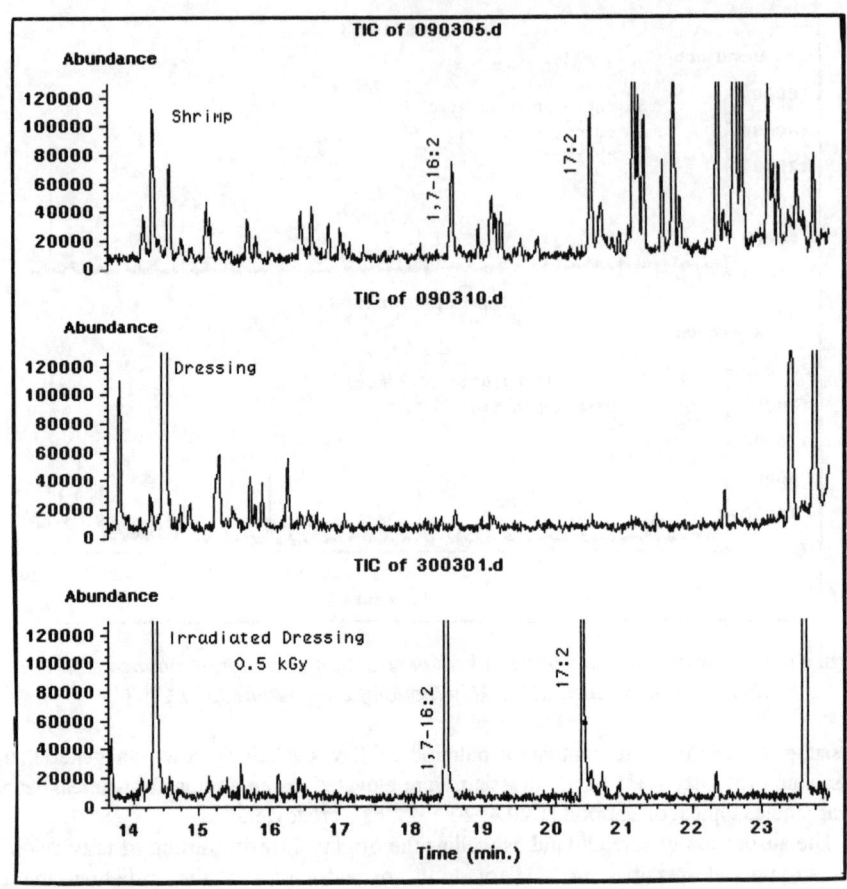

Figure 7 *LC-LC-GC analysis of a shrimp product*
Gas chromatograms of the alkadiene fraction from the isolated shrimp (top), the isolated dressing (centre) and the dressing after irradiation with 0.5 kGy (bottom)

of the isolated shrimp, 1,7-16:2 and 17:2 could be identified in very similar amounts. (Figure 7). From the approximate ratio of the precursor fatty acids, the amount of 17:2 was expected to be lower than that of 1,7-16:2. Again, there was no agreement between the hydrocarbon concentration and the fatty acid composition.

No peaks corresponding to radiation-induced hydrocarbons were found in the chromatogram of the oily dressing (Figure 7). The latter finding means, that only the shrimps had been treated with ionising radiation prior to preparation of the final product. In the chromatogram of the dressing which was irradiated prior to examination, the alkadienes 1,7-16:2 and 6,9-17:2 were clearly detectable even after irradiation with 0.5 kGy (Figure 7).

4 DISCUSSION

On-line coupled LC-GC proved to be a highly efficient method for the determination of the radiation-induced hydrocarbons in different foods. Compared with the well established Florisil method, nearly identical qualitative as well as quantitative results were obtained. However, on-line coupled LC-GC is not only an equivalent method to Florisil chromatography but there are some advantages which could help to solve difficult problems in hydrocarbon analysis.

The quantity of sample material required for LC-GC determination is drastically reduced, to a few milligrams. Food containing less than 1% lipid, like the described fish species and shrimps, can be analysed without any problems. The detection limit given by the Florisil method can be lowered by increasing the quantity of lipid injected into the LC column. This feature was very important for the analysis of shrimps in dressing and in the case of fish samples for detection at low doses. In the cheese experiment, samples irradiated with a dose of 0.25 kGy could be clearly identified. The detection of doses below 0.1 kGy will be possible with increased lipid amounts.

For analysis of the more complex fish and shrimp samples, an extension of the LC instrumentation to a two-column system (LC-LC) was necessary to select the alkadienes from the hydrocarbon fraction. According to the fatty acid composition, 1,7-16:2 can be expected as a marker in the alkadiene fraction. Since it contains an internal double bond it is considered as an unequivocal confirmation of radiation treatment. Further work has to be carried out to elucidate the formation and structure of 17:2.

Despite the difficulties with the lipid of the shrimps in dressing, the results of TL analysis could be confirmed by gas chromatographic analysis. Even further information about the product was obtained by analysis of the dressing as it could be proved that the shrimps had been irradiated prior to further processing. Providing the whole product would have been irradiated, a selective identification of the shrimp would not have been possible, since the alkadienes generated in the dressing oil would have masked those formed in the shrimp fat. Such a problem may be solved by looking for hydrocarbons originating from the higher unsaturated fatty acids which are typical of fish and seafood.

References

1. J. R. Kavalam. and W. W. Nawar, *JAOCS,* 1969, **46**, 387.
2. J. R. Champagne and W. W. Nawar, *J. Food Sci.*, 1969, **34**, 335.
3. M. F. Dubravcic and W. W. Nawar, *J. Agric. Food Chem.*, 1969, **17**, 639.
4. W. W. Nawar, *Rad. Res. Rev.*, 1972, **3**, 327.
5. W. W.Nawar, Z. R. Zhu and Y. J. Yoo, "Food Irradiation and the Chemist," D. E. Johnston and M. H. Stevenson, Eds., Royal Society of Chemistry, Cambridge, 1990, p. 13.
6. K. M. Morehouse and Y. Ku, *Rad. Phys. Chem.*, 1990, **35**, 337.
7. K. M. Morehouse and Y. Ku, *J. Agric. Food Chem.*, 1992, **40**, 1963.
8. A-M. Sjoberg, J. P. Tuominen, T. Kiutamo and S. M. Luukkonen, *J. Sci. Food Agric.*, 1992, **59**, 65.
9. N. Helle, G. Schulzki, B. Linke, A. Spiegelberg, K. W. Bögl, G. A. Schreiber, H. U. Grabowski von, J. Pfordt, U. Mauermann, S. Julicher, C. Bischoff, V. Vater and M. Heitmann, *Z. Lebens. Unters. Forsch.*, 1993, **197**, 1.

10. A. Spiegelberg, G. Schulzki, N. Helle, K. W. Bögl and G. A. Schreiber, *Rad. Phys. Chem.,* 1994, **43**, 433.
11. G. A. Schreiber, G. Schulzki, A. Spiegelberg, N. Helle and K. W. Bögl, *JAOAC,* 1994 **77**, 1202.
12. G. A. Schreiber, N. Helle, G. Schulzki, B. Linke, A. Spiegelberg, M. Mager and K. W. Bögl, K.W., ADMIT, 1994, this publication.
13. G. A. Schreiber, G. Schulzki, A. Spiegelberg, J. Ammon, U. Banziger, P. Baumann, R. Brockmann, C. Droz, P. Fey, K. Fuchs, C. Gemperle, T. Gollner, C. Hees, D. Jahr, K. Jonas, W. Krolls, M. Langer, H. Lohse, G. Mildau, F. Parsch, J. Pfordt, B. Ronnefahrt, J. Rumenapp, W. Ruge, H. Stemmer, B. Studer, C. Trapp, N. Vreden, R. Wohlfarth and K. W. Bögl, "An Interlaboratory Study on the Detection of Irradiated Camembert, Avocadoes, Papayas and Mangoes by Gas Chromatographic Analysis of Radiation-Induced Hydrocarbons," Bundesinstitut für gesundheitlichen Verbraucherschutz und Veterinarmedizin (BgVV), Berlin, BgVV-Heft 6, 1995.
14. M Biedermann, K. Grob, D. Frohlich and W. Meier, *Z. Lebensm. Unters. Forsch.,* 1992, **195**, 409.
15. K. Grob, "On-Line Coupled LC-GC," Huthig, Heidelberg, 1991.
16. D. C. Malins, "Fish Oils - Their Chemistry, Technology, Stability, Nutritional Properties, and Uses," M. E. Stansby, Ed., The AVI Publishing Company, Inc. Westport, Connecticut, 1967, p. 31.
17. K. Grob, M. Biedermann, A. Artho and J. Egli, *Z. Lebens. Unters. Forsch.,* 1991, **193**, 213.
18. "Deutsche Einheitmethoden zur Untersuchung von Fetten, Fettprodukten, Tensiden und verwandten Stoffen," Deutsche Gesellschaft fur Fettwissenschaft e.V. Munster, wissenschaftliche Verlagsgesellschaft mbH Stuttgart, Grundwerk, 1. Auflage.
19. G. Schulzki, A. Spiegelberg, N. Helle, K. W. Bögl and G. A. Schreiber, "New Developments in Food, Feed and Waste Irradiation," G. A. Schreiber, N. Helle and K. W. Bögl, Eds., Bundesgesundheitsamt (BGA), Berlin, SozEp-Heft 16, 1993, p. 55.

VALIDATION OF THE CYCLOBUTANONE PROTOCOL FOR DETECTION OF IRRADIATED LIPID CONTAINING FOODS BY INTERLABORATORY TRIALS

M. H. Stevenson

Food Science Division, The Department of Agriculture for Northern Ireland and
The Queen's University of Belfast
Newforge Lane
Belfast BT9 5PX
U.K.

1 INTRODUCTION

Following the positive conclusion of the Food and Agricultural Organisation (FAO)/ International Atomic Energy Agency (IAEA)/World Health Organisation (WHO) Joint Expert Committee on the Wholesomeness of Irradiated Food (JECFI)[1] in 1981 that "the irradiation treatment of any food commodity up to an overall average dose of 10 kGy presents no toxicological hazard and introduces no special nutritional or microbiological problems" there was renewed interest in the use of irradiation for the preservation of food. The last decade has witnessed significant advancement of food irradiation processing and at present 37 countries have approved one or more food items for human consumption and 25 countries have commercialised the technology.[2] As a result of the progress made in commercialisation of the process, greater international trade in irradiated foods, differing regulations relating to the use of the technology in many countries and consumer demand for clear labelling of the treated food, the need arose for reliable and routine tests to confirm that food had been irradiated. Although, not essential for management of the process, it was envisaged that the availability of such tests would encourage world-wide acceptance of food irradiation and may help in the enforcement of labelling regulations.[3]

While the formal requirement for general methods of identification of irradiated food was only recognised in 1988[4] the search for such tests had commenced somewhat earlier in anticipation of this need materialising. When the work was initiated it was evident that before a method could become applicable it had to fulfill a number of requirements. For example, the test should (a) be specific and not influenced by other processes or storage, (b) be accurate and reproducible, (c) have a detection limit below the minimum dose likely to be applied to the food, (d) be applicable to as wide a range of foods and food products as possible, and (e) be capable of withstanding legal scrutiny should the validity of the method be challenged in a court of law. The criteria which an ideal detection method should meet were further developed during the first meeting of the IAEA co-ordinated research programme on Analytical Detection Methods for Irradiation Treatment of food (ADMIT)[5] and allowed the scrutiny of possible detection methods and their applicability to be assessed.

The use of 2-alkylcyclobutanones as potential markers for the detection of irradiated foods was based on work carried out in the early 1970's by LeTellier and Nawar.[6,7] In a detailed investigation of the major classes of compounds formed from fats on irradiation, these workers, using simple triglycerides irradiated at high doses (60 kGy) under vacuum isolated a series of cyclic compounds known as the 2-alkylcyclobutanones. These compounds are not formed by normal degradative processes and are known to have the same number of carbon atoms as the parent fatty acids from which they are formed with an alkyl group located in ring position 2 (Figure 1). Formation of the 2-alkylcyclobutanones is believed to be initiated by the loss of an electron from the oxygen on the carbonyl of a fatty acid or triglyceride, followed by a rearrangement process to produce 2-alkylcyclobutanones specific to their parent fatty acids (Table 1). This work provided an important lead and a theoretical basis for the use of these compounds as markers for irradiation but it was far from being a technique which could be used routinely to identify irradiated food. It was necessary to develop an analytical procedure which could cope with the use of different irradiation doses and a range of lipid contents and also have adequate sensitivity and specificity.

The potential for using 2-alkylcyclobutanones as markers for irradiated food was first realised[8] in 1990 when, using a sensitive gas chromatography/mass spectrometry (GC/MS) technique, it was possible to unambiguously demonstrate the presence of 2-dodecylcyclobutanone (2-DCB) in minced chicken meat irradiated at 5 kGy. However, an essential requirement for this work was the availability of an authentic standard of 2-DCB in order that the extraction, subsequent chromatographic separation and detection procedures could be optimised. Once synthesis of this standard was completed[9] and a decision taken to use chicken as a general model system for a lipid containing food, work commenced on systematically evaluating a wide range of variables which could possibly affect formation of the 2-alkylcyclobutanones and hence their ability to form the basis of a definitive detection method.

To date 2-DCB has never been detected in any non-irradiated or microbiologically spoiled samples and has always been found in irradiated samples thereby confirming the specificity of this compound as a marker of irradiation treatment. Detailed experimental work has clearly demonstrated that the compound is not produced by cooking,[10] by packaging in air, vacuum or carbon dioxide[11] or during storage for prolonged periods of time.[12] There is also no difficulty in detecting 2-DCB in cooked, irradiated samples although the amount present has been found to be slightly reduced.[12] The compound has also been shown to be present following freeze-drying of irradiated material.[12]

The concentration of 2-DCB formed during the irradiation process follows a linear response to increasing dose up to 10 kGy for both fresh and frozen chicken meat, although slightly reduced amounts were observed for samples irradiated in the frozen state.[10-13] Extensive studies carried out on 2-DCB following storage under a variety

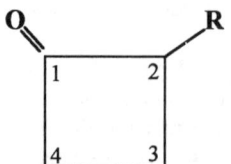

R is an alkyl group which contains 4 carbon atoms less than the parent fatty acid

Figure 1 *Generalised structure of 2-alkylcyclobutanones*

Table 1 *2-Alkylcyclobutanones Produced from their Parent Fatty Acids*

Fatty Acid		2-Alkylcyclobutanone Produced
Myristic	(C14:0)	2-Decylcyclobutanone
Palmitic	(C16:0)	2-Dodecylcyclobutanone
Stearic	(C18:0)	2-Tetradecylcyclobutanone
Oleic	(C18:1)	2-Tetradecenylcyclobutanone
Linoleic	(C18:2)	2-Tetradecadienylcyclobutanone

of conditions have also clearly shown that the compound is remarkably stable.[9-12] Initial studies by Boyd and co-workers[9] demonstrated the presence of 2-DCB not only in samples immediately following irradiation but also in irradiated samples stored for 20 days at 4°C. Further evidence of the usefulness of 2-DCB as a reliable qualitative marker was provided in yet another study where the compound was detected in chicken meat which had been sterilised at -40°C by either gamma or electron-beam irradiation and stored for 12 years at room temperature. The compound was not present in chicken meat sterilised by thermal processing 13 years previously.[12]

Whilst initial studies focused on the detection of 2-DCB in chicken meat, the synthesis of an authentic 2-tetradecylcyclobutanone (2-TCB) standard allowed the method to be modified so that both these 2-alkylcyclobutanones could be employed as markers for the irradiation process.[14] As with 2-DCB, the amount of 2-TCB formed in chicken meat increased with increasing irradiation dose, although the amounts present were lower and reflected the lower concentrations of stearic acid in chicken meat.

The methodology was further extended to other irradiated meats such as pork, lamb, beef and mechanically recovered meat (MRM) and, moreover, to detect the inclusion of irradiated foods in secondary and tertiary food products providing that the level of inclusion of irradiated product is high enough.[15] Except for pork, the amounts of 2-DCB and 2-TCB were found in a ratio similar to that of the precursor fatty acids of the meat.[14] The amount of 2-DCB present in pork was lower than expected relative to the amount of 2-TCB produced and the effect was further exaggerated when the pork was irradiated frozen. Both 2-DCB and 2-TCB have also been successfully detected and quantified in irradiated liquid whole egg.[16] Since this product also contains substantial quantities of oleic acid an investigation was carried out to determine the presence of the predicted cyclobutanone, 2-tetradecenylcyclobutanone (2-TDCB). Although an authentic standard of 2-TCCB has yet to be synthesised the presence of this 2-alkylcyclobutanone in irradiated egg has been established unambiguously, using a variety of analytical techniques.[16] In the future, a variety of 2-alkylcyclobutanones may be used to detect and/or confirm irradiation treatment.

The sensitivity of the detection technique based on 2-alkylcyclobutanones is, however, dependent on the concentration of the originating parent fatty acid in the food product. Normally with high fat foods such as chicken meat, beef and pork, 2-DCB can be easily detected at irradiation doses as low as 0.5 kGy. Unfortunately, this is not the case for low fat food such as prawn meat where the low levels of lipid present a particular challenge.

Further development of the existing methodology for such foods is currently being pursued.[17]

The 2-alkylcyclobutanone method has immense potential as a reliable test for the detection of irradiated foods. However, further application of the technique requires validation, preferably in the form of interlaboratory blind trials. Such trials would permit the repeatability and reproducibility of the method to be determined in the hands of other analysts who may need to use this methodology for food control purposes.

This paper will summarise the results of two interlaboratory blind trials carried out under the auspices of the European Union's, Community Bureau of Reference (BCR)[18] and the Joint FAO/IAEA Division of Nuclear Techniques in Food and Agriculture under their co-ordinated ADMIT programme.[19]

2 BCR INTERLABORATORY COLLABORATIVE BLIND TRIAL[18]

The objective of this trial was to examine the potential of using 2-DCB for the identification of irradiated chicken meat. Also, in order to establish if storage at -20°C affected the results, samples were analysed upon receipt and then after a 6 month storage period at frozen temperatures.

2.1 Sample Preparation

Each of 5 participating laboratories were required to analyse samples of chicken for the concentration of 2-DCB present. Whole chickens were obtained from a local chicken processor in Northern Ireland. Breast and thigh meat together with the attached skin was taken, homogenised in a food processor, bulked and mixed. Approximately 50 g samples were weighed into plastic bags, heat sealed and stored at 3°C prior to irradiation.

2.2. Irradiation and Dosimetry

Samples were given irradiation doses of either 0.5, 3 and 5 kGy (dose rate, 2.0 kGy h^{-1}, environmental temperature, 4° ± 1°C) or were non-irradiated. Cobalt 60 was used as the source of ionising radiation (Gammabeam 650, Nordion International Inc., Kanata, Canada). In order to measure the actual irradiation dose received by each sample, a perspex gammachrome YR dosimeter (AEA Technology, Harwell, UK) was attached to each of the samples given an irradiation dose of 0.5 kGy while an amber perspex dosimeter (Type 3042C, AEA Technology, Harwell, UK) was used for those receiving 3 and 5 kGy. The change in absorbance of the gammachrome and amber dosimeters was measured spectrophotometrically at 530 and 603 nm, respectively. The corresponding doses of irradiation were calculated using calibration graphs provided by the National Physical Laboratory, Teddington, UK. In order to further check dosimetry, 12 alanine dosimeters (National Physical Laboratory, Teddington, UK) were also included with the samples at each irradiation dose and the dose received by these dosimeters was measured by the National Physical Laboratory.

Following irradiation the samples were frozen at -20°C, packed in polystyrene boxes containing dry ice and sent to each of the participating laboratories by express post.

2.3 Samples for Analysis

Each participating laboratory received 48 samples, 24 of which were to be analysed immediately and the remainder to be stored frozen for a 6 month period prior to analysis. Within an analytical batch, each laboratory was expected to analyse 3 known non-irradiated samples, 6 samples irradiated at 3 kGy and 15 blind coded samples. The samples were to be further divided into 3 batches each to include 1 known non-irradiated sample, 2 known irradiated samples and 5 unknown samples. Analysis of each batch was to be carried out on 3 separate occasions in order to provide an estimate of both the within and between days variation on the analytical result. The laboratories were also supplied with the cyclobutanone working protocol[18] and an authentic 2-DCB standard for calibration purposes.

2.4 Statistical Analysis

The concentrations of 2-DCB at each of the irradiation doses were subjected to analysis of variance in order to test for differences between the laboratories at each of the storage times. The repeatability (r) of the results was calculated for each laboratory and dose, as the average difference between analyses carried out on different samples on the same day. In a few instances a laboratory did not carry out more than one analysis of a particular dose on the same day and hence r could not be calculated. The reproducibility (R) was calculated as the average difference between the mean results on different days.

2.5 Results

Analysis of the known irradiated (approximately 3.0 kGy) samples indicated that there was considerable variation in the amounts of 2-DCB determined both within and between laboratories (Figure 2) although there was no difficulty in detecting irradiated samples.

All of the samples analysed were correctly identified after the first storage period of 1 month and even after 6 months of storage only 2 false negatives were reported. Both of these samples were associated with 1 laboratory and were at the lowest dose level of 0.5 kGy (Table 2).

There was considerable variation both between and within laboratories in the concentration of 2-DCB (Figures 3 to 5; Table 3). As expected, the concentrations of 2-DCB increased with increasing irradiation dose although the concentration of 2-DCB was significantly reduced after storage for 6 months at -20°C (Table 3). Substantial differences were also observed in the repeatability and reproducibility of the 2-DCB analyses both within and between laboratories (Table 4).

In absolute terms the repeatability and reproducibility values increased with increasing irradiation dose but, when expressed relative to the mean concentrations, they were similar except for the lowest irradiation dose 0.5 kGy where they were higher.

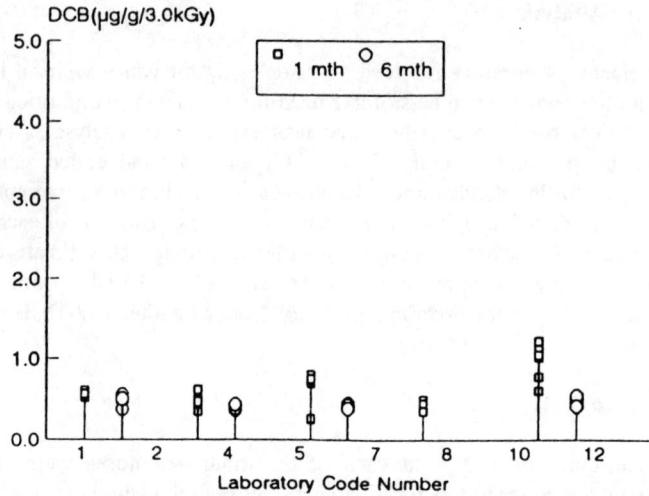

Figure 2 *Concentrations ($\mu g\ g^{-1}$ lipid) of 2-DCB for each known irradiated (3.0 kGy) sample of homogenised chicken meat analysed by each laboratory one and 6 months after irradiation.*

Table 2 *Total Number of Samples Analysed for 2-DCB at Each Time Period Together with the Number of Correct Identifications*

Dose (kGy)	No. of Samples	Storage Time (Months)	Correct Identifications
0.0	21	1	21
	14	6	14
0.5	16	1	14
	16	6	16
3.0	18	1	18
	16	6	16
5.0	19	1	19
	14	6	14
Total	134		132

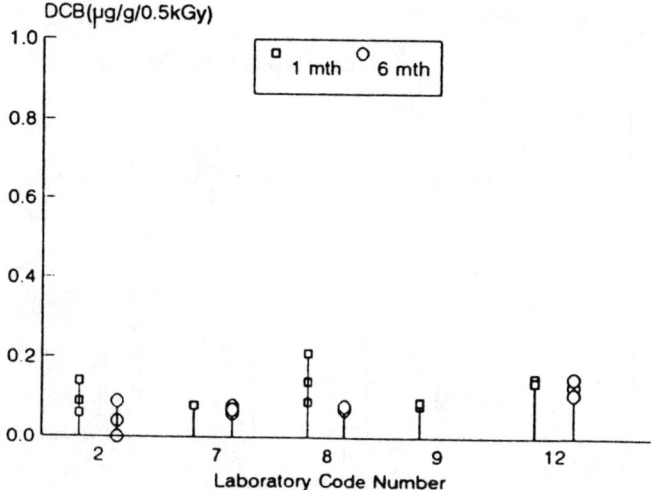

Figure 3 *Concentrations (μg g⁻¹ lipid) of 2-DCB for each blind irradiated (0.5 kGy) sample of homogenised chicken meat analysed by each laboratory one and 6 months after irradiation*

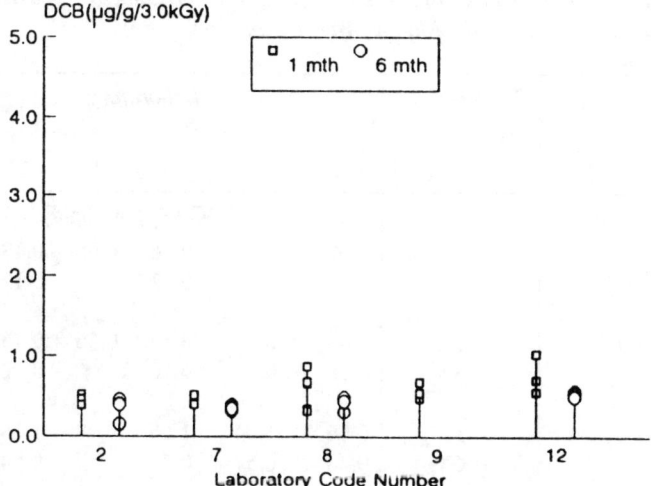

Figure 4 *Concentrations (μg g⁻¹ lipid) of 2-DCB for each blind irradiated (3.0 kGy) sample of homogenised chicken meat analysed by each laboratory one and 6 months after irradiation*

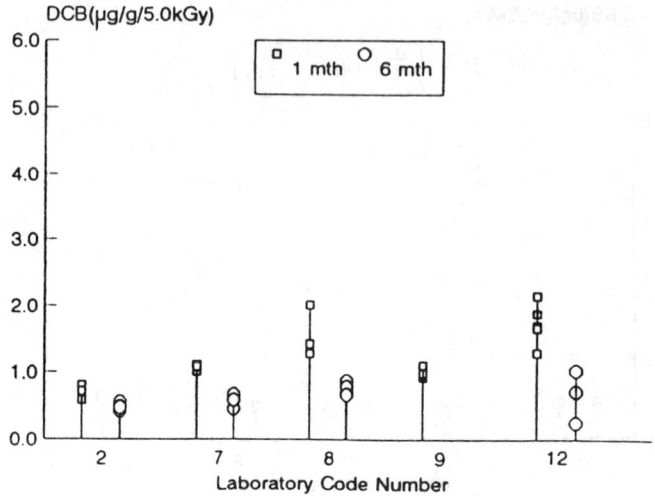

Figure 5 *Concentrations ($\mu g\ g^{-1}$ lipid) of 2-DCB for each blind irradiated (5.0 kGy) sample of homogenised chicken meat analysed by each laboratory one and 6 months after irradiation*

Table 3 *Mean Concentration of 2-DCB Detected in Irradiated Chicken by Each Laboratory One and 6 Months After Irradiation*

Dose (kGy)	Storage Time (Months)	Laboratory					
		2	7	8	9	12	SED
Blind Samples		2-DCB ($\mu g\ g^{-1}$ lipid)					
0.5	1	0.09	0.07	0.16	0.08	0.13	0.02
	6	0.03	0.07	0.72	**	0.12	
3.0	1	0.44	0.46	0.57	0.56	0.75	0.13
	6	0.28	0.36	0.40	**	0.52	
5.0	1	0.66	1.04	1.66	0.98	1.78	0.25
	6	0.48	0.57	0.74	**	0.64	
Known Samples							
3.0	1	0.55	0.50	0.58	0.42	0.97	0.11
	6	0.49	0.39	0.42	**	0.52	

SED = Standard Error of the Difference; ** No analyses carried out

Table 4 *Repeatability and Reproducibility of the 2-DCB Analysis of Irradiated Chicken Carried Out by Each Laboratory One and 6 Months After Irradiation*

Laboratory	Storage Time (Months)	Dose (kGy)							
		0.5†		3.0†		5.0†		3.0‡	
		r	R	r	R	r	R	r	R
		2-DCB (µg g^{-1} lipid)							
2	1	0.04	0.05	0.09	0.00	0.13	0.18	0.05	0.04
	6	0.02	0.06	0.26	0.03	0.06	0.07	0.09	0.04
7	1	0.00	0.00	*	0.12	0.07	0.04	0.14	0.14
	6	0.02	0.00	0.02	0.04	0.12	0.11	0.02	0.05
8	1	0.07	*	0.52	0.10	0.12	0.70	0.04	0.52
	6	0.01	0.01	*	0.17	0.03	0.17	0.03	0.05
9	1	0.01	0.01	*	0.15	0.10	0.02	0.05	0.09
	6	**	**	**	**	**	**	**	**
12	1	0.01	0.01	0.51	0.12	0.24	0.68	0.14	0.44
	6	0.01	0.03	0.02	0.02	*	0.74	0.07	0.06
Mean	1	0.03	0.02	0.37	0.10	0.13	0.32	0.08	0.25
	6	0.01	0.02	0.10	0.06	0.07	0.27	0.05	0.05
Overall Mean		0.02	0.02	0.24	0.08	0.10	0.30	0.07	0.15

† Blind samples r = repeatability
‡ Known irradiated samples R = reproducibility
* Only one sample analysed on each day so repeatability cannot be calculated
** No analysis carried out

3 FAO/IAEA INTERLABORATORY COLLABORATIVE BLIND TRIAL[19]

The objective of this collaborative trial was to provide additional information on the feasibility of using both 2-DCB and 2-TCB as specific markers for the identification of chicken, pork and liquid whole egg irradiated at commercial doses. The 11 laboratories who participated in this trial were required to determine the amounts of 2-DCB and 2-TCB present in each sample.

3.1 Sample Preparation

Chicken samples were prepared as described previously for the BCR interlaboratory trial. Pork shoulder was prepared in a similar manner to the chicken samples. Pasteurised liquid whole egg was obtained from a local processing factory and 100 g well mixed samples placed in glass bottles prior to irradiation.

3.2 Irradiation and Dosimetry

Samples were given doses of approximately 1 and 3 kGy (dose rate, 2.0 kGy h^{-1}, environmental temperature, 4° ± 1°C). The dosimetry system employed was similar to that used for the BCR trial with perspex gammachrome YR dosimeters being used for the 1 kGy samples. Following irradiation samples were frozen at -20°C prior to being dispatched to the participating laboratories.

3.3 Samples for Analysis

Each laboratory received a total of 33 samples, 11 of each food product. Within each batch of 33 samples, 3 were identified as non-irradiated (1 for each product), 3 identified as having received approximately a 3 kGy dose of irradiation (1 for each product) and the remaining 27 were blind coded. For the unidentified samples, participants received at least 2 and not more than 4 samples given each dose. Over all the participants, there was an equal number of samples at each dose. A protocol for measuring both 2-DCB and 2-TCB along with the appropriate authentic standards were also supplied.

3.4 Statistical Analysis

The concentrations of 2-DCB and 2-TCB at each of the irradiation doses were subjected to analysis of variance in order to test for differences between the laboratories. The repeatability (r) of the results was calculated for each laboratory and dose as the difference between analyses carried out on different samples on the same day.

3.5 Results

All of the samples analysed were correctly identified by the 11 participating laboratories with the exception of one sample of chicken which had received a dose level of 3.0 kGy (Table 5). This was an unexpected result and as further analyses of the sample using volatile hydrocarbons[18] also indicated that this sample had not been treated with ionising radiation it was concluded that it had been miscoded. As not all laboratories had time to perform the requested analyses the numbers of pork samples actually analysed were lower than those for chicken and egg.

Although all the samples analysed (both irradiated and non-irradiated) were correctly identified, except for the one sample of chicken mentioned previously, considerable variation both between and within laboratories in the determined concentrations of 2-DCB and 2-TCB was observed (Tables 6 to 8). The concentrations of the marker compounds were also found to increase with increasing irradiation dose as expected.

In the case of 2-DCB the repeatabilities for the egg samples were poorer than those for samples of chicken or pork while those of 2-TCB for all three products were similar.

As for the BCR trial the repeatability values increased with increasing irradiation dose but, similar values were observed when expressed relative to the mean concentrations

Table 5 *Total Number of Samples Analysed for 2-DCB and 2-TCB Together With the Number of Correct Identifications*

Product	Dose (kGy)	Number of Samples	Correct Identifications
Chicken	0.0	34	34
	1.0	33	33
	3.0	32	31*
Egg	0.0	32	32
	1.0	34	34
	3.0	33	33
Pork	0.0	25	25
	1.0	23	23
	3.0	24	24

* Volatile hydrocarbons were also not detectable in this sample

There was no evidence of any difference in relative variability between the two 2-alkylcyclobutanone compounds.

4 DISCUSSION

In the BCR trial, although there was considerable variation in the amount of 2-DCB found in both the known irradiated and the blind samples, the number of correct identifications was high. All samples receiving irradiation doses of 3 and 5 kGy, which is within the dose range likely to be used commercially, and the vast majority of samples given a low dose of 0.5 kGy, were correctly identified. A 99 % correct identification rate was achieved. It is also noteworthy that although only one of the 5 laboratories participating in this trial had substantial experience in the use of the methodology for the detection of 2-DCB.

Considerable variation in the concentration of 2-DCB and 2-TCB was observed in blind samples analysed in the FAO/IAEA trial. Nevertheless, all samples were correctly identified qualitatively except for one sample which was shown to be miscoded. As expected, neither 2-DCB or 2-TCB was detectable in non-irradiated chicken, pork or liquid whole egg thus confirming the specificity of these compounds as markers for irradiated high lipid fat containing foods. Whilst the ratio of 2-DCB to 2-TCB in chicken and egg reflected the fatty acid composition of these foods, the ratio of 2-DCB to 2-TCB in irradiated pork varied from that of the precursor fatty acids. These findings were not unexpected since a similar phenomenon was reported previously.[14]

As the irradiation dose increased, the concentrations of 2-DCB and 2-TCB were also found to increase although there was variability in the reported concentrations both between and within laboratories. It would, therefore, prove difficult to accurately estimate the dose received by samples of unknown origin unless an irradiated sample, subjected to the same processing and storage conditions, was available for analysis at the same time as the unidentified samples which in practice would be unlikely. Nevertheless, it would be possible to indicate whether samples had received a low, medium or high dose of irradiation.

Table 6 *Mean Concentrations of 2-DCB and 2-TCB in Irradiated Chicken Together with the Repeatability of the Values for Each Laboratory*

Laboratory	Dose (kGy)	2-DCB	r	2-TCB	r
		$\mu g\, g^{-1}$ lipid			
1	1.0	0.313	0.01	0.160	0.06
	3.0	0.760	0.16	0.420	0.12
2	1.0	0.225	0.05	0.055	0.01
	3.0	0.350	0.30	0.085	0.03
3	1.0	0.259	0.05	0.088	0.02
	3.0	0.557	0.50	0.224	0.16
5	1.0	0.080	0.13	0.033	0.05
	3.0	0.168	0.14	0.050	0.00
6	1.0	0.070	0.04	0.015	0.01
	3.0	0.247	0.23	0.053	0.07
7	1.0	0.215	0.06	0.063	0.02
	3.0	0.686	0.03	0.216	0.02
8	1.0	0.217	0.02	0.056	0.01
	3.0	0.679	0.07	0.218	0.05
10	1.0	0.233	0.01	0.071	0.00
	3.0	0.651	0.05	0.218	0.02
11	1.0	0.322	0.01	0.120	0.01
	3.0	0.812	0.05	0.321	0.03
12	1.0	0.153	0.12	0.040	0.03
	3.0	0.225	0.05	0.060	0.02
14	1.0	0.427	0.03	0.135	0.01
	3.0	1.097	0.07	0.417	0.05
Overall mean	1.0	0.228	0.05	0.076	0.021
	3.0	0.566	0.15	0.207	0.052
‡Overall mean	1.0	0.276	0.03	0.094	0.017
	3.0	0.699	0.15	0.265	0.060

‡ Excludes laboratories 5, 6 and 12
r = repeatability

Table 7 *Mean Concentrations of 2-DCB and 2-TCB in Irradiated Liquid Whole Egg Together with the Repeatability of the Values for Each Laboratory*

Laboratory	Dose (kGy)	2-DCB	r	2-TCB	r
		\multicolumn{4}{c}{$\mu g\ g^{-1}$ lipid}			
1	1.0	0.157	0.03	0.097	0.03
	3.0	0.357	0.14	0.220	0.07
2	1.0	0.150	0.30	0.032	0.07
	3.0	0.600	0.60	0.145	0.11
3	1.0	0.147	0.04	0.064	0.12
	3.0	0.730	0.21	0.263	0.06
5	1.0	0.208	0.19	0.052	0.04
	3.0	0.580	0.27	0.137	0.06
6	1.0	0.045	0.05	0.010	0.00
	3.0	0.090	0.00	0.033	0.02
7	1.0	0.393	0.12	0.129	0.03
	3.0	1.248	0.33	0.386	0.02
8	1.0	1.177	0.01	0.031	0.01
	3.0	0.568	0.01	0.115	0.02
10	1.0	0.173	0.11	0.047	0.02
	3.0	0.512	0.31	0.148	0.08
11	1.0	0.274	0.00	0.094	0.00
	3.0	0.748	0.08	0.258	0.03
12	1.0	0.167	0.13	0.047	0.08
	3.0	0.150	0.02	0.065	0.01
14	1.0	0.321	0.02	0.097	0.02
	3.0	0.698	0.13	0.247	0.06
Overall mean	1.0	0.201	0.09	0.064	0.03
	3.0	0.224	0.19	0.183	0.05
‡Overall mean	1.0	0.571	0.08	0.074	0.05
	3.0	0.683	0.23	0.223	0.06

† Excludes laboratories 5, 6 and 12
r = repeatability

Table 8 *Mean Concentrations of 2-DCB and 2-TCB in Irradiated Pork Together with the Repeatability of the Values of Each Laboratory.*

Laboratory	Dose (kGy)	2-DCB	r	2-TCB	r
		μg g⁻¹ lipid			
1	1.0	0.188		0.257	0.12
	3.0	0.420		0.727	0.15
3	1.0	0.109		0.123	0.04
	3.0	0.264		0.359	0.00
6	1.0	0.037		0.040	0.04
	3.0	0.095		0.110	0.00
7	1.0	0.115		0.117	0.00
	3.0	0.360		0.351	0.04
8	1.0	0.062		0.066	0.00
	3.0	0.165		0.155	0.07
10	1.0	0.124		0.119	0.02
	3.0	0.319		0.331	0.03
11	1.0	0.155		0.176	0.01
	3.0	0.447		0.504	0.03
14	1.0	0.275		0.217	0.02
	3.0	0.544		0.517	0.05
Overall mean	1.0	0.133	0.02	0.139	0.03
	3.0	0.328	0.06	0.382	0.05
‡Overall mean	1.0	0.147	0.02	0.154	0.03
	3.0	0.360	0.07	0.421	0.05

‡ Excludes laboratory 6
r = repeatability

5 CONCLUSIONS

The two trials clearly demonstrate that the cyclobutanone methodology can be used to distinguish between irradiated and non-irradiated samples of chicken, pork or liquid whole egg. They have also demonstrated that the procedure is a robust and efficient technique even when used by inexperienced analysts. Also, as a result of these trials, the

2-alkylcyclobutanone methodology is currently being adopted by the European Committee for Standardization (CEN) as a standard reference method for the detection of irradiated lipid containing foodstuffs.

The existing protocol could be used for the detection of irradiated foods containing greater than 1% lipid and investigations are in progress to evaluate the potential of extending the applicability of the technique to low lipid containing foods.

Acknowledgments

Thanks are due to the Ministry of Agriculture, Fisheries and Food (MAFF) for partial funding of the 2-alkylcyclobutanone method development programme carried out in Belfast, and to BCR and the FAO/IAEA Joint Irradiation Division for financing of the interlaboratory trials.

The author also wishes to thank Dr R. Gray and Mr W. D. Graham for sample preparation, Dr D. J. Kilpatrick for statistical analysis and the various laboratories who participated in the trials.

References

1. Joint FAO/IAEA/WHO Expert Committee on the Wholesomeness of Irradiated Food (JECFI), Report of the Working Party on Irradiation of Food, WHO Technical Report Series 659, WHO, Geneva, 1981.
2. M. Ahmed, *Radiat. Phys. Chem.*, 1993, **42**, 245.
3. Advisory Committee on Irradiated and Novel Foods (ACINF), "Report on the Safety and Wholesomeness of Irradiated Foods," HMSO, London, 1986.
4. Anon., "Acceptance, Control of and Trade In Irradiated Food," Conference Proceedings, Geneva, IAEA, Vienna, 1989.
5. Anon, "Co-ordinated Research Programme on Analytical Detection Methods for Irradiation Treatment of Foods (ADMIT)," First Research Co-ordination Meeting, Poland, IAEA, Vienna, 1990.
6. P. R. LeTellier and W. W. Nawar, *Lipids*, 1972, **7**, 75.
7. W. W. Nawar, "Health Impact and Control Methods of Irradiated Food," K. W. Bögl, D. F. Regulla and M. J. Suess, Eds., Neuherberg/Munich, WHO, Copenhagen, 1988, p. 287.
8. M. H. Stevenson, A. V. J. Crone and J. T. G. Hamilton, *Nature*, 1990, **334**, 202.
9. D. R. Boyd, A. V. J. Crone, J. T. G. Hamilton, M. V. Hand, M. H. Stevenson and P. J. Stevenson, *J. Agric. Food Chem.*, 1991, **39**, 789.
10. A. V. J. Crone, J. T. G. Hamilton and M. H. Stevenson, *J. Sci. Food Agric.*, 1992a, **58**, 249.
11. M. H. Stevenson, A. V. J. Crone, J. T. G. Hamilton and C. H. McMurray, *Radiat. Phys. Chem.*, 1993, **42**, 363.
12. A. V. J. Crone, J. T. G. Hamilton and M. H. Stevenson, *Int. J. Food Sci. Technol.*, 1992b, **27**, 691.
13. M. H. Stevenson, *Trends Food Sci. Technol.*, 1992, **3**, 257.
14. M. H. Stevenson, *Food Technol.*, 1994, **48**, 141.

15. A. V. J. Crone, "The Use of 2-Alkylcyclobutanones as Markers for the Identification of Irradiated Lipid Containing Foods," PhD Thesis, The Queen's University of Belfast, 1992.
16. A. V. J. Crone, M. V. Hand, J. T. G. Hamilton, N. D. Sharma, D. R. Boyd and M. H. Stevenson, *J. Sci. Food Agric.*, 1993, **62**, 361.
17. B. T. McMurray, W. C. McRoberts, J. T. G. Hamilton and M. H. Stevenson, In: "Detection Methods for Irradiated Food - Current Status," ADMIT, this publication.
18. M. H. Stevenson, W. Meier and D. J. Kilpatrick, "A European Collaborative Blind Trial Using Volatile Hydrocarbons and 2-Dodecylcyclobutanone to Detect Irradiated Chicken Meat," Commission of the European Communities (BCR), Brussels, Luxembourg, EUR 15969 EN, 1994.
19. M. H. Stevenson, D. J. Kilpatrick and C. H. McMurray, "Report on the FAO/IAEA Collaborative Blind Trial Using 2-Dodecylcyclobutanone and 2-Tetradecylcyclobutanone to Detect Irradiated Chicken, Pork and Liquid Whole Egg," FAO/IAEA, Vienna, 1994.

THE USE OF 2-SUBSTITUTED CYCLOBUTANONES IN THE DEVELOPMENT OF AN ENZYME-LINKED IMMUNOSORBENT ASSAY (ELISA) FOR THE DETECTION OF IRRADIATED FOODS

L. Hamilton,[1] C. T. Elliott,[3] D. R. Boyd,[2] W. J. McCaughey,[3] and M. H. Stevenson.[1,4]

[1]Department of Food Science and [2]School of Chemistry
The Queen's University of Belfast, Belfast BT9 5AG
[3]Veterinary Science Division and [4]Food Science Division
The Department of Agriculture for Northern Ireland
Newforge Lane, Belfast, BT9 5PX
U.K.

1 INTRODUCTION

When lipid-containing foods are irradiated a series of cyclic compounds, 2-substituted cyclobutanones, are formed from the corresponding fatty acids. Upon irradiation the four major fatty acids present in most foods, namely palmitic, stearic, oleic and linoleic acid are converted into 2-dodecyl-, 2-tetradecyl-, 2-tetradecenyl- and 2-tetradecadienylcyclobutanone (Table 1), respectively. The use of 2-substituted cyclobutanones as a detection method for lipid-containing irradiated foods has previously been reported.[1-5] The original method was based upon gas chromatography/mass spectroscopy (GC/MS) identification and quantification of 2-dodecyl- and 2-tetradecylcyclobutanone markers. This method requires the use of relatively sophisticated analytical equipment and takes several hours to complete. The aim of the present work is to develop an enzyme-linked immunosorbent assay (ELISA) for the detection of a range of 2-substituted cyclobutanones. This should provide a rapid, simple and sensitive screening method for the detection of these radiolytic products in both high and low lipid-containing foods. In addition, it should offer the potential for on site testing of irradiated foods, which is presently impossible using the existing detection methods.

Generally the development of an ELISA involves:-
(1) preparation of a hapten,
(2) conjugation of the hapten to a carrier protein,
(3) generation of antibodies, and
(4) validation of the ELISA.

Two approaches to the development of an ELISA for 2-substituted cyclobutanones were considered:-
(1) Generation of antibodies to conjugates of the actual 2-substituted cyclobutanones formed upon irradiation of lipid-containing foods *i.e.* to cyclobutanones containing long side-chains of 10, 12 and 14 carbon atoms.
(2) Generation of antibodies to conjugates of 2-substituted cyclobutanones containing short side-chains *i.e.* 3, 4 and 5 carbon atoms.

Table 1 *Fatty Acid Composition of Chicken*

Fatty Acid		% Composition in Chicken	2-Substituted cyclobutanone Formed Upon Irradiation of Fatty Acid	
			Side-chain	Structure
Oleic	$C_{18:1}$	45.5	$C_{14:1}$	cyclobutanone-$(CH_2)_4CH=CH(CH_2)_7CH_3$
Palmitic	$C_{16:0}$	24.0	$C_{12:0}$	cyclobutanone-$(CH_2)_{11}CH_3$
Linoleic	$C_{18:2}$	11.9	$C_{14:2}$	cyclobutanone-$(CH_2)_4CH=CHCH_2CH=CH_2(CH_2)_4CH_3$
Stearic	$C_{18:0}$	6.3	$C_{14:0}$	cyclobutanone-$(CH_2)_{13}CH_3$
Myristic	$C_{14:0}$	1.1	$C_{10:0}$	cyclobutanone-$(CH_2)_9CH_3$

Initially the second option was pursued as it was anticipated that the short chain 2-substituted cyclobutanone derivatives would provide better antigens from which antibodies could be raised specifically to the cyclobutanone ring portion of the molecules. On this basis it was hoped that such antibodies would detect the range of 2-substituted cyclobutanones formed upon irradiation of lipid-containing foods. In addition, it was anticipated that the synthesis of the short chain 2-substituted cyclobutanones would be more convenient.

Derivatives of such 2-substituted cyclobutanones (C_3, C_4 and C_5 side-chains) were synthesised and polyclonal antibodies were raised against these haptens in rabbits. A longer chain 2-substituted cyclobutanone derivative (C_{10} side-chain) was later synthesised[6] and polyclonal antibodies were subsequently raised to this hapten.

2 MATERIALS AND METHODS

2.1 Synthesis of 2-Substituted Cyclobutanone Derivatives

A series of 2-substituted cyclobutanones (2-CB) containing a terminal carboxyl group (1) were synthesised by a multi-step route.[6] This was achieved by an adaption of the procedure described by Miller and Gadwood[7] for the synthesis of 2-alkyl substituted cyclobutanones.

$$n = 3,4,5,10$$

(1) $(CH_2)_nCOOH$

2.2 Conjugation of 2-Substituted Cyclobutanones

The short chain (C_3, C_4, C_5) 2-CB derivatives were each conjugated to the carrier proteins human serum albumin (HSA) and transferrin (Tf), and to the enzyme horse radish peroxidase (HRPO) by the carbodiimide method.[8] An identical method was used for the conjugation of the longer C_{10} 2-CB derivative to bovine thyroglobulin (BTG), Tf and HRPO.

2.3 Immunisation

Priming immunogens were freshly prepared for each of the 2-CB-carrier protein conjugates. In all cases these were prepared by combining the 2-CB-carrier protein conjugate with complete Freund's adjuvant in a ratio of 1:1. For each conjugate prepared, two New Zealand white rabbits were inoculated *via* multiple subcutaneous injections along both flanks of the animal. Succeeding rabbit inoculations were prepared using incomplete Freund's adjuvant. These booster inoculations took place at fortnightly intervals. Test bleeds from the rabbits were taken *via* the marginal ear vein at weekly intervals and antibody responses monitored. After 12 weeks an intravenous dose of the immunogen was given to each rabbit and 7 days later all were exsanguinated.

2.4 Monitoring

In all cases antisera monitoring was carried out by competitive ELISAs. In the case of the short chain 2-CBs (C_3, C_4, C_5) the 2-CB-carrier protein conjugate was bound to a flat bottom microtitre plate. 2-CB antibody was added to one well, while the antibody plus a 2-CB standard was added to the other. The absorbance in each well was measured at 450 and 630 nm in a microtitre plate reader.

In the case of the C_{10} 2-CB, the antisera was bound directly to a microtitre plate. The HRPO-2-CB conjugate was added to one well, while the HRPO-2-CB conjugate plus a 2-CB standard was added to the other. The absorbance in each well was measured after addition of a chromogen solution.

2.5 Antibody Cross Reactivity Evaluation

The specificity of the antisera produced was determined by the preparation of dose response curves to a range of 2-CBs and non-related compounds (Tables 2 and 3). These curves were compared directly to those produced by the 2-CB used in the immunization. The cross reactivity 50% value (CR_{50}) was defined as:- the amount (ng) of the 2-CB (used in the immunization) required to reduce colour production by 50%, divided by the amount (ng) of the compound tested required to reduce colour production by 50%, multiplied by 100.

2.6 Sample Pre-treatment, Method Development and Extraction

2.6.1 Experiment 1. Five non-irradiated samples of minced chicken meat (breast, thigh and skin, 20 g) were extracted according to the method of Crone *et al.*[4] A portion of the lipid extracts (200 mg) were purified by column chromatography on deactivated Florisil using hexane (140 ml), followed by 1% diethyl ether/hexane (150 ml) and 2% diethyl ether/hexane (75%) as eluant. The 1% diethyl ether/hexane fractions were taken to dryness and spiked with 2-dodecylcyclobutanone (1 mg) in methanol (20 ml). Aqueous buffer (230 ml) was added and each of the samples were tested in an ELISA beside the 2-dodecylcyclobutanone standard in aqueous medium.

2.6.1 Experiment 2. Five non-irradiated samples of chicken meat (breast, thigh and skin, 20 g) were spiked with C_{12} 2-CB (200 ng) in hexane (200 ml). Five non-irradiated samples of chicken meat (breast, thigh and skin, 20 g) were spiked with C_{12} lactone. The lipid was extracted from each of the samples and was purified on a Florisil column (as above) and the extracts were analysed by GC/MS, using identical conditions to those previously reported.[2]

3 RESULTS AND DISCUSSION

3.1 Generation of Antibodies to 2-Substituted Cyclobutanones

Low molecular weight compounds such as 2-CBs are not antigenic, *i.e.* they do not stimulate a humoral immune response. To raise antibodies to such compounds they must be conjugated to large molecular weight carriers before immunization. In order to facilitate such conjugations 2-CB derivatives containing a terminal carboxyl group were synthesised. Initially a series of short chain 2-CB derivatives were prepared containing 3, 4 and 5 carbon atoms in the side-chain (not including the carboxylate carbon, coded C_3, C_4 and C_5, respectively). Each was conjugated to HSA and Tf and polyclonal antibodies were obtained in workable titres for each of the 2-CB-Tf conjugates, while titres for the analogous 2-CB-HSA conjugates were very low.

The cross reactivities of the antibodies raised to the 2-CB-Tf conjugates are reported in Table 2. The C_3 antibody showed 100% cross reaction with the C_3 2-CB. However, cross reactivities of 42 and 30% were observed for the C_5 and C_6 2-CBs, respectively. When the longer chain C_{10} and C_{12} 2-CBs were tested the cross reactivity was found to be < 0.1%. Thus cross reactivity decreased rapidly with increasing chain length. A similar

Table 2 Cross Reactivities of C_3, C_4 and C_5 2-Substituted Cyclobutanone Antibodies

Compound	CR_{50}		
	C_3 Antibody	C_4 Antibody	C_5 Antibody
2-CB-$(CH_2)_3$COOH	100	14	5
2-CB-$(CH_2)_4$COOH	–	100	–
2-CB-$(CH_2)_5$COOH	42	22	100
2-CB-$(CH_2)_6$OH	30	–	60
2-CB-$(CH_2)_9$CH$_3$	< 0.1	< 0.1	< 0.1
2-CB-$(CH_2)_{11}$CH$_3$	< 0.1	< 0.1	< 0.1
$CH_3(CH_2)_2$COOH	< 0.1	–	–
$CH_3(CH_2)_3$COOH	–	< 0.1	–
$CH_3(CH_2)_4$COOH	–	–	< 0.1

Table 3 Cross Reactivities (CR_{50}) of C_{10} 2-Substituted Cyclobutanone Antibodies

Compound	CR_{50}
2-CB-$(CH_2)_9$CH$_3$	100
2-CB-$(CH_2)_{11}$CH$_3$	73
2-CB-$(CH_2)_{13}$CH$_3$	13
2-CB-$(CH_2)_5$COOH	<0.1
γ-butyrolactone-$(CH_2)_{11}$CH$_3$	72
ascorbic acid derivative	< 0.1

pattern was observed for the C_4 and C_5 antibodies. These results suggested that the short chain 2-CB antibodies were very specific to the length of the side-chain.

In order to determine if the short chain 2-CB antibodies had been raised solely against the side-chain of the 2-CB, the cross reactivities of the analogous carboxylic acids were tested. None of the antibodies cross reacted with the acids. Thus both the CB ring and side-chain must be involved in the immune recognition process. It was concluded that although good immune responses to the short chain 2-CB-Tf conjugates were achieved, the antibodies were very specific to short chain 2-CBs and did not cross react with the 2-CB irradiation products, as required. Consequently, an alternative approach was adopted, *i.e.* generation of antibodies to conjugates of the actual 2-CBs formed upon irradiation of lipid-containing foods. Considering the fatty acid composition of chicken (Table 1) it would appear that antibodies to the 2-CB formed upon irradiation of oleic acid (the most abundant fatty acid) would be the most useful for development of the ELISA. However, this would require a lengthy and difficult synthesis of a 2-CB with a side-chain containing both a terminal carboxyl group and a localised double bond.

Considering only the composition of the saturated fatty acids in chicken meat, the most appropriate 2-CB to target appeared to be that derived from palmitic acid, that is a 2-CB with a saturated 12 carbon side-chain. Solubility problems were encountered in the synthesis of the C_{12} 2-CB derivative bearing a terminal carboxyl group, so the more soluble C_{10} derivative was synthesised. On the basis of the cross reactivity results from the short chain 2-CB antibodies, it was anticipated that antibodies raised to the C_{10} derivative would also detect the C_{12} 2-CB.

The C_{10} 2-CB derivative containing a terminal carboxyl group (1, n = 10) was synthesised, conjugates were produced to Tf and BTG and polyclonal antibodies were raised in rabbits. A very positive response was observed for the BTG conjugates, while the Tf conjugates showed a poor response. This is in contrast to the earlier results obtained for the short chain 2-CB conjugates. The observation that the optimum carriers for the long and short chain 2-CBs are different is not unusual and is possibly due to animal to animal variation or how the hapten is presented with different carriers.

Cross reactivity studies of the C_{10} antibodies indicated that they cross reacted well with 2-decyl- and 2-dodecylcyclobutanone (Table 3) as anticipated. A value of 73% cross reaction was observed for the C_{12} derivative, while a low cross reaction of 13% was observed for the longer chain 2-tetradecylcyclobutanone which was also expected. This is further evidence that the 2-CB side-chain is involved in the immune recognition process. As initial problems associated with the synthesis of the C_{10} 2-CB derivative have now been overcome it is anticipated that synthesis of the analogous C_{12} derivative will be possible. Antibodies from such a derivative should detect the CB radiolytic products from palmitic, stearic and possibly oleic acid.

An additional observation from the cross reactivity data is that the C_{10} 2-CB antibodies were found to give a good cross reaction with the C_{12} lactone, in fact the cross reactivity with the C_{12} 2-CB (73%) and the C_{12} lactone (72%) were almost identical. Five-membered lactones may be formed by oxidation of the analogous 2-CB. It has previously been found in these laboratories that 2-dodecylcyclobutanone stored in hexane in the fridge for a few months undergoes partial autoxidation to the analogous lactone (2). It is possible that the C_{10} 2-CB conjugate was partially metabolised to the analogous lactone upon immunization. Thus polyclonal antibodies raised to such a conjugate may

(2)

[structure: cyclobutanone with (CH₂)₁₁CH₃ side chain] → [O] → [structure: γ-lactone (5-membered) with (CH₂)₁₁CH₃ side chain]

contain a mixture of anti-2-CB and anti-lactone antibodies. This may explain why the C_{10} antisera cross reacted with both the C_{12} 2-CB and C_{12} lactone.

Lactones occur naturally, *e.g.* vitamin C (3). The fact that the C_{10} antisera detected both the C_{12} 2-CB and the C_{12} lactone presented a potential problem as the ELISA requires an antibody specific to 2-CBs formed in irradiated foods. It is essential that the antibody does not detect naturally occurring lactones and thus give a false positive result. However, considering the very specific nature of the 2-CB antisera previously raised, it was anticipated that the C_{10} antisera would only detect lactones of similar structure to the 2-CBs, *i.e.* with side-chains of 8 to 12 carbon atoms, which do not occur widely in nature. As anticipated no cross reaction was observed between the C_{10} antisera and vitamin C (Table 3).

3.2 Sample Pre-treatment, Method Development and Extraction

Having obtained antisera to the C_{10} 2-CB two questions arose:-

(1) Would the existing extraction procedure developed by Crone *et al.*[4] be compatible with an ELISA ?
(2) Would lactones such as the C_{12} lactone interfere with the test ?

To answer these questions two experiments were conducted. In the first, chicken samples were extracted and the lipid extracts were purified and spiked with 2-dodecyl-cyclobutanone. Each of the spiked lipid extracts were tested using the C_{10} 2-CB antibodies in an ELISA and the results compared directly with those obtained from an analogous 2-dodecylcyclobutanone standard in an aqueous medium. The ELISA was found to detect the 2-CB in all cases, indicating that the extraction procedure was compatible with the ELISA.

In the second experiment chicken samples were spiked with 2-dodecylcyclobutanone and the analogous C_{12} lactone. The lipid was extracted, purified and analysed by GC/MS. The CB was extracted into the 1% diethyl ether/hexane fraction, while the more polar lactone was detected only in the 2% diethyl ether/hexane fraction. Thus, if the lipid extract is purified by chromatography prior to the ELISA, any lactone present should be separated from the analogous CB due to the difference in polarity of the compounds.

(3) [structure of vitamin C / ascorbic acid]

4 CONCLUSIONS

The syntheses of short chain 2-CB derivatives $(1, n = 3,4,5)$ and their subsequent conjugation to carrier proteins were achieved. Good immune responses were observed for the 2-CB-Tf conjugates. However, the antibodies were found to be very specific to short chain 2-CBs and did not cross react with the 2-CB irradiation products. Thus they were unsuitable for use in an ELISA for the detection of irradiated lipid-containing foods.

The C_{10} 2-CB derivative $(1, n = 10)$ was synthesised and conjugated to Tf and BTG. Antibodies were raised and BTG proved to be the optimum carrier in this case. The C_{10} antibodies showed a good cross reaction with 2-dodecyl- and 2-decylcyclobutanone and cross reacted with 2-tetradecylcyclobutanone to a limited extent. These antibodies have been shown to detect a 2-dodecylcyclobutanone standard in an ELISA and they have great potential for use in the development of an ELISA for the detection of irradiation of lipid-containing foods.

References

1. M. H. Stevenson, A. V. J. Crone and J. T. G. Hamilton, *Nature*, 1990, **334**, 202.
2. D. R. Boyd, A. V. J. Crone, J. T. G. Hamilton, M. V. Hand, M. H. Stevenson and P. J. Stevenson, *J. Agric. Food Chem.*, 1991, **39**, 789.
3. A. V. J. Crone, J. T. G. Hamilton and M. H. Stevenson, *J. Sci. Food Agric.*, 1992, **58**, 249.
4. A. V. J. Crone, J. T. G. Hamilton and M. H. Stevenson, *Int. J. Food Sci. Technol.*, 1992, **27**, 691.
5. M. H. Stevenson, A. V. J. Crone, J. T. G. Hamilton and C. H. McMurray, *Radiat. Phys. Chem.*, 1993, **42**, 363.
6. L. Hamilton, M. H. Stevenson (deceased), D. R. Boyd, I. N. Brannigan, A. B. Treacy, J. T. G. Hamilton, W. C. McRoberts and C. T. Elliott, *J. Chem. Soc., Perkin Trans. 1*, 1996, 139.
7. S. A. Millar and R. C. Gadwood, *J. Org. Synth.*, 1988, **67**, 210.
8. T. Chard, "An Introduction To Radioimmunoassay and Related Techniques," Elsevier Biomedical Press, Amsterdam, 1982, Appendix VI, p. 264.

APPLICATION OF DCI TO THE LIPID METHOD

J. Raffi,* G. Lesgards,[+] I. Pouliquen,[+] P. Giamarchi[+] and A. Fakirian[+]*

[+]Laboratoire de Chimie des produits Naturels
*Laboratoire de Recherche sur la Qualite des Aliments
Faculte de Saint-Jerome, Avenue Escadrille Normandie Niémen
F-13397 Marseille cedex 20
France

1 INTRODUCTION

At the end of the sixties Nawar[1] proposed a cleavage point on the triglycerides which can produce alkanes and alkenes with one or two carbons less, aldehydes and free fatty acids. The first results of work on pork[2] were extended to chicken and poultry meats.[3] The methodology used by Nawar involved extraction of the lipid fraction followed by vacuum distillation and analysis by gas chromatography (GC). Other extraction and fractionation procedures have been investigated by ADMIT and BCR groups which are more appropriate for the routine examination of large numbers of sample. For instance, in the protocol[4] discussed at present by the European Committee for Standardization (CEN), the hydrocarbons are extracted together with the fat from the meat, separated from the fat using florisil columns and then detected with GC/FID or GC/MS.

In the present study, the radio-induced volatile compounds were analysed with a DI200 chromatograph, used with a head-space system,[5,6] also called the DCI system (Desorption, Concentration, Injection). The main advantage of the method is that it avoids the soxhlet extraction of the lipid fraction from the foodstuffs. However, the sensitivity may be decreased and in each case, it must be checked that the "direct sensitivity" is sufficient enough to allow good identifications. Several products were studied; oils, poultry meat, avocado pear. It appears that the DCI is a good and fast method provided that the temperature of the oven is controlled, which is not the case with the commercial apparatus used.

Firstly, this study commenced by working on vegetable, sunflower and peanut oils irradiated at various doses up to 50 kGy. These oils and this high irradiation dose were chosen as model system in order to perfect the method. Sunflower oil was chosen because of it's high level of unsaturation (64% of linoleic acid), and peanut oil for it's low level of unsaturation level (57% of oleic acid). In fact vegetable oils can be irradiated at low doses in order to avoid aflotoxine contamination, or as a component of fresh ready prepared salads. Secondly, the results obtained for the first study were compared with those obtained for a study on the thermolysis effects on the same oils.

Lastly, the concrete case was studied: effect of irradiation on poultry meat, avocado-pear lipids, and on fresh pilchards lipids, in order to make a test to detect irradiated foods.

2 MATERIALS AND METHODS

2.1 Reagents

The standards used for chromatography were as follows: oleic acid, nonadecylic acid, methyl oleate, methyl nonadecylate, methyl heneicosanoate were obtained from Alltech Associates (Deerfield, Illinois); Triolein was obtained from Fluka Chemie (Bucks, Switzerland); Butyl Hydroxy Toluene (BHT, 2,6-di-terbutyl-4-methylphenol) was from Janssen Chimica (Geel, Belgium). Other chemical products and solvents (heptane, octane, diethyl oxide, chloroform, methanol *etc.*), were obtained from SDS (Marseille, France).

2.2 Extraction of Lipids

Eight grams of avocado-pear pulp, pilchard fillet, poultry meat *etc.* were crushed with the same quantity of extra pure sea sand and anhydrous sodium sulphate in order to obtain a fine powder. The mixture was put in extraction thimbles and extracted with 100 ml of diethyl oxide in a soxhlet extractor for 4 h. The diethyl oxide, from the extract obtained, was then evaporated in a vacuum extractor at room temperature in order to obtain pure avocado-pear or other lipids.

2.3 Sample Irradiation

Samples were irradiated by γ-rays from ^{137}Cs (Cadarache source of 18600 Ci, and approximately 60 Gy min^{-1}), at room temperature and kept at -20°C in a domestic freezer.

2.4 Gas Chromatography of Volatile Compounds

A GC DI200 from Delsi (Argenteuil, France), with helium carrier gas, split injector, and flame ionisation detector (FID), was used with an head space dynamic system, for desorption and concentration of the sample volatile compounds: a small oven containing the sample was heated under carrier gas flow in order to evaporate the volatile compounds. These compounds were then cryo-concentrated on a tenax trap (tenax is a porous polymeric of 2,6-diphenyl-P-phenylene oxide), for a few minutes and afterwards they were injected into the GC by flash heating of the trap.

The following conditions were used: 30 mg of oil sample to be desorbed; desorbtion at 150°C for 10 min and trapping on tenax at -40°C; injection by trap flash heating at 200°C for 1 min; inlet pressure, 1 bar; detector temperature, 260°C; fused silica column, length 50 m; external diameter, 0.45 mm, internal diameter 0.32 mm, stationary bonded phase SE30, film thickness 0.2 μm; temperature programmed from 100°C to 250°C at 5°C min.$^{-1}$

3 EXPERIMENTAL RESULTS AND DISCUSSION

3.1 Volatile Compound Analysis

Volatile compounds were analysed in sunflower, peanut, avocado and poultry oils extracted from the original products (non-irradiated or irradiated at different doses) or directly from the foodstuff itself.

The results obtained[6] are summarised in Table 1. The names of the lipids studied and related radio-induced hydrocarbons along with the values of the minimal detectable doses obtained by the two methods are given.

Moreover, it should also be noted that the amount of these compounds in oils, was not significantly modified by addition of BHT antioxidant to the oil.

3.2 Heated Oils Test Quality

In the volatile hydrocarbons fraction only alkanes and alkenes with 17 carbons were present and there was no formation of 16 carbon alkenes by the heat treatment

Table 1 *Foodstuffs Studied by the DCI Technique*

Foodstuff	Main Lipids	Main Hydrocarbons Studied	Minimal Detectable Dose	
			With Lipid Extraction	With Direct Injection
Oils	C16:0	C15:0	-	0.15 kGy
		C14:1		
	C18:1	C17:1		
		C16:2		
	C18:2	C17:2		
		C16:3		
Avocado-pear	C16:0	C15:0	0.25 kGy	0.5 kGy
		C14:1		
	C16:1	C15:1		
		C14:2		
	C18:1	C17:1		
		C16:2		
	C18:2	C17:2		
		C16:3		
Poultry meat	C16:0	C15:0	0.25 kGy	0.5 kGy
		C14:1		
	C18:0	C17:0		
		C16:1		
	C18:1	C17:1		
		C16:2		
	C18:2	C17:2		
		C16:3		

The thermolytic and radiolytic reactions were different, as were the mechanisms involved. It is possible to show the difference between an irradiated and a heated oil by analysis of the 16 carbon volatile hydrocarbons. In fact it is possible to detect an irradiated oil, even after a heat treatment, by the presence of 16 carbon hydrocarbons in the volatile carbohydrate analysis.

This technique could also be used to control the quality of cooking oils in restaurants or small retailers of French fried potatoes.

4 CONCLUSION

The results obtained for DCI in this study[7] were promising. However, further experiments on food containing lipids are necessary. The detection limits must be defined for different foodstuffs in relation to several food industry applications and by comparison to the future "official" method, the "DCI" method being an alternative test.

Acknowledgments

We are indebted to the French Research Ministry (Grant N° 88.G.1014), to the Community Bureau of Reference (BCR, Brussels) (Agreements 5348/1/5/340/ 90/4-BCR F[10] and 5415/1/5/340/90/11-BCR-F[10]) and to IAEA (Agreement N°5154/CF) for financial support and helpful discussions during the meetings they organised.

References

1. W. Nawar and J. Balboni, *JAOAC*, 1970, **53**, 726.
2. M. Dubravcic and W. Nawar, *JAOCS*, 1966, **17**, 639.
3. W. Nawar, Z. Zhu and Y. Yoo, "Food Irradiation and the Chemist," D. E. Johnston and M. H. Stevenson, Eds., Royal Society of Chemistry, Special Publication No. 86, 1990, p. 13.
4. J. Raffi, H. Delincée, E. Marchioni, C. Hasselmann, A-M. Sjöberg, M. Leonardi, M. Kent, K-W. Bögl, G. Schreiber, H. Stevenson, W. Meier, "Methods of Identification of Irradiated Foods," Commission of the European Communities (BCR), Brussels, Luxembourg, EUR 15261 EN, 1993/94.
5. G. Lesgards, P. Giamarchi, M. Prost, M. Michel and J. Raffi, *Ann. Fals. Exp. Chim.*, Paris, 1990, **83**, 299.
6. G. Lesgards, J. Raffi, I. Pouliquen, A-A. Chaouch, P. Giamarchi and M. Prost, *JAOCS*, 1993, **70**, 179.
7. A. Fakirian, D.Sc. thesis, Université d'Aix-Marseille, faculté de Saint-Jérôme, April 1994.

Other Chemical Methods

THE STATUS OF DETECTION METHODS BASED ON RADIOLYTIC PRODUCTS

P. B. Roberts

Institute of Geological and Nuclear Sciences
Lower Hutt
New Zealand

1 INTRODUCTION

In principle, quantification of any product of radiolysis is a potential method by which the irradiation of food can be detected. Recent advances in chemically-based detection techniques have concentrated on two types of radiolytic products, as discussed elsewhere in this volume. Cyclobutanones are attractive because present analytical techniques have not found detectable amounts in non-irradiated foods (Stevenson, this volume). Volatile hydrocarbons produced by lipid cleavage are also well studied (Nawar, this volume). Although not uniquely present in irradiated foods, their quantitative distribution is significantly changed upon irradiation, and several marker molecules can be used in a single test to increase reliability.

A number of other radiolytic products have also been studied. Generally they suffer from the disadvantage of being based on increases in the concentration of a single compound over variable baseline levels in non-irradiated foods. Such measurement methods are not as well advanced as the cyclobutanone and hydrocarbon methods. Three radiolytic products have had significant attention recently. They are based on the radiolytic production of o-tyrosine, lipid peroxides and gases such as hydrogen (H_2) and carbon monoxide (CO).

2 METHODOLOGIES

2.1 o-Tyrosine

o-Tyrosine is produced by OH radical attack on phenylalanine-containing proteins, and was first suggested as a detection method for meats by Karam and Simic.[1] The original analysis proposed used gas chromatography/mass spectrometry (GC/MS) with selective ion monitoring (SIM). Initial enthusiasm for the method waned as conflicting results emerged concerning the presence or virtual absence of o-tyrosine in non-irradiated meat.[2,3] Practical difficulties in the analysis also became evident both for the GC/MS method and for the HPLC method proposed more recently.[4]

These difficulties have delayed development of the o-tyrosine method, which basically could be a relatively cheap and fast method for analytical food laboratories. Meier and co-workers have continued to develop the methodology (Meier, this volume). Quantitation of o-tyrosine is now carried out by HPLC with fluorescence detection, with a modified procedure to improve separation of the o-tyrosine peak. Background levels have been established in non-irradiated samples of chicken, fish, shrimps, mussels and frogs' legs. Doses of radiation in the commercially relevant dose range cause substantial increases in the o-tyrosine concentration.

The method requires validation by a large-scale interlaboratory comparison and this is now in progress.

2.2 Lipid Peroxides

Radiolysis of aerated unsaturated fatty acids leads to the production of lipid hydroperoxides (LPH). Analysis by iodometric titration, for example, is simple which is an attractive feature for laboratories which cannot afford sophisticated equipment.

However, there are several serious drawbacks to LHP as markers for irradiated foods. They are produced by autoxidation and are unstable; anti-oxidants and oxygen levels in the food, as well as its composition, will influence yields. The method may be restricted to frozen foods.

At the time of the second ADMIT meeting in 1992 three laboratories were studying the method.[5] However, although LPH may be under investigation by many researchers from a radiation chemistry viewpoint, only one laboratory appears to be continuing development of LPH as a detection method (Wu, this volume). Recent research, limited to pork, shows that autoxidation ceases for several months at -18°C in non-irradiated pork. After irradiation, however, an increase in LPH occurs. Baseline concentrations in non-irradiated pork have been measured, and irradiation up to 1 kGy has been shown to increase LPH content 3-7 fold shortly after treatment.

A blind trial of LPH as a detection method for pork has been completed successfully. However, LPH measurement as a detection method must be regarded as a research project at present, with many difficulties to be overcome by greater collaboration between different laboratories.

2.3 Gas Evolution

Radiolysis of water-containing foods yields several low molecular weight gases.[6] Two of these, H_2 and CO, have been investigated as potential detection methods. The advantage of the proposed method is it's extreme simplicity and speed. Radiolytic gases trapped in the food are released by brief microwave heating or grinding, and detected by gas chromatography or other gas sensors. Gas sensors which are robust, cheap and simple are under investigation (Hitchcock; Delincée, this volume). Analyses can be completed within about 5 min.

The disadvantages of the method arise principally from the likelihood of low molecular weight gases diffusing out of the food prior to detection. This limits the method to frozen foods and foods, in which gas is trapped within the food matrix. The limited data available suggest that diffusion will result in highly variable results between samples and food types and be dependent on the storage time and temperature.

Furuta and co-workers[7] first showed that H_2 could be a useful marker for irradiated pepper during several months of storage. They extended the method to frozen beef, pork and chicken, showing that H_2 was only useful for a few weeks after irradiation.[8] However, CO remained a useful marker for up to one year. Roberts (this volume) confirmed these results in chicken but also showed that, in frozen shrimps, the evolved gas concentration fell far more rapidly to values indistinguishable from the concentration evolved from unirradiated samples. In frozen shrimp H_2 does not appear to be a useful marker, while CO is useful for only a few weeks and at doses near the upper practical range. Roberts (this volume) also showed that the results were dependent on storage temperature. Other papers in this volume (Hitchcock; Delincée) report results obtained with poultry using novel H_2 and multiple gas sensors. The impact of these sensors on the sensitivity and robustness of the method remains to be fully assessed. Diffusion of gas out of the food seems likely to severely restrict the usefulness of gas evolution as a detection method for irradiated foods. The sensitivity of the amount of gas released to the nature of the food and the conditions of storage and measurement will probably preclude interlaboratory comparisons of the type conducted for other detection methods. Nevertheless, research is likely to continue to define the optimum parameters for the method, stimulated by its speed and simplicity. The method may well find use within analytical laboratories as a rapid preliminary screen for the irradiation of a limited range of foodstuffs.

3 CONCLUSIONS

The number of potential radiolytic product markers for irradiated foods is large. All require the increase in concentration above irradiation levels to be significant, well defined and consistent.

Three methods are under investigation. Lipid peroxides suffer further constraints as a result of their instability and production by autoxidation; the method is not being widely researched. Gas evolution suffers severe constraints in its usefulness since it is susceptible to diffusion out of the food. This can be highly dependent upon conditions. Nevertheless, there is interest in the method as a rapid screening method. Research on o-tyrosine as a marker has made significant progress. More work must be done before the data available are as extensive as that for cyclobutanones and lipid-derived hydrocarbons, methods which may remain more robust than the o-tyrosine method. Results from the interlaboratory comparisons will be viewed with interest and may well establish o-tyrosine as a further valuable marker for irradiated foods.

References

1. L. R. Karam and M. G. Simic, "Health Impact, Identification and Dosimetry of Irradiated Foods," WHO, Copenhagen, Denmark, 1988, p. 297.
2. L. R. Karam and M. G. Simic, *Anal. Chem.*, 1988, **60**, 1117A.
3. R. J. Hart, J. A. White and W. J. Reid, *Int. J. Food Sci. Technol.*, 1988, **23**, 643.
4. N. Chuaqui-Offermans and T. McDougall, *J. Agric. Food Chem.*, 1991, **39**, 300.
5. International Atomic Energy Agency, Report of the 2nd Research Co-ordination Meeting on Analytical Detection Methods for Irradiation Treatment of Foods (ADMIT), 15-19 June, 1992, Budapest, Hungary, IAEA, Vienna, p. 10.

6. C. von Sonntag, "The Chemical Basis of Radiation Biology," Taylor and Francis, London, 1987, Chapter 15, p. 458.
7. T. Dohmaru, M. Furuta, T. Katayama, H. Toratani, and A. Takeda, *Radiat. Res.*, 1989, **120,** 552.
8. M. Furuta, T. Dohmaru, T. Katayama, H. Toratani and A. Takeda, *J. Agric. Food Chem.*, 1992, **40,** 1099.

DETERMINATION OF o-TYROSINE IN SHRIMPS, FISH, MUSSELS, FROG LEGS AND EGG-WHITE

W. Meier, H. Hediger and A. Artho

Kantonales Laboratorium Zurich
8030 Zurich
Switzerland

1 INTRODUCTION

o-Tyrosine was proposed as a marker in post-irradiation dosimetry and different methods for the determination of this marker have been published:
Gas Chromatography/Mass Spectroscopy (GC/MS) after derivatisation[1-4]
High Pressure Liquid Chromatography (HPLC) with fluorescence detection[5-7]

However, there are two main problems which still need to be solved. The background levels of o-tyrosine need to be established in non-irradiated samples and the poor separation of the o-tyrosine peaks in the chromatogram using the HPLC methodology.

Using a newly developed procedure these problems can be overcome. The HPLC chromatograms are now run with a very low concentrated buffered system and all the chromatograms show a very good separation of the o-tyrosine peak. If the separation is not sufficient or if the results require confirmation, the o-tyrosine fraction is collected. The o-tyrosine can be determined using GC/MS after derivatisation with chloroformic acid methyl ester. The latter step can be carried out in a buffered system, so that there is no need to evaporate the system.

In an earlier publication[8] the use of a cation-exchange column was proposed to confirm an uncertain amount of o-tyrosine, but the retention times of o-tyrosine are not consistent and, therefore, this procedure has been superseded.

2 EXPERIMENTAL

1.1 Sample Preparation

Samples were homogenised and approximately 250 mg were hydrolysed with 900 µl 6 N hydrochloric acid under vacuum for 24 h at 110°C. The hydrolysate was filtered (microfilter 0.45 µm) and stored, if necessary, at -20°C. In contrast to earlier work, the freeze-drying step is no longer necessary making the method more simple and rapid.

2.2 HPLC (Reversed Phase HPLC Column)

Stainless steel was used for the injector and all capillaries. To preserve the column, the system was backflushed after the o-tyrosine eluted.

Gradient pump L-6200, Merck/Hitachi
Fluorescence-detector, Merck/Hitachi (EX 275 nm, EM 305 nm)
Column: Hypersil-5-ODS, 250 x 4.6 mm with a Precolumn Hypersil-5-ODS, 20 x 4.6 mm (both Stagroma, Wallisellen, Switzerland)
Standard-Eluent solution: 14.1 g of Trilithium-citrate, 4.2 g of citric acid, 9 ml of concentrated HCl in 1 litre of double distilled water
Eluents:
Eluent A: 10 ml of standard-eluent solution, 10 ml of ethanol in 1 litre of double distilled water, pH = 4.3
Eluent B: ethanol : double distilled water, (4:1)

Concentration of the standards (o-, m- and p-tyrosine): 100 ng ml^{-1} 6 N HCl each

Chromatographic conditions:
    0 - 14 min Eluent A, normal flush
    14 - 30 min Eluent B, back flush
    30 - 50 min Eluent A, normal flush

Injection: 100 µl

Retention times:
    p-Tyrosine   6.5 min
    m-Tyrosine   8.3 min
    o-Tyrosine   11.3 min

Chromatograms of non-irradiated and irradiated shrimps are shown in Figure 1.

If the separation is not sufficient or if some results require confirmation, the o-tyrosine fraction (approximately 2 ml) is collected and the o-tyrosine determined by GC/MS after derivatisation.

3 GC/MS

3.1 Reagents

Chloroform
Solvent mixture: double distilled water : methanol : pyridin = 50:30:20
Chloroformic acid methyl ester (MCF), Merck
MCF-solution: 1% chloroformic acid methyl ester in chloroform

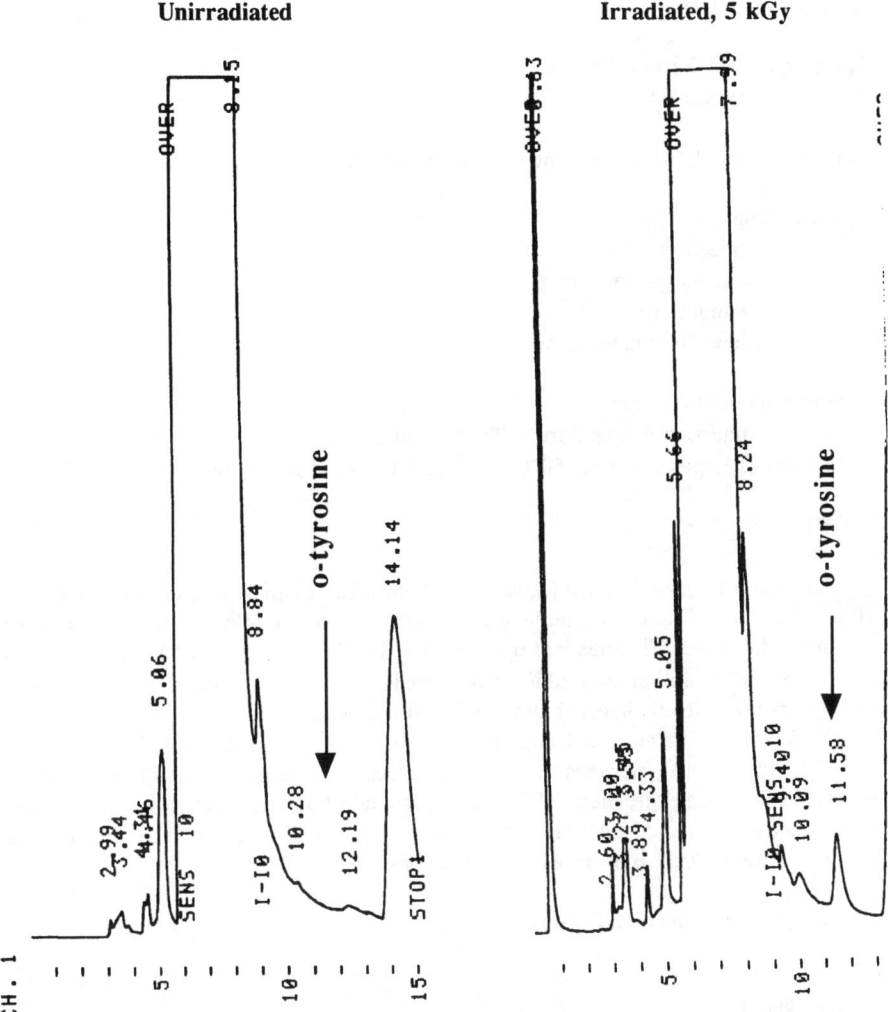

Figure 1 *Determination of o-tyrosine in shrimps*

3.2 GC/MS Sytem

Apparatus: QMD 1000, Finnigan
             Mode: CI$^+$, NH$_3$

Column: DB 1, 30 m x 0.32 mm i.d., 0.25 µm thickness

Operating conditions:
    Injector: 250°C
    Carrier gas: helium
    Sample size: 2 µl
    Injection mode: on column

Column temperature program:
    Injection temperature: 40°C for 2 min
    Temperature rate: 50°C min^{-1} until 190°C, than 5°C min^{-1} until 250°C

3.3 Derivatisation

The total o-tyrosine fraction (2 ml) was mixed with 100 µl of the solvent mixture in a 10 ml tube (Note: The pH should be greater than 7). Ten microlitre of MCF were added and mixed for 5 sec. The tube was opened, closed and mixed again for 5 sec. Then 100 µl of the MCF-solution was added and mixed again. The chloroform-phase (lower phase) was then directly injected into the GC/MS system.

The MS after chemical ionisation (NH$_3$) showed only 3 main mass-signals (M + 1, M + 18 and M - 14), whereas the MS after electrical ionisation showed a number of mass signals. Using the chemical ionisation, it could be shown that the non-irradiated shrimp sample contained less than 0.05 mg kg^{-1} o-tyrosine, but using electrical ionisation this was more difficult to prove as it was less sensitive.[9]

The MS (CI$^+$) are shown in Figures 2 and 3

4 RESULTS

The amount of o-tyrosine in 57 non-irradiated samples (30 shrimp, 7 fish, 10 chicken, 7 froglegs and 3 mussels) and 20 irradiated samples (6 shrimp, 5 chicken, 3 fish, 5 froglegs and 1 mussel) was as follows:

    Non-irradiated: 0.03 - 0.08 mg kg^{-1}
    Irradiated (5 kGy): 0.55 - 0.70 mg kg^{-1}

5 CONCLUSIONS

With this new HPLC system the o-tyrosine in irradiated shrimps, fish, mussels, froglegs and egg-white can be well separated from other peaks in the chromatogram. It is not

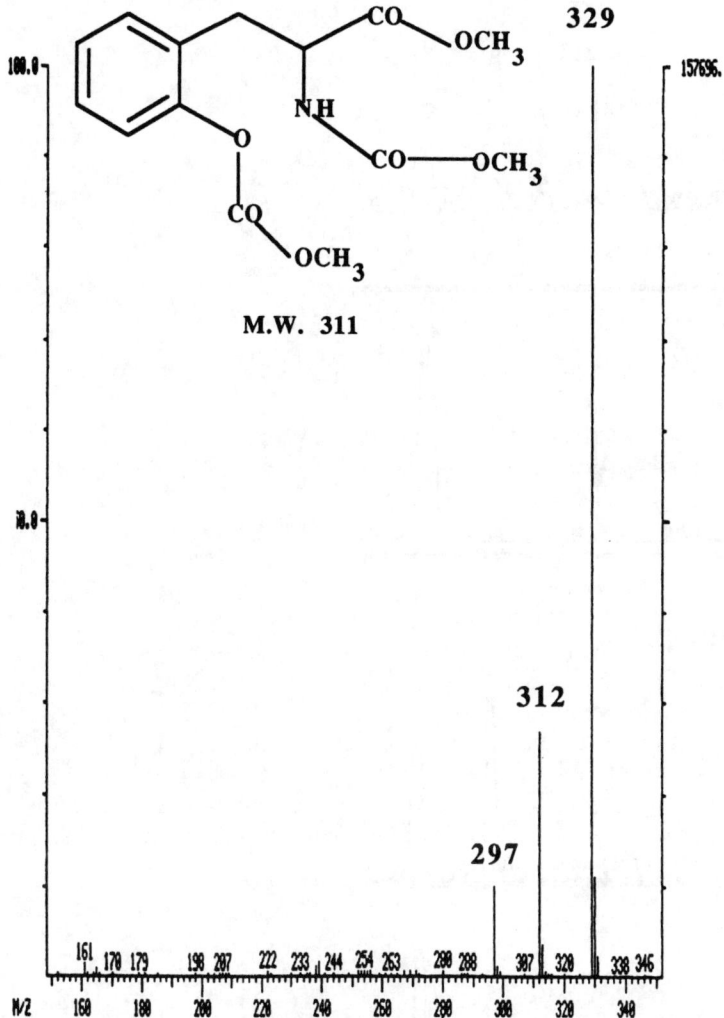

Figure 2 *GC-MS of o-tyrosine after derivatisation mode:* CI^+, NH_3

Unirradiated

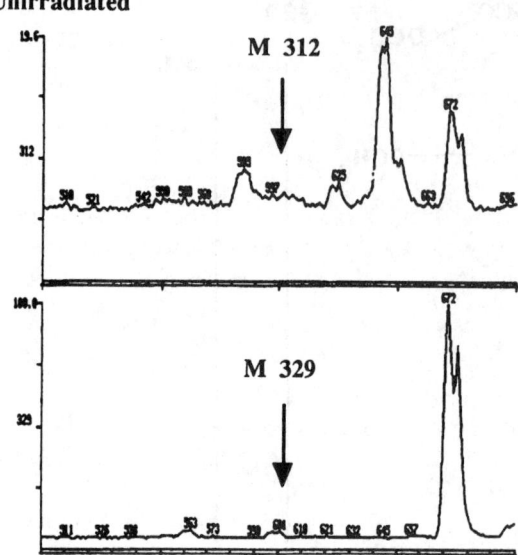

Irradiated, 5 kGy

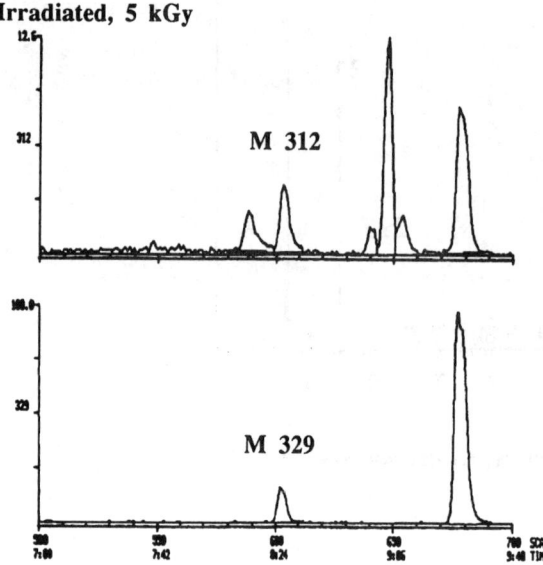

Figure 3 *GC-MS of the HPLC-fractions after derivatisation*

now necessary to freeze-dry the samples. Samples with a concentration of o-tyrosine greater than 0.1 mg kg^{-1} are suspected as having been irradiated. To confirm such results, the o-tyrosine fraction can be collected and the o-tyrosine can be determined by GC/MS after derivatisation with chloroformic acid methyl ester.

Acknowledgements

Gratitude for financial support is expressed to the Swiss Federal Office of Public Health, Division of Food Control, Bern. This project is part of a co-ordinated programme of research under the sponsorship of the IAEA, Vienna and BCR, Brussels.

References

1. L. R. Karam and M. G. Simic, *Anal. Chem.*, 1988, **60**, 1117A.
2. R. J. Hart, J. A. White and W. J. Reid, *Int. J. Food Sci. Technol.*, 1988, **23**, 643.
3. C. T. Pedersen and R. Fuhlendorff, Commission of the European Communities (BCR), Brussels, Luxembourg, EUR 13331 EN, 1991, p. 213.
4. R. Fuhlendorff and T. C. Pedersen, *Jysk Teknologisk, Teknologiparken*, 1989, 8000 Arhus C.
5. W. Meier, R. Bürgin R. and D. Fröhlich, *Mitt. Geb. Lebensm. Hyg.*, 1989, **80**, 22.
6. O. Zoller, D. Schöni and B. Zimmerli, Commission of the European Communities (BCR), Brussels, Luxembourg, EUR 13331 EN, 1991, p. 217.
7. N. Chuaqui-Offermanns and T. McDougall, *J. Agric. Food Chem.*, 1991, **39**, 300.
8. W. Meier, H. Hediger, A. Artho and E. J. M. Meier, BGA, Berlin, SozEp-Heft, 1993, **16**, 88.
9. W. Meier and E. J. M. Meier, IAEA, Second Research Co-ordination Meeting, 15-19 June, 1992, Budapest.

IDENTIFICATION OF IRRADIATED FOODS BY AN IMMUNOCHEMICAL METHOD

T. Kume and T. Matsuda*

Takasaki Radiation Chemistry Research Establishment, Japan Atomic Energy Research Institute, Takasaki, Gunma 370-12, Japan

* Department of Applied Biological Sciences, School of Agricultural Sciences, Nagoya University, Chikusa-ku, Nagoya 464-01, Japan

1 INTRODUCTION

Recent developments of analytical detection methods; physical methods including thermoluminescence (TL) and electron spin resonance (ESR), chemical methods including the detection of hydrocarbon from lipid, biological methods including seed germination or change in microbial flora, have been summarised by Delincée.[1,2] However, the immunochemical detection method has not been described in detail. In this paper, the specific identification method using immunochemical detection mainly in irradiated chicken eggs is introduced.[3]

2 METHODS FOR IMMUNOCHEMICAL DETECTION

2.1 Preparation of Antisera

Rabbit antisera to egg white proteins, ovalbumin (OVA), ovomucoid (OM) and ovotransferrin (OT) were prepared as described previously.[4,5] The proteins were dissolved in phosphate buffered saline (PBS), emulsified with an equal volume of Freund's complete adjuvant (Difco), and injected subcutaneously on the back of a rabbit. The rabbit was given two booster injections of the same antigen emulsified with Freund's incomplete adjuvant (Difco) 14 and 35 days after the first immunization. Bleeding was performed 10 days after the last injection. The sera were separated by centrifugation and stored at -80°C before use. Commercial antiserum to OVA (Nordic Immunology) was also used.

2.2 Polyacrylamide Gel Electrophoresis

Sodium dodecyl sulfate (SDS) polyacrylamide gel electrophoresis (SDS-PAGE) (10-15% acrylamide) was performed according to the method of Laemmli.[6] Gel sheets were stained with a solution of 0.2% Coomassie Brilliant Blue R-250 in water : 2-propanol:acetic acid (5:5:1 v/v/v) and destained with 7% acetic acid containing 10% methanol.

2.3 Protein Blotting

The proteins separated by SDS-PAGE were transferred electrophoretically onto a nitrocellulose (0.45 μm, Advantec Toyo) or polyvinylidene difluoride (PVDF, 0.45 μm, Millipore) sheet by the method of Towbin et al.[7] The nitrocellulose sheet was incubated at 4°C overnight in 3% bovine serum albumin (BSA) in PBS for blocking. After being washed with PBS, the sheet was incubated with rabbit antisera at 37°C for 2 h with gentle agitation. After being washed with PBS containing 0.02% Tween-20 (PBST), the sheet was incubated with peroxidase-coupled anti-rabbit IgG at 37°C for 1 h. After washing with PBST, the protein bands with reactivity to the specific antibody were visualised by activity staining for peroxidase using 4-chloro-1-naphthol (Bio-Rad).

3 CHANGES IN PROTEINS BY IRRADIATION

Investigations have been carried on the changes in proteins as a result of irradiation[8-11] and their applicability for the identification of irradiated foods. Radiation causes the denaturation of proteins resulting in changes in molecular weight, electrophoretic patterns and inactivation. Figure 1 shows the HPLC of OVA irradiated in N_2 and in O_2. Biological activities of lysozyme (LY) and OVA were decreased by irradiation but the inactivation in egg was small. Therefore, the change in biological activity is not applicable for identification purposes as the necessary dose for inactivation is higher than the practical dose level. Changes in molecular weight and electrophoretic patterns are also difficult to distinguish between changes caused by irradiation and those caused by heat denaturation.

SDS-PAGE of irradiated OVA showed that some components with lower molecular weight were newly produced by irradiation (Figure 2). The production of these low molecular weight components suggested that radiation-induced peptide bond cleavage produced the degraded fragments of OVA. The bands with small electrophoretic mobilities were observed at a higher dose, suggesting the production of aggregated

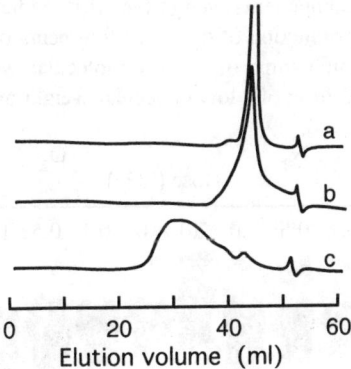

Figure 1 *Changes in HPLC elution curves of OVA a) non-irradiated, b) 5 kGy in O_2 and c) 5 kGy in N_2*

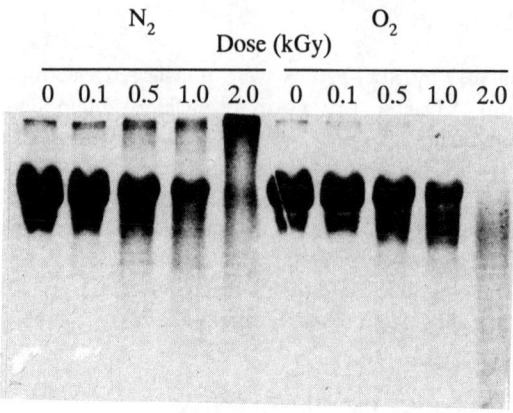

Figure 2 *SDS-PAGE of OVA irradiated in N_2 and O_2*

proteins. Such protein aggregation was caused more strongly in N_2 than in O_2. As shown in Figure 3, some bands corresponding to the degraded fragments were visualised more clearly on the membrane by immunoblotting using anti-OVA antibody. These results show that cleavage of the peptide bond occurs by irradiation and that the induced fragments have reactivity to the specific antibody.

4 IDENTIFICATION OF IRRADIATED CHICKEN EGGS

4.1 SDS-PAGE of Irradiated Egg White Proteins

SDS-PAGE of irradiated egg white showed that many minor components with molecular weight less than 30 kDa were newly produced by irradiation and that the production of such low molecular weight components was more remarkable for the 10 kGy irradiated egg than the 2 kGy irradiated egg (Figure 4). The production of these low molecular weight components suggested that radiation-induced peptide bond cleavage resulting in the production of degraded fragments of egg white proteins such as OVA, OT and OM, although some of the low molecular weight components produced might be in the aggregated form of a low molecular weight protein, LY.

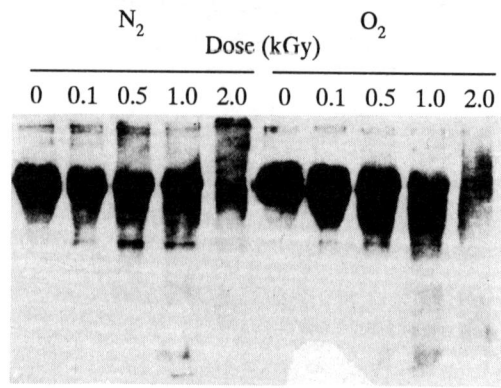

Figure 3 *Immunoblotting of OVA irradiated in N_2 and O_2*

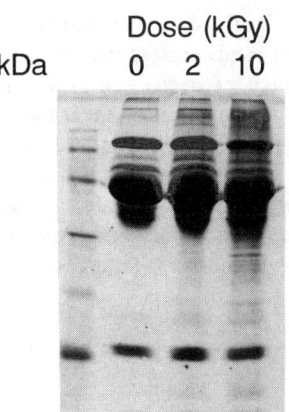

Figure 4 *SDS-PAGE patterns of irradiated egg white
Molecular weights of the standard protein markers are indicated on the left
OT; OVA, OM and LY indicate ovotransferrin, ovalbumin, ovomucoid and
lysozyme, respectively*[3]

4.2 Immunoblotting of Irradiated Egg White Proteins

The radiation-induced fragmentation of egg white proteins was confirmed by a specific and sensitive detection method, known as immunoblotting. This method used rabbit antibodies raised against the major egg white proteins, OVA, OT and OM. As shown in Figure 5, some major bands as well as many minor bands, corresponding to the

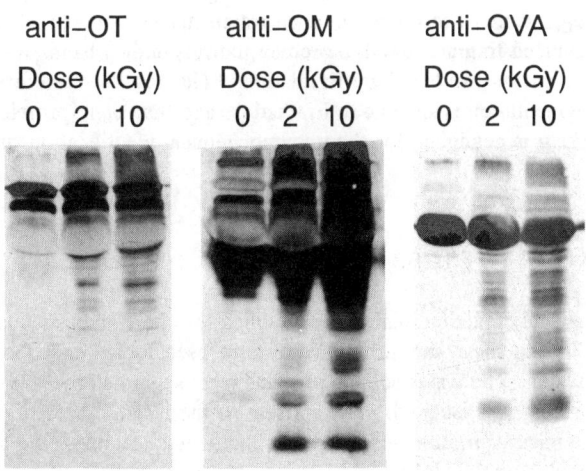

Figure 5 *Immunoblotting analysis of irradiated egg white for anti-OT(left), OM (middle) and OVA (right)*[3]

degraded fragments were clearly visualised on the membrane by immunoblotting using specific antibodies. Some indistinct bands corresponding to the aggregated proteins with lower mobility were also detected, especially in the case of OM. Thus, the immunoblotting analysis using specific antibodies was far more sensitive and specific than the simple electrophoretic analysis for the detection of radiation-induced fragmentation of egg white proteins. Although the colour intensity of bands was changed by the amount of sample loaded on, the same electrophoretic and immunoblotting patterns were obtained. The same patterns were obtained for eggs purchased at different markets and irradiated at different times thereby demonstrating the good reproducibility.

5 DISTINCTION BETWEEN IRRADIATION AND HEAT TREATMENT

Both irradiation and heat treatments cause the aggregation of proteins. To determine whether the immunoblotting was applicable to identification of irradiated egg after cooking, the degraded fragments of proteins were examined in egg white heated at 100°C for 15 min following irradiation at 2 or 10 kGy. An OVA/anti-OVA system was used because OVA is the most abundant egg white protein and radiation-induced fragmentation was the most remarkable of the three major proteins tested. There were no major differences in the SDS-PAGE pattern between the cooked and uncooked egg whites, except that the band intensity of OT was slightly decreased and a small amount of protein aggregates were formed by cooking in both irradiated and non-irradiated egg whites.

After the transfer on membrane, OVA, its aggregates and degraded fragments were detected by the anti-OVA antibody (Figure 6). The OVA degraded fragments in the irradiated and cooked egg white were also visualised by the specific antibody, indicating that the immunochemical detection method was applicable to heat-denatured irradiated egg white proteins. Furthermore, few degraded fragments were detected in the cooked non-irradiated egg white. These heat-induced fragments were much smaller in quantity than radiation-induced fragments and, moreover, corresponding bands were not found in the degraded fragments produced by irradiation. The results, therefore, suggest that such protein fragmentation is specific to irradiation, and that slight protein fragmentation induced by cooking is negligible for the immunochemical identification of irradiated egg white.

6 APPLICATION TO OTHER FOODS

The immunochemical detection method was applied to other foods such as chicken meat and shrimp. Chicken meat and shrimp were purchased locally and the proteins were extracted with water. The water soluble proteins were separated by SDS-PAGE and the main band of protein was isolated. The antisera to the isolated proteins were prepared from mice. The mice were immunised by subcutaneous injection of the antigen and the sera was separated from the blood.

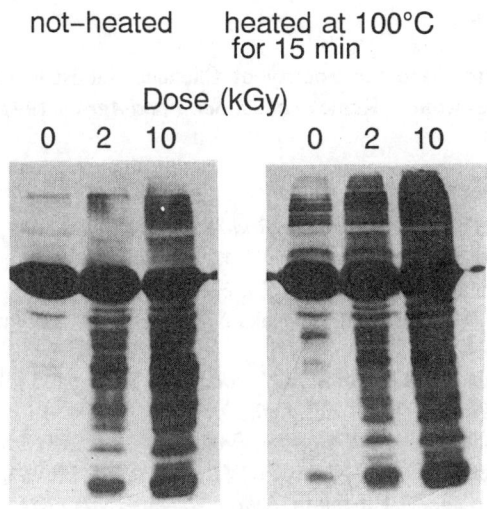

Figure 6 *Immunoblotting for anti-OVA of egg white boiled after irradiation*[3]

Chicken meat was irradiated at 2, 5 and 10 kGy in ice. Shrimp was irradiated at 2 and 10 kGy in ice or under frozen conditions with dry ice, respectively. The water soluble proteins were extracted and analyzed by SDS-PAGE. Many protein bands were detected on SDS-PAGE but the change in electrophoretic patterns of proteins was not significant. Immunoblotting using the antibodies for the proteins isolated from chicken meat and shrimp did not show clear results, as the specificity and reaction activities of these antibodies were not high enough for detection. It is not clear whether the changes of proteins in these foods irradiated up to 10 kGy were too small to induce the detectable fragments or if the specificity and activity of antisera were not high enough to detect the degraded fragments. Since these experimental conditions are at a preliminary stage, further study is necessary to find the most suitable protein changed at these dose levels and to prepare the specific antibody for the purified protein.

7 CONCLUSION

Irradiated egg could be identified immunochemically by the specific detection of radiation-induced degraded fragments of egg white proteins. Protein and peptide antigens can be generally detected by immunoblotting using specific antibodies even in the presence of other co-existing proteins and peptides. It is, therefore, considered that the immunochemical method for the identification of irradiated egg is quite effective and would be applicable not only to irradiated egg but also to irradiated egg white proteins added in foods.

Acknowledgments

The authors wish to thank the Society of Chemical Industry for the permission to reproduce the figures from T. Kume et al., J. Sci. Food Agric.. 1994, **65**, 1.

References

1. H. Delincée and D. A. E. Ehlermann, Radiat. Phys. Chem., 1989, **34**, 877.
2. H. Delincée, Radiat. Phys. Chem., 1992, **42**, 351.
3. T. Kume, T. Ishii and T. Matsuda, J. Sci. Food Agric., 1994, **65**, 1.
4. T. Matsuda, K. Watanabe and R. Nakamura, Biochim. Biophys. Acta, 1982, **707**, 121.
5. J. Gu, T. Matsuda and R. Nakamura, J. Food Sci., 1986, **51**, 1448.
6. U. K. Laemmli, Nature, 1970, **227**, 680.
7. H. Towbin, T. Stachelin and J. Gordon, Proc. Natl. Acad. Sci USA, 1979, **76**, 4350.
8. Y. Sato, Y. Umemoto and T. Kume, Food Irrad. Japan, 1969, **4**, 42.
9. T. Kume, Y. Sato and Y. Umemoto, Nihon Nougei-kagaku Kaishi, 1973, **47**, 549.
10. T. Kume and I. Ishigaki, Biochim. Biophys. Acta, 1987, **914**, 101.
11. T. Kume and T. Matsuda, Radiat. Phys. Chem., 1995, **46**, 225.

USE OF THE PEROXIDE METHOD FOR IDENTIFYING IRRADIATED FOOD

S. C. Qi and J. L. Wu

Department of Technical Physics
Peking University
Beijing 100871
China

1 INTRODUCTION

The peroxide method is important not only for detecting irradiated food but also for the control of the quality of food. Recent research showed that DNA damage induced by lipid peroxidation may be involved in processes of aging and tumorigenesis.

The main difficulty of applying the peroxide method in identifying irradiated food is the lack of knowledge about factors which influence the levels of peroxide in non-irradiated foods. Studies have shown that the rate of peroxidation is dependent on the concentrations of radicals and oxygen which exist in foods.[1] Ionising radiation can accelerate peroxidation of lipid-containing food. The peroxidation rate induced by radiation is much faster than that of autoxidation.

This paper presents new information about γ-irradiation peroxidation and autoxidation in pork, liquor and docosahexaenoic acid aqueous solution systems as well as the feasibility of detecting irradiated foods by the peroxide method. The blind test for identifying irradiated pork by the peroxide method is also included in this work.

2 EXPERIMENTAL

2.1 Materials

The fresh pork and liquor used in these experiments were purchased from the local market. The samples were irradiated by γ-rays from a ^{60}Co source at a dose rate of 0.64-4.2 Gy s^{-1} for pork, and 0.14-1.8 Gy s^{-1} for liquor in the presence of air at room temperature. The pork cut off 5-10 mm in thickness from the four sides of a lump of pork is referred to as the outer part of pork and the remainder as the inner part of the pork.

Cis-4,7,10,13,16,19-docosahexaenoic acid (DHA) obtained from Sigma Chemical Company was used without further purification. Irradiation was carried out by using a ^{60}Co γ source and a 2.8 Mev van de Graaff generator. The electron pulses were of 10^{-6}s duration with doses of 0.4-29 Gy. For pulse radiolysis, DHA aqueous solutions were saturated with N_2O and N_2O/O_2 (4/1 v/v), respectively and for γ-irradiation, the solutions saturated with NO_2/O_2 were used.

2.2 Determination of Peroxides

The organic peroxides in pork were determined by an iodometric method described elsewhere.[1] Hydrogen peroxide and total peroxides (H_2O_2 and organic peroxides) in liquors were analysed by Ce^{+4} titrimetry and by a spectrophotometric method at 350 nm, respectively.[2,3] The iodine liberated from the reaction of peroxides with KI appeared to give maximum absorption. The conjugated diene peroxides and malondialdehyde formed in γ-radiolysis of DHA aqueous solution can be measured spectrophotometrically at 234 nm (with ε = 28000 dm^3 mol^{-1} cm^{-1})[4] and 267 nm (with ε = 31506 dm^3 mol^{-1} cm^{-1}),[5] respectively. Generally, the peroxides were determined immediately after the samples had been irradiated.

2.3 Blind Test

The samples were supplied by an engineer of the Gammaster at Peking University under supervision.

3 RESULT AND DISCUSSION

3.1 Pork

3.1.1 Unusual Features in Irradiated Pork. Figure 1 displays the variation in peroxide content in irradiated and non-irradiated pork for the duration of storage at -18°C in the presence of air. It can be seen from Figure 1 that the level of peroxides in the non-irradiated sample is essentially unchanged within the storage time of about 7

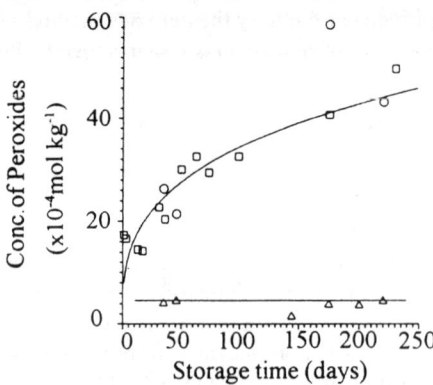

Figure 1 *The difference in the formation of peroxides between irradiated and non-irradiated pork at -18°C*
   Δ non-irradiated pork; □ irradiated pork (dose = 2.18 kGy); O irradiated pork (dose = 3.12 kGy)

Other Chemical Methods

months under the experimental conditions. It shows that the typical background average value in the control can be used as the parameter to identify whether or not the pork has been irradiated by comparing the peroxide concentration between irradiated and non-irradiated samples.

The probability distribution of peroxide content measured by 80 non-irradiated samples is given in Figure 2. From Figure 2, the average peroxide content is taken as $5.4 \pm 3.0 \times 10^{-5}$ mol kg^{-1}.

Figure 3 shows the dependence of the formation of peroxides on absorbed dose in irradiated pork. It can be found from Figure 3 that G(peroxides) is equal to 4.2 over the range of doses used and the content of peroxides in the pork irradiated at dose of approximately 4 kGy is about 1.7×10^{-3} mol. kg^{-1}, which approaches 31 times greater than the background average value of the control. It is well known that γ-irradiation can be used to eliminate salmonella in pork with doses of 2-7 kGy and to treat pork products for trichinae control with doses of 0.5-1 kGy. According to this study, even in irradiated pork treated with 0.5 kGy, the peroxide content is still about 3 times higher than the background level.

An examination similar to that in Figure 1 has also been made for the inner part and surface part of a lump of pork and it has been found that in the inner part of pork, whether irradiated or not, the peroxide levels are unchanged within a storage time of about 7 months. In the outer part, it varies with the storage time for irradiated pork and remains constant for non-irradiated pork (Figure 4). The results clearly show that γ-irradiation can accelerate autoxidation on the surface of irradiated pork. From the above mentioned results it is, therefore, possible to confirm whether the pork has been irradiated by analysing the peroxide content.

3.1.2 Criteria for Testing Irradiated Pork. As shown in Figure 1, the formation of peroxides is very slow in non-irradiated pork, so the average background value ($5.4 \pm 3.0 \times 10^{-5}$ mol kg^{-1}) might be used as the parameter to judge if the pork has been irradiated. For example, if the peroxide concentration in a sample is abnormally high by comparison with the background value, the pork will be considered to be irradiated. Generally, for pork treated with medium dose (such as ~2 kGy), the peroxide level produced by irradiation is about 10^{-3} mol kg^{-1} (Figure 3), which is about 18 and 5 times greater, respectively, than the average background value ($5.4 \pm 3.0 \times 10^{-5}$ mol kg^{-1}) and

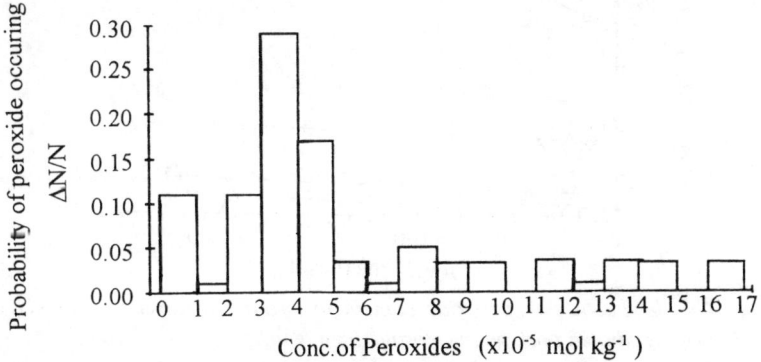

Figure 2 *The plot of probability distribution of peroxide contents in random sampling of 80 samples of fresh pork*

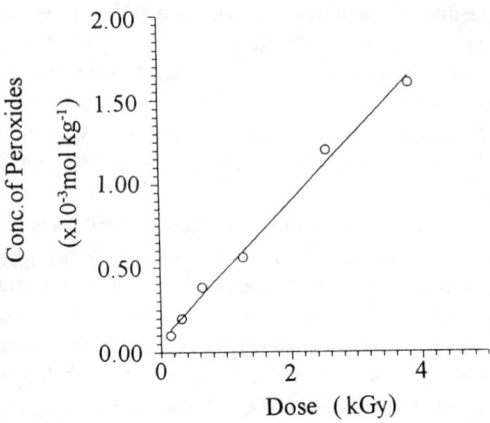

Figure 3 *The dependence of formation of peroxides on absorbed dose*

the maximum peroxide content (17×10^{-5} mol kg^{-1}, Figure 2) determined in 80 random samplings of non-irradiated fresh pork. According to the plot of the distribution of peroxide content (Figure 2), it is improbable for the unirradiated fresh pork with peroxide value of 10^{-3} mol kg^{-1}.

Therefore, it seems to be acceptable to take $(0.5 \pm 0.17) \times 10^{-3}$ mol kg^{-1} as the upper

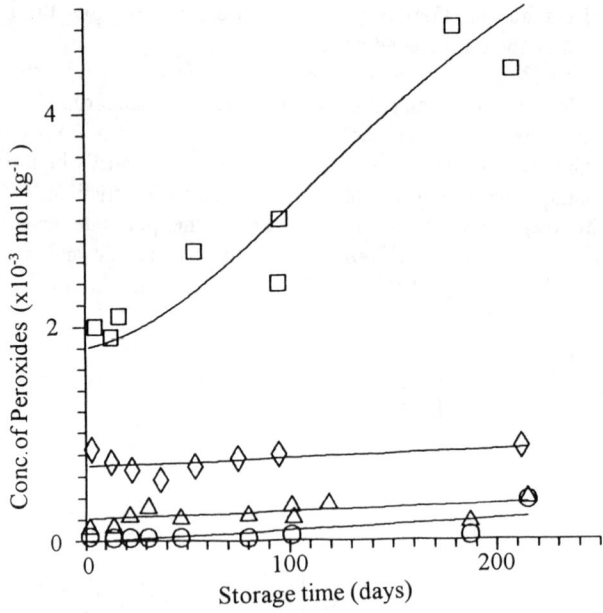

Figure 4 *The dependence of peroxide content in irradiated (dose = 1.74 kGy) and non-irradiated pork on the storage time at -18°C*
  □ *outer part of irradiated pork;* ◇ *inner part of irradiated pork;* Δ *outer part of non-irradiated pork;* O *inner part of unirradiated pork*

Table 1 *The Results of Blind Test for Irradiated Pork by LHP Method*

Sample Number	Storage Time (Days)	Peroxide Content (mol kg^{-1})		Conclusion[1]	Remarks[2]
		Inner	Outer		
1	2	9.0×10^{-4}	1.4×10^{-3}	irradiated	2.2 kGy
2	4	9.5×10^{-4}	1.7×10^{-3}	irradiated	2.0 kGy
3	8	7.5×10^{-4}	1.5×10^{-3}	irradiated	irradiated[3]
4	9	7.3×10^{-4}	1.2×10^{-3}	irradiated	1.0 kGy
5	11	1.6×10^{-4}	4.5×10^{-4}	uncertain	non-irradiated
6	33	6.6×10^{-4}	2.8×10^{-3}	irradiated	1.0 kGy
7	39	1.6×10^{-3}	4.2×10^{-3}	irradiated	2.2 kGy
8	44	1.3×10^{-4}	3.9×10^{-4}	uncertain	non-irradiated
9	100	1.5×10^{-3}	5.6×10^{-3}	irradiated	2.9 kGy
10	247	8.0×10^{-4}	1.1×10^{-2}	irradiated	irradiated[3]
11	261	7.1×10^{-4}	1.4×10^{-3}	irradiated	2.2 kGy
12	275	6.0×10^{-4}	9.1×10^{-3}	irradiated	1.6 kGy
13	278	8.7×10^{-4}	1.5×10^{-2}	irradiated	2.9 kGy
14	296	7.3×10^{-4}	5.4×10^{-3}	irradiated	2.0 kGy
15	297	6.2×10^{-4}	4.8×10^{-3}	irradiated	2.0 kGy
16	303	5.6×10^{-4}	8.5×10^{-3}	irradiated	2.2 kGy
17	310	4.0×10^{-4}	3.4×10^{-4}	uncertain	non-irradiated
18	311	1.4×10^{-4}	2.9×10^{-4}	uncertain	non-irradiated
19	312	1.8×10^{-4}	3.3×10^{-4}	uncertain	non-irradiated

[1] Given by the analysts
[2] Given by the engineer of the Gammaster
[3] The dose value was not given

limit of peroxide content for pork irradiated at < 1 kGy. The method can be used to identify the pork irradiated with doses ≥ 1 kGy accurately by taking approximately 0.7×10^{-3} mol kg^{-1} as the peroxide reference value.

As shown in Figures 1 and 4, the peroxide concentration in a non-irradiated sample is essentially constant within the storage time and increases gradually in irradiated samples. Therefore, irradiated pork can be detected by determining the variation of peroxide content on standing. The procedure is significant for the identification of pork irradiated with a lower dose (such as < 1 kGy). Table 1 lists the preliminary results of the blind test. It is obvious that the accuracy of detection is very high when the dose is ≥ 1 kGy by using the reference standard of 0.7×10^{-3} mol kg^{-1} peroxide. For the samples which are given as uncertain in Table 1, they can be further examined by ESR spectroscopy or by determining the variation of peroxide content on standing. Recent research shows that the ESR signal difference exists also in bone irradiated with a dose of 0.5 kGy.

3.2 Liquor

3.2.1 The Formation of Peroxides by γ-Irradiating Liquor. Gamma-irradiation can accelerate aging of wine and improve the quality of some low-grade liquors. For example, it can mellow the liquors. Peroxides are the main radiolytic product on the irradiation of oxygen containing liquor. It has been found that oxygen plays an important role in the formation of peroxides. Under air, the level of peroxides increased linearly with an increase of absorbed dose and then reached a maximum. Subsequently, it decreased with further increase in absorbed dose and depletion of dissolved oxygen in the liquor.[6] At steady state, peroxide concentration is about 0.1 ppm (by H_2O_2). In irradiated liquor, peroxides occur mainly in two forms, namely 92% H_2O_2 and 8% organic peroxides.[6] It is noteworthy that the peroxides produced by irradiation are quite stable in some liquors.[7]

3.2.2 D-2,3-Butanediol. This is the main radiolytic product from the irradiation of liquor in the absence of oxygen. When the liquor is irradiated in the presence of oxygen, there is an induction period in the formation of 2,3-butanediol following which the concentration of 2,3-butanediol increases with the increase in absorbed dose.[8] It has been demonstrated by gas chromatography (GC) and polarimetry that the 2,3-butanediol formed on the irradiation of liquor is an equimolar mixture of racemoid and mesomeride. 2,3-Butanediol is also a natural component and exists widely in a many commercial liquors (or wines). However, the 2,3-butanediol isomers produced by the action of microorganism are a mixture of l- and meso-isomers. When a liquor sample is irradiated by ionising radiation, d-2,3-butanediol is a unique radiolytic product. Therefore, it can be used to determine whether or not the liquor has been irradiated.

3.2.3 Conclusions. When the liquor was irradiated in the presence of air, the sample could be identified by the peroxide method and when irradiated in the absence of air, the detection of d-2,3-butanediol was used to indicate irradiation treatment. If an oxygen containing liquor was processed by ionising radiation without oxygen being present then the available detection method was dependent on the dose used. When the sample was irradiated with doses less than 1.5×10^3 Gy,[8] the peroxide method could be used, and when irradiated with doses more than 1.5×10^3 Gy, the d-2,3-butanediol method was used.

3.3 Cis-4,7,10,13,16,1 9-Docosahexaenoic Acid

3.3.1 Yields of Biallylic Radicals and Conjugated Diene Hydroperoxides. As a model system of unsaturated lipid, a DHA aqueous solution was studied with pulse and γ-irradiation techniques respectively. Figure 5 shows the absorbance change at 288 nm in a N_2O saturated DHA aqueous solution after a 10^{-6} s pulse of irradiation. A rapid increase in absorbance was observed after the pulse. The increase is due to the formation of DHA biallylic radicals. The OH (or O^-) radical can abstract a H atom from biallylic sites in DHA molecule to form the biallylic radical, which can also be formed by intramolecular H transfer processes. The G values of biallylic radicals are given in Table 2. It can be seen from Table 2 that at a higher dose (> 5.0 Gy), G (biallylic radical) is dose dependent and at a low dose (< 5.0 Gy), the G values are invariable at pH 10.8, G=2.3 and pH 12.0, G=2.72.

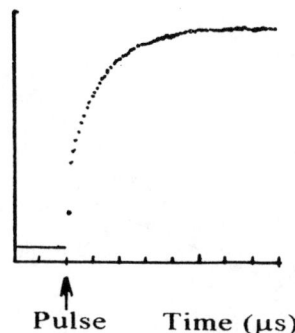

Figure 5 *The absorbance change at 228 nm in a N_2O saturated 4.3×10^{-4} mol dm^{-3} DHA aqueous solution (pH 11.6) irradiated with a 10^{-6} s electron pulse of 4.5 Gy (2.0×10^{-5} s/division)*

Table 3 lists the G (conjugated diene peroxides) values obtained from γ-radiolysis of 8.6×10^{-4} mol dm^{-3} DHA aqueous solution saturated with N_2O/O_2. It can be seen from Table 3 that G (conjugated diene peroxides) values are also pH dependent. G (conjugated diene peroxides) are consistent with G(biallylic radical) at the same pH. The results seem to show that the biallylic radical is the precursor of the peroxides. The effect of SOD on the yields of conjugated diene peroxides is also included in Table 3.

3.3.2 *Malondialdehyde.* The formation of malondialdehyde is related to pH and absorbed dose under γ-irradiation of N_2O/O_2 saturated DHA aqueous solution (Figure 6). It has also been found that the G(malondialdehyde) increases from 0.61 to 0.88 in the presence of SOD at pH 10.8. Therefore, malondialdehyde seems to be a secondary end product of the peroxidative degradation process.

Table 2 *The Yields of Biallylic Radical Determined in N_2O Saturated 8.6×10^{-4} mol dm^{-3} DHA Solution Irradiated with a 10^{-6}s Electron Pulse*

pH 10.8		pH 12.0	
Dose (Gy)	G (Biallylic Radical)[1]	Dose (Gy)	G (Biallylic Radical)
5.5	2.3	19.1	1.98
3.6	2.3	15.1	2.10
2.2	2.3	11.3	2.24
		7.1	2.42
		4.5	2.51
		4.4	2.68
		2.6	2.72
		1.7	2.76
		1.1	2.72

[1] The G values of biallylic radical are calculated by using the value of $\varepsilon = 3.0 \times 10^4$ dm^3 mol^{-1} cm^{-1} at 288 nm[4]

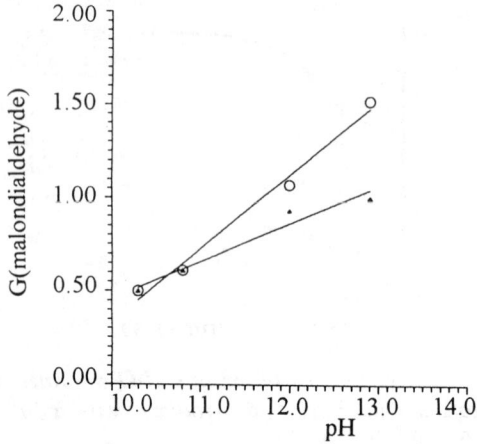

Figure 6 *Dependence of G (malondialdehyde) value on pH* : ○ < 36 G; △ > 36 Gy

Table 3 *The Yields of Conjugated Diene Peroxides*

	pH				Remarks
	10.3	10.8	12.0	12.9	
G(biallylic radical)		2.30	2.72		dose < 5 Gy
G(conjugated diene peroxides)[1]	1.9 2.68	2.30 3.40	2.64	3.00	dose < 36 Gy 2.1-7 mg SOD 100 ml^{-1}

[1] G values are calculated by taking $\varepsilon = 28000$ dm^3 mol^{-1} cm^{-1} at 234 nm[4]

3.3.3 Conclusions. In a N_2O (or N_2O/O_2) saturated aqueous solution system, the hydrated electron is converted into an OH radical. In the system, $G(OH+O^-) = 5.4$ the OH radical is a reactive species. It can attack on any site of the DHA molecule. For example, the OH radical can abstract H atoms from biallylic, allylic and non-allylic sites of an unsaturated fatty acid and can add to double bonds in the molecule. It can be seen from Table 3 that about 2/3 OH radicals (5.4) generated in the radiolysis of water could be expected to add to double bonds of the DHA molecule or to extract allylic and non-allylic H atoms from the molecule at pH 10.3 to produce a wide variety of lipid hydroperoxides. The difference in the reactivity of radicals produced between irradiation peroxidation and autoxidation will bring on the disparity in peroxidative product, which may be useful to indicate the irradiation of DHA containing foods. Generally, the rate of irradiation peroxidation is greater than that of autoxidation, *i.e.* the content of conjugated diene peroxides and malondialdehyde in irradiated food containing DHA will be higher than that in non-irradiated food. These differences may also used for identifying irradiated foods containing DHA.

Acknowledgments

The study was supported by the State Science and Technology Commission (PRC). We are very grateful to IAEA Food and Agriculture Organisation of the United Nations Food Preservation Section and the Volkswagen-Stiftung for their support. The work on DHA was carried out under IAEA support in Max-Planck Institut fuer Strahlenchemie. In particular we wish to thank Prof. Dr. C. von Sonntag for his kind help throughout the experiments.

Co-workers - S. H. Yuan, Z. Z. Li, Y. R. Zhou, L. B. Qu, X. M. Pan and H. F. Ha

References

1. S. C. Qi, J. L. Wu and Y. Zhu, *Radiat. Phys. Chem.*, 1993, **42**, 371.
2. Z. X. Zhang "Shiyong Youji Dingliang Fenxi," Shanghai Kexue Chubanshe, Shanghai, 1965.
3. C. J. Hochanadel, *J. Phys. Chem*, 1952, **56**, 587.
4. C. von Sonntag, "The Chemical Basis of Radiation Biology," Taylor and Francis, London, New York, Philadelphia, 1987.
5. A. S. Csallany, M. D. Guan, J. D. Manwaring and P. B. Addis, *Anal. Biochem.*, 1984, **142**, 277.
6. S. C. Qi, S. H. Yuan and J. L. Wu, *J. Radiat. Res. Radiat. Process.*, 1991, **9**, 150.
7. J. L. Wu, X. M. Pan, S. C. Qi, L. B. Qu and H. F. Ha, *J. Radiat. Res. Radiat. Process.*, 1989, **7**, 61.
8. S. C. Qi, J. L. Wu and R. Y. Yuan, *Radiat. Phys. Chem.*, 1990, **35**, 329.

A RAPID AND SIMPLE SCREENING TEST TO IDENTIFY IRRADIATED FOOD USING MULTIPLE GAS SENSORS

H. Delincée

Federal Research Centre for Nutrition
Karlsruhe
Germany

1 INTRODUCTION

A number of analytical detection methods for the irradiation treatment of foods have been developed in recent years by international co-operation.[1] Most of these methods require relatively expensive equipment and/or extended sample preparation time. It would be desirable to have quick and simple screening tests which immediately give some indication whether a food product has been irradiated or not. A simple method was described by Furuta and co-workers,[2] who estimated the amount of carbon monoxide (CO) in irradiated frozen meat, *i.e.* chicken, pork and beef, by expelling the trapped gas by quick microwave heating and detecting released CO in the head space using gas chromatography (GC). This work was followed up by Roberts and co-workers,[3] who showed promising results with frozen chicken meat and shellfish. In order to speed up the time of analysis, the present laboratory used an electrochemical CO sensor in previous work[4,5] to estimate the CO content in the released gas from irradiated food. Similar equipment is used to estimate CO in ambient air in workplaces or in an underground garage.

In the mean time, Hitchcock[6] reported the use of a novel hydrogen-specific electronic sensor to detect the evolved hydrogen gas from irradiated samples.

This paper describes the simultaneous use of multiple gas sensors, *e.g.* for carbon monoxide (CO), hydrogen (H_2), hydrogen sulfide (H_2S) and ammonia (NH_3), to increase the reliability of the gas evolution method in detecting irradiated food.

2 EXPERIMENTAL

Foods were purchased in local shops, and irradiated either with ^{60}Co-γ-rays (Gammacell 220, AECL, dose rate ~0.1 Gy s^{-1}) or 10 MeV electrons (Circe III linear accelerator, BFE). Non-irradiated as well as irradiated (maximum dose 5 kGy) mechanically deboned poultry meat was kindly provided by a French company.[7]

Meat was coarsely chopped, transferred to a 1 litre glass flask and a similar weight of water was added to ensure even heating. The air-tight flask was heated in a microwave oven for about 2 min (visible boiling) and the expelled vapor collected through silicone tubing pierced through the lid of the flask. In initial experiments using only a CO sensor, the released gas was collected in a plastic bag and, after cooling, the gas was directed over the sensor. In later experiments the circuit design developed by Hitchcock[6] was used, and 1 litre of evolved gas was led through the tubing in a gas circuit with a total volume of 1.3 litre (gas volume measurement by water displacement). The gas in the circuit was pumped around using a flow rate of 1 litre min^{-1} and the gas stream was directed over all gas sensors. After a constant measuring value had been achieved (approximately 2.5 min), the sensor values were recorded (Figure 1).

The CO and H_2S "Pac II" sensors were supplied by 'Dräger' (Lübeck), and the NH_3 and H_2 sensors by 'Gesellschaft für Gerätebau' (Dortmund). It should be emphasised that sensor values in ppm (parts per million) represent the part of the detected gas in the gas stream and not the amount of gas per weight of meat, although values will be related.

3 RESULTS AND DISCUSSION

First successful experiments were carried out with chicken meat using commercial gas diffusion tubes for carbon monoxide analysis (Dräger CO 2/a) which show a colour change from white to blue. The use of an electrochemical CO sensor, however, simplified the procedure. At the beginning approximately 250 g of sample was used to obtain a reliable signal from the quick measuring CO sensor, which is much more material than used by Furuta and co-workers[2] who only applied 1-5 g of meat. In later experiments 50 g of meat was used. With commercial meat samples, however, this amount creates no problem. To counteract scattering of individual values, which may be rather large, 6 to 10 samples were measured. Promising results were obtained for chicken meat, whole liquid egg and soup mix.[5] Since a preliminary test with lean shrimps showed no difference upon irradiation, it was believed that mainly the lipid fraction was responsible for the CO evolution. Later experiments showed, however, that these shrimps had not in fact been irradiated as expected, and also that low-fat meat released similar amounts of CO to high-fat meat, thus the idea that only the fat fraction is responsible for CO evolution could not be upheld. In experiments varying the fat content of chicken meat (by mixing lean meat

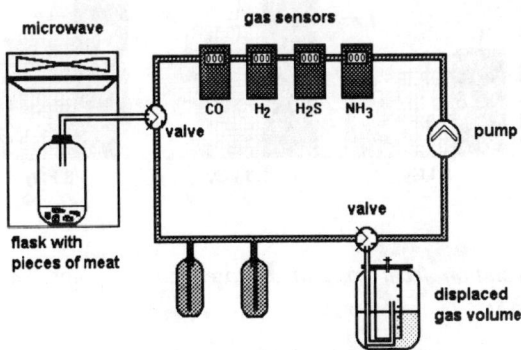

Figure 1 *Gas circuit for measurement of gas evolution from irradiated food*

with 0, 20, or 40% skin), even slightly decreasing CO amounts with increasing fat content were found, indicating that the protein content may be more important. In fact, experiments more than 20 years ago,[8] demonstrated the formation of headspace gases in canned radiation-sterilized foods. The major food constituents, carbohydrates, proteins and lipids, all gave rise to the development of headspace gases after irradiation.[9]

Recent experiments with shrimps, which actually were irradiated, also show a large increase in CO evolution after irradiation.[10]

Further experiments using three gas sensors with frozen chicken legs showed again large increases in CO after irradiation with 1.5 and 3 kGy. Increases in H_2S and NH_3 contents were also observed upon irradiation, however, not so markedly as for CO (Figure 2).

With mechanically deboned poultry meat, detection of the irradiation treatment using all 4 gas sensors is still possible after several months of frozen storage. The CO content was greatly increased in the irradiated sample. Hydrogen, which was virtually undetectable in the non-irradiated sample, could definitely be detected in the irradiated meat. An increase in H_2S was also noted, whereas the change in NH_3 content was negligible (Figure 3).

If products were not frozen but only chilled, such as in an experiment with chicken legs, a large increase in gas content is observed immediately after irradiation, but not after a 14 day storage period. The gases formed obviously diffuse out of the non-frozen food, and cannot be used as a radiation marker. Similar results were also obtained by measurement of released H_2 from irradiated pepper.[11]

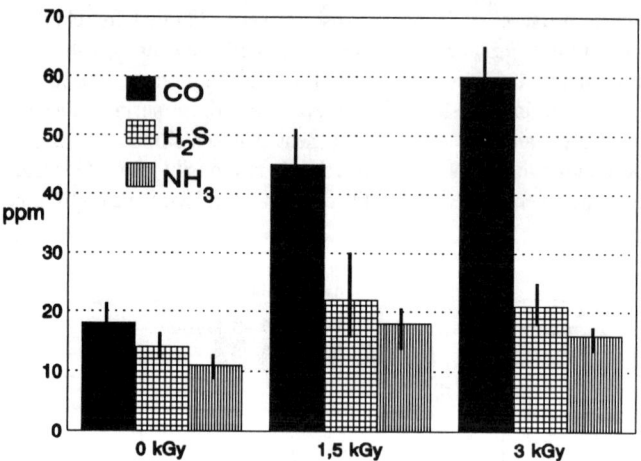

Figure 2 *Gas evolution from irradiated frozen chicken legs (mean values of 10 replicates)*

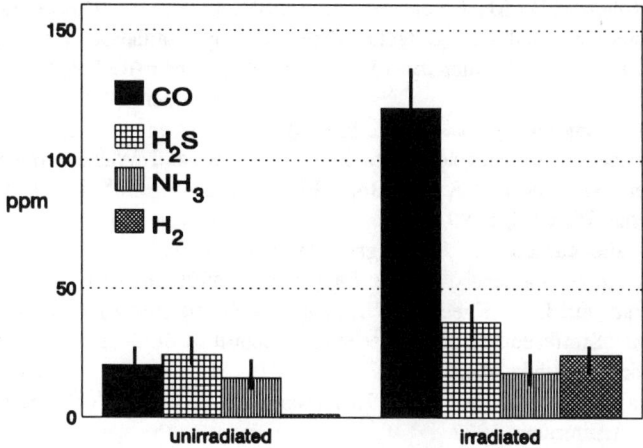

Figure 3 *Gas evolution from irradiated frozen mechanically deboned poultry meat (mean values of 6 - 10 replicates)*

More experiments are still needed to check the cross-reactivity and interference factors of the various gas sensors, but the method seems to offer some potential as a rapid and simple pre-screening test. By applying several gas sensors, a "gas fingerprint" for different foods can be established. More work is also needed to establish background values for various foods, and to study the influence of radiation parameters.

4 CONCLUSION

Measurement of gas evolution is a promising screening test to detect irradiated meat. The use of multiple gas sensors increase the reliability of the test, which is cheap and rapid and easy to perform. Suspected samples can subsequently be analysed by established techniques which offer unequivocal identification.

Acknowledgements

The author is most grateful for the support of the Joint FAO/IAEA Division of Nuclear Techniques in Food and Agriculture, Food Preservation Section, particularly to L. Ladomery, for the opportunity to participate in the ADMIT programme. The skilful experimental work of Miss P. Funk is highly appreciated.

References

1. H. Delincée, *Z. Lebensm. Unters. Forsch.*, 1993, **197**, 217.
2. M. Furuta, T. Dohmaru, T. Katayama, H. Toratani, and A. Takeda, *J. Agric. Food Chem.*, 1992, **40**, 1099.

3. P. B. Roberts, D. M. Chambers and G. W. Brailford, "Co-ordinated Research Programme on Analytical Detection Methods for Irradiation Treatment of Foods (ADMIT)," Budapest, Hungary, 15-19 June 1992, Joint FAO/IAEA Division, Vienna, 1994, p. 78.
4. H. Delincée, *Bundesgesundheitsbl.*, 1993, **36**, 378.
5. H. Delincée, "New Developments in Food, Feed and Waste Irradiation," G. A. Schreiber, N. Helle and K. W. Bögl, Eds., Bundesgesundheitsamt (BGA), Berlin, SozEp Heft 16, 1993, p. 99.
6. C. H. S. Hitchcock, *J. Sci. Food Agric.*, 1993, **62**, 301.
7. T. Sadat and M. Vassenaix, *Radiat. Phys. Chem.*, 1990, **36**, 661.
8. G. B. Pratt and L. E. Kneeland, "Irradiation Induced Headspace Gases in Packaged Radiation Sterilized Food," Technical Report 72-55-FL, US Army Natick Laboratories, Natick, Mass., 1972
9. P. S. Elias and A. J. Cohen Eds., "Radiation Chemistry of Major Food Components," Elsevier, Amsterdam, 1977.
10. H. Delincée, P. Funk and D. Roig, "4. Deutsche Tagung Lebensmittelbestrahlung," Hamm, 6.-7. April 1994, Bundesgesundheitsamt (BGA), Berlin, SozEp Heft 5, 1994, p. 149.
11. T. Dohmaru, M. Furuta, T. Katayama, H. Toratani and A. Takeda, *Radiat. Res.*, 1989, **120**, 552.

GAS EVOLUTION AS A RAPID SCREENING METHOD FOR DETECTION OF IRRADIATED FOODS

P. B. Roberts, D. M. Chambers and G. W. Brailsford*

Institute of Geological Nuclear Sciences and
*National Institute of Water and Atmospheric Research
Lower Hutt
New Zealand

1 INTRODUCTION

A number of detection methods for irradiated foods are in an advanced state of development.[1] No single method is likely to be universally applicable but a battery of tests such as thermoluminescence, electron spin resonance and analysis of lipid radiolytic products may soon be available for most foods and technical uses of irradiation. Most of these proposed tests require relatively sophisticated equipment or technical skills and are often time consuming and costly. There would be value in relatively simple tests which could be used as a rapid screening system or confirmatory method.

The detection of gases produced radiolytically and trapped within the food may be such a simple test. Dohmaru et al.[2] considered hydrogen (H_2) release from irradiated peppercorns to be a potential detection method. Hydrogen is highly diffusible and rapid loss from foods may render it less useful as a long term marker for irradiation. However, Hitchcock[3] has suggested that a novel gas sensing module warrants investigation. Carbon monoxide (CO) trapped within frozen meat and poultry can be detected for several months after irradiation in a rapid, simple test.[4]

The literature on the use of radiolytic gases as a detection method is limited and this paper extends the above studies. In particular, it extends the work to frozen shellfish, for which irradiation has been used as a commercial decontamination technique for many years, and considers the effect of storage temperature. Work on poultry is also reported as a cross-reference to earlier work and because irradiated poultry has recently been released into the US retail trade.

2 MATERIALS AND METHODS

2.1 Samples

Boneless chicken and de-shelled shrimps were obtained from local retailers, cut into approximately 1 g cubes and frozen at -25°C before use.

2.2 Irradiation and Storage

For each dose to be tested, frozen samples were placed in a 30 ml polyethylene vial and lightly capped to permit free diffusion of gas. Samples were irradiated in a Gammacell 220 ^{60}Co irradiator at approximately 0.3 Gy s^{-1}. Irradiations took between 1 to 4 h. Sample temperatures were maintained at between -25°C and -35°C during irradiation by means of an insulating box in contact with dry ice. Post-irradiation storage was at -15°C or -25 °C.

2.3 Gas Analysis

At least 3, and usually 5, replicates were analysed for each irradiation dose, storage temperature and time. Non-irradiated controls were incorporated in all experiments.

A sample was placed while still frozen in a 5 ml glass vial and 1 ml of distilled water was added. The screw cap for the vial was tightly closed; it contained a re-sealing septum through which head-space gas could be sampled using a non-coring hypodermic needle and gas-tight syringe.

The sample was rapidly thawed in a microwave oven (15 s at 700W) and about 3-4 ml of head-space was drawn off into a syringe. The H_2 and CO concentrations were measured in a single analysis using a HP5890 Series II Gas Chromatograph with a 72 x 1/8 inch column packed with a 5A molecular sieve at 100°C. Samples to be measured were injected into a 1 ml sample loop and dried via passage through 1/16 in. tubing immersed in dry ice-ethanol at 70°C.

The carrier gas was zero air, free of H_2 and CO. An RGD2 mercuric oxide reduction gas detector was used and calibrated against a laboratory standard (R1.1) for CO and against outside air for H_2. The limits of detection were about 50 ppb (parts per billion) and 20 ppb by volume for H_2 and CO, respectively. Measurement errors on individual samples are considered to be about 5% and 2% for H_2 and CO respectively. Results are reported as the mean ± standard error of the mean for the replicate samples.

3 RESULTS AND DISCUSSION

3.1 Chicken

Figure 1 shows the amount of gas evolution detected after a 5 kGy dose and storage up to 40 days. Evolution from non-irradiated samples was 15 and 25 times greater than the detection limits for H_2 and CO respectively. A 5 kGy dose further increased gas evolution within 24 h by about 20-fold for H_2 and 10-fold for CO. During storage at -15°C detectable H_2 decreased to near control levels within 40 days. Apart from a small decrease in the first few days, detectable CO remained almost constant throughout 40 days storage.

The dose response for gas evolution after 10 days storage is show in Figure 2. The increase with dose is not large for H_2 since considerable loss by diffusion occurs over 10 days. A clear dose dependence is evident for CO evolution which apparently saturates above 3 kGy. The results indicate that CO evolution could be a useful detection method for frozen poultry meat down to about 0.5 kGy which is well below the 1.5 kGy minimum dose required in US legislation.[5] Since H_2 and CO are lost from the samples at a different rate, there may be some potential for estimating the approximate time prior to testing of the irradiation treatment.

The results are broadly in agreement with those reported by Furuta et al. for CO evolution from chicken, beef and pork.[4] It is of interest that results from the present study were obtained under conditions more likely to favour diffusion loss of gas. In our work the storage temperature was higher (-15°C versus -20°C) and irradiated samples had a far higher surface to volume ratio since 1 g rather than 50 g samples were irradiated and stored.

3.2 Shrimps

The main purpose of this work was to extend the gas evolution method to frozen shellfish and examine the storage temperature effects. Evolution of H_2 and CO from non-irradiated samples was essentially as found for chicken. However, the increase induced by irradiation was far less than that observed with chicken. For H_2 the increase was barely significant even within 24 h of irradiation and H_2 is not further considered.

Table 1 summarises the dose response for CO evolution after various storage times and the effect of 6 weeks storage at either -15°C or -25 °C. With 25°C storage for 24 hours post-irradiation, an increase of 5-fold over controls (Cf. 10-fold for chicken) was observed in the dose range of commercial relevance. No dose dependency was seen between 1 to 3.5 kGy. A saturation effect at a rather higher dose range was apparent with chicken (Figure 2). A similar lack of dose dependence can be inferred from the results of Furuta et al for beef, less clearly for chicken but not for pork.[4] There was no explanation at this time for such a result which may be of interest in terms of the radiation chemistry involved.

Table 1 also shows that, unlike chicken, detectable CO decreases markedly with storage time within a 6 week period. Even after -25°C storage, CO evolution was only above control levels at the highest dose tested. With higher temperature storage (-15°C) even this increase above controls was no longer evident. At -25°C an increase in CO levels above controls was just measurable at 5 kGy following 95 days storage. Given the lack of dose dependence after 24 h storage, it is perhaps surprising that after 6 weeks storage at -25°C only the highest dose resulted in significant radiation-induced gas evolution. This may indicate that radiolytic CO is produced in more than one type of "trapping site" or over more than a single time period.

Table 1 *CO Evolution from Frozen Shrimps*

Dose (kGy)	CO (ppb by volume)			
	-25°C			-15°C
	1 Day	42 Days	95 Days	42 Days
0	308 ± 67	181 ± 241	275 ± 162	46 ± 79
1.4	1666 ± 338	-	703 ± 238	-
2.7	1629 ± 340	532 ± 567	498 ± 244	-
3.8	1644 ± 289	475 ± 370	328 ± 77	406 ± 272
5.0	1240 ± 166	957 ± 446	581 ± 152	311 ± 143

The lesser increase in radiation-induced CO evolution for shrimps compared with poultry and, particularly, its relatively rapid loss with storage is disappointing. The difference may be due to the different structure and chemical composition of the meats.

4 CONCLUSIONS

This study confirmed that detection of trapped radiolytic H_2 and, particularly, CO can be a rapid (less than 5 min per sample) test for poultry over the dose range relevant commercially and for long storage periods at temperatures up to -15°C.

The new results show that the method would only be valid for shellfish with short storage times at very low temperatures. Storage temperature, as would be expected, also influences the rate at which trapped H_2 diffuses out of the irradiated products. Commercial storage temperatures can vary considerably. Even for a single batch, and assuming no temperature abuse occurs, different temperatures may be encountered between the irradiation plant and the final retail outlet.

However, the results reported here indicate that lack of such an increase could not constitute proof of no irradiation treatment. The situation is more favourable for other meats but it is recommended that further studies of the effects of storage temperature before the method could be considered to provide evidence of irradiation or non-irradiation of food.

In summary, evolution of radiolytic CO is potentially a rapid and relatively cheap method for the detection of irradiated frozen meats and poultry. It requires equipment found in most analytical laboratories and minimal other technical skills. However, its usefulness may be restricted to a limited number of frozen and possibly dry foods and it is also dependent on any temperature fluctuation which may have occurred during storage.

References

1. International Atomic Energy Agency (1992), "Report of the Second Research Coordination Meeting on Analytical Detection Methods for Irradiation Treatment of Foods," Budapest, 15 June 1992, IAEA, Vienna.
2. T. Dohmaru, M. Furuta, T. Katayama, H. Tortani and A. Takeda, *Radiat. Res.*, 1989, **120** 552.
3. C. H. S. Hitchcock, *J. Sci. Food Agric.*, 1993, **62** 301.
4. M. Furuta, T. Dohmaru, T. Katayama, H. Tortani and A. Takeda, *J. Agric. Food Chem.*, 1992, **40**, 1099.
5. USDA, United States Department of Agriculture 1992, "Irradiation of Poultry Products," Federal Register 57, No 183, p. 43588.

DETERMINATION OF HYDROGEN IN ICE AND IN IRRADIATED FROZEN CHICKEN

C. H. S. Hitchcock

Food Safety Research Group
School of Biological Sciences
University of Surrey
Guildford GU2 5XH
U.K.

1 INTRODUCTION

The irradiation of food is a process which can improve the quality and safety of the diet; it complements more familiar methods of food preservation such as cooking and freezing. The Food Safety Act 1990 allows the controlled production and sale of irradiated food in the UK, but the market is being initially inhibited by consumer resistance; this is partly based on lack of confidence. While a detection test is not officially considered to be a prerequisite for control of irradiated food, it would provide additional safeguards valued by the consumer. Two types of test are needed firstly, a precise reference method that can be undertaken in a specialised laboratory, preferably leading to an estimation of the original radiation dose, and secondly, a rapid screening method that can be used more widely. In the second category, a simple cheap test for the presence of the hydrogen generated during irradiation might provide a suitable basis.[1]

Current approaches to the detection of irradiated food are meeting with some success. Preferred methods are detailed in these Proceedings, and in the Report of the previous FAO/IAEA research co-ordination meeting on Analytical Detection Methods for Irradiation Treatment of Foods (ADMIT).[2] They have been comprehensively reviewed,[3] and include physical methods (*e.g.* luminescence or electron spin resonance (ESR) spectroscopy), microbiological analysis and the chemical determination of radiolysis products. In particular, photostimulated luminescence (PSL) is the basis of an instrument developed at the Scottish Universities Research and Reactor Centre, and described in these Proceedings: it automatically measures the luminescence of solid mineral (*e.g.* silicate) in the sample, allowing confident screening (*e.g.* of spices).

The more general chemical determination of stable markers in irradiated food requires the application of sophisticated procedures to detect the very small changes that occur during irradiation. The success of a chemical approach will have two opposite effects on consumers: while being reassured by the fact that food can be tested for irradiation, they will be dismayed by the demonstrable presence of "unnatural" chemicals, in, however, low concentrations. This fear must be allayed by the generally favorable comparisons of the acceptability of irradiated food with that of stored non-irradiated food, of heated food containing "unnatural" thermal products, of food containing "natural" microorganisms or pests, and even of fresh unprocessed food containing "natural" toxic chemicals.

The rapid diffusion of hydrogen has hitherto inhibited research on its potential use as a marker for irradiated food. Unpublished observations by Furuta et al. led to a preliminary investigation by Roberts et al. reported at the previous ADMIT meeting:[2] this showed that rapid thawing of frozen irradiated meat (0.7-5.0 kGy) released hydrogen that was detectable by gas-liquid chromatography. The hydrogen was continuously lost on storage at -15°C, observed levels decreasing to those of non-irradiated controls after 40 days. It was concluded that hydrogen would be a useful marker for only a week or two following irradiation of frozen meat.

An initial investigation[1] has demonstrated that the irradiation of water (>0.1 kGy) generates hydrogen gas that can be quantified by headspace analysis using a novel electronic sensor. Since the hydrogen diffused quickly away from unsealed samples, it was concluded that the potential application of the hydrogen-specific detector for monitoring irradiated food is limited to packaged or solid foods which retain the characteristic hydrogen marker. Frozen food contains solid water, which could provide an appropriate matrix for two possible reasons: firstly, irradiation could produce free radicals that are stabilised in the solid matrix until melting allows the reactions that give rise to hydrogen; and secondly, hydrogen could be produced during irradiation, and trapped in the solid matrix until melting occurs. This preliminary report covers the application of the specific sensor to the analysis of ice and of irradiated frozen chicken.

2 MATERIALS AND METHODS

2.1 Materials

Ice cubes containing hydrogen were prepared by filling a 500 ml screw-cap bottle with water, and displacing half with hydrogen from a commercial cylinder. After storing the sealed inverted bottle at ambient temperatures for several days with occasional shaking, 10 ml aliquots of the aqueous hydrogen solution were pipetted into an ice tray, and frozen at -18°C for 3 h. The resulting ice "cubes" measured approximately 2.5 x 2.5 x 1.6 cm (surface area, about 30 cm^2); they were wrapped individually in gas-permeable plastic bags and stored at -18°C.

Samples of frozen irradiated chicken were supplied by Dr. M. H. Stevenson (Food and Agricultural Research Division, Department of Agriculture for Northern Ireland, Belfast BT9 5PX, UK). Samples (10 g) of minced chicken were sealed in gas-permeable plastic and shaped into thin square plates of approximate dimensions 5 x 5 x 0.3 cm (surface area, about 50 cm^2). They were kept at temperatures lower than -18°C in dry ice during irradiation at 5 kGy and transport (1 day), and then stored at -18°C. Whole chicken legs (about 100 g) were wrapped, frozen, irradiated at 5 kGy and stored in the same way.

2.2 Methods

A procedure for the determination of hydrogen with a specific sensor has been described,[1] and was modified somewhat for the analysis of solid frozen samples. The whole sample (at -18°C) was removed from its packaging and immediately placed in the sample vessel (capacity 180 ml or 360 ml); if necessary, the headspace volume of this vessel was quickly adjusted to 160 ml with water to maintain consistent calibration. The sample vessel was then sealed (at zero time), with the pump recirculating the headspace gases through the detector vessel. The sample was then allowed to thaw, aided by a warm (40°C) water bath until no ice remained. The sensor output (V mV) was recorded at zero time and at appropriate intervals thereafter. The ice cubes and chicken mince thawed within 5 min and after a further 5 min a constant final reading could be recorded. The whole chicken legs had thawed after 2 h. The decrease of V due to the hydrogen released after thawing was calculated, and the corresponding volume of hydrogen estimated from calibration curves.[1] This volume was corrected by subtracting blank readings from non-irradiated samples which had been stored over the same periods of time.

Occasionally, when the sample contained more than 10 µl of hydrogen (which might overload the sensor), it was thawed in a separate sample vessel. Aliquots of the headspace were then injected directly into the detection system, which also contained water to ensure the presence of water vapour. The volume of hydrogen in the whole sample was then calculated from the known dilution factor.

3 RESULTS

3.1 Ice Cubes

Two trays of ice cubes were prepared from the aqueous solution of hydrogen on two different occasions. The solubility of hydrogen in water suggests that a 10 ml ice cube saturated with hydrogen would contain about 200 µl; however, the freshly frozen cubes from each tray were found to contain only 144 µl and 110 µl respectively. Evidently, some uncontrolled loss of hydrogen occurred during the freezing of the hydrogen solution.

After a given time in storage at -18°C, hydrogen was detected at the same level whether or not the plastic bag surrounding each ice cube was removed before thawing in the detector; the plastic was therefore gas-permeable, and did not impede release of hydrogen from the frozen cube.

From the 10 ml ice cubes, more than 9.1 ml of water were always recovered by syringe, demonstrating little loss of water during storage up to 5 months; no correction was made for such loss. The plastic wrapping effectively prevented an observed continuous loss of water from unwrapped cubes.

Both trays demonstrated that hydrogen was slowly lost from the ice cubes during storage at -18°C; after a month, the hydrogen level had fallen to about one-tenth of the initial value. After 6 months, residual hydrogen could no longer be determined with confidence. However, hydrogen is lost from liquid water much more rapidly (*e.g.* total loss in 12 h from 21 cm^2 of undisturbed water/air surface).[1] This retention of hydrogen in ice encouraged the analysis of frozen chicken samples.

3.2 Chicken Mince

Freshly irradiated 10 g frozen samples contained 5 µl of hydrogen after 2 days delivery time; after a further 30 days storage at -18°C, residual hydrogen had fallen to 0.2 µl. After this time, the hydrogen in a single 10 g sample became undetectable (signal less than twice blank). Non-irradiated samples always gave blank readings corresponding to about 0.1 µl hydrogen.

3.3 Chicken Legs

The signal from the sensor was recorded every 10 min or less as the sample slowly thawed inside the detection system; the signal from both the irradiated samples and the non-irradiated blanks decreased with time. The frozen chicken legs were initially examined 11 days after irradiation at 5 kGy. After thawing for 1 h, 11.1 µl of hydrogen had been released; after 2 h, 17.2 µl; and after 3 h, 17.2 µl. Evidently all the hydrogen had escaped from the sample after thawing for 2 h, and observations at this time were selected for quantification of the hydrogen.

The observed levels of hydrogen in whole chicken legs (approximately 100 g) after storage at -18°C are illustrated in Figure 1. The irradiated frozen chicken legs contained 17 µl of hydrogen after 11 days; after 212 days, 1.0 µl of hydrogen could still be

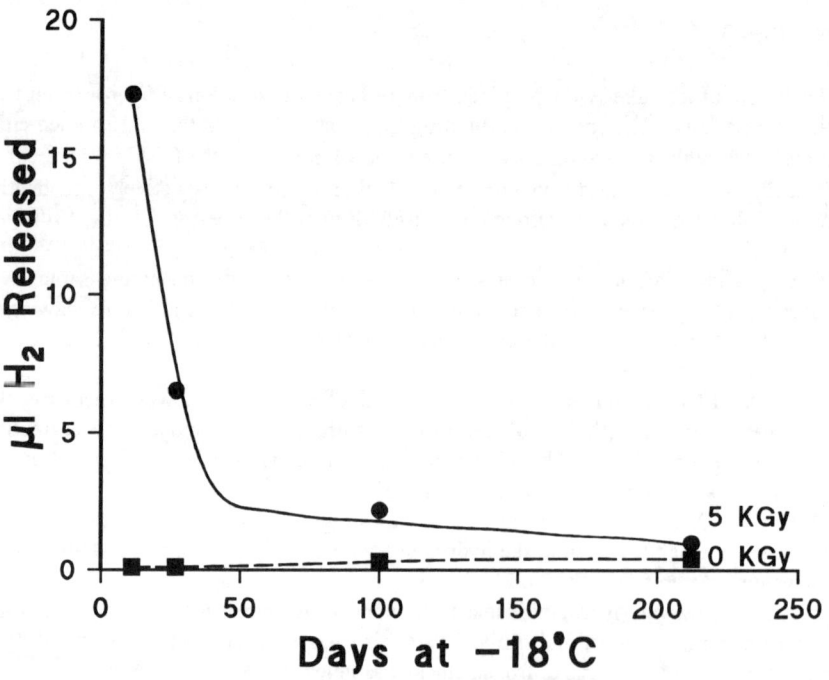

Figure 1 *Volume of hydrogen (µl) released by irradiated (5 kGy) and non-irradiated (0 kGy) chicken legs after storage at -18°C for various periods of time*

detected, a significantly higher level than the corresponding blank (0.4 µl). It was, therefore, possible to distinguish between irradiated and non-irradiated chicken after storage at -18°C for over 6 months.

The apparent level of hydrogen in the non-irradiated chicken was initially less than 0.1 µl, rising to 0.4 µl after 212 days storage. This was found to be due to two factors, both of which were eliminated by correcting the observed level in the irradiated chicken (*i.e.* subtracting the corresponding blanks). Firstly, a few volatiles normally present in food interfere somewhat with the sensor (unpublished observations); the increase from 0.1 µl to 0.4 µl may be due to the formation of similar volatiles during storage. Secondly, the electronic drift in the signal from the sensor is significant at these levels; variation in drift (even within a working day) may not be consistent, so that the correction introduces some uncertainty. Drift is an important factor, since it proved impossible to regain the original signal from irradiated chicken by flushing out the hydrogen and vapours after 3 h recirculation. Moreover, when the volatiles from non-irradiated samples were flushed out, the signal remained largely unchanged, suggesting that the decrease in signal ascribed to apparent hydrogen was due mainly to drift.

4 DISCUSSION

The analysis of the frozen aqueous solutions of hydrogen indicated that irradiated frozen food could be usefully examined by the same method after storage at -18°C. The chicken mince (10 g) contained 5 µl of hydrogen 2 days after irradiation at 3 kGy; this was somewhat less than expected, since 100 ml of water irradiated at 2 kGy released 8 µl of hydrogen per 10 ml.[1] Although the chicken contained roughly 8 g of water, its radiolysis to (and the diffusion of) hydrogen was probably affected by the matrix. Moreover, the geometry of the chicken sample (3 mm thickness) facilitated loss of hydrogen in the two days transit before analysis, and indeed thereafter. While the chicken mince could be successfully analysed after one month in storage, a target of at least one year is reasonable. This result suggested that larger samples of chicken could be successfully analysed after lengthier storage, provided that they could be accommodated inside the headspace detector, and that the blank readings did not interfere. The analysis of the chicken legs confirmed this suggestion.

4.1 Polar Ice

The immobilisation of hydrogen in solid ice is inevitably thwarted by the high diffusivity of the lightest molecule in Chemistry through and between the ice crystals. The speed of this diffusion limits the demonstrated usefulness of hydrogen as a marker for irradiated food. Such hydrogen diffusion has not been extensively studied, though the lack of diffusion of atmospheric gases in ice is the basis of the analysis of ice cores drilled out from the polar ice-caps in order to determine the composition of ancient atmospheres. As the surface snow is compacted into ice, air is trapped as closed bubbles and carried downwards by the weight above. As the pressure increases beyond 100 atmospheres, the bubbles disappear due to the formation of clathrates (hydrates, *e.g.* $O_2 \cdot 6H_2O$), which immobilise the gases in the ice layers. When an ice core is brought to the surface, the pressure falls below the dissociation pressure of the clathrates, and bubbles slowly re-form; nevertheless, the cores can be stored at polar temperatures (*e.g.* -20°C), and cut into discs (typically 1 cm thick) which are separately analysed for trapped gases. Each disc corresponds to a specific date, depending on its depth in the core: gases such as oxygen, nitrogen, argon, methane, nitrous oxide and carbon dioxide are immobilised in polar ice for at least 100,000 years. Evidently, the formation of clathrates involves interaction between host (water) and guest (gas) molecules which prevents diffusion at high pressures. When brought to the surface, there may be some delay before the guest gas is released from the clathrate; but even after this, diffusion is also impeded by the ice structure itself.

The relevant procedure during ice core analysis is the storage of the frozen cores at ambient pressures after drilling. Such storage apparently does not allow significant diffusion of the gases under investigation; for instance, no significant change of carbon dioxide level was observed between ice core samples stored for a week, 21 weeks or 49 weeks.[4] It was concluded[5] that the small permeability excludes the possibility of a loss of carbon dioxide during storage of ice cores, provided the ice is not fractured. Other atmospheric gases routinely monitored are less soluble, but are similarly immobilised.

4.2 Frozen Chicken

Hydrogen would doubtless also form a clathrate at high pressures, but not in frozen food; at ambient pressures gases form simpler solutions. Ice has an open structure, allowing space for interstitial molecules at a relatively low cost of energy; in general the diffusion of molecules through the crystal is made possible by lattice defects.[6] Thus, molecules as large as water itself can move between vacant sites in the crystal lattice, or as interstitial molecules within the lattice. However, small molecules like helium and neon can take advantage of open channels (diameter 2.4Å) within the ice crystal and parallel to its c-axis;[7,8] similar channels are aligned perpendicular to this axis. They lead to far greater permeabilities for smaller molecules and would allow rapid diffusion of hydrogen as well as helium through ice.[9] The diffusion coefficients of helium and water through ice are therefore very different, and have been estimated as 10^{-6} and 10^{-11} $cm^2\ sec^{-1}$ respectively.[8] These values imply that helium would be released from the ice in frozen food roughly 10^5 times faster than most other gases (*e.g.* oxygen), and suggest that hydrogen would diffuse almost as rapidly. Moreover, discontinuities in the ice and the presence of soluble and insoluble food components would facilitate this diffusion.

It therefore appears that the determination of hydrogen for the detection of irradiated frozen food is feasible but limited by the diffusivity properties of hydrogen itself, which it shares only with a few other small molecules. This discussion suggests that the same approach using carbon monoxide as the diagnostic gas would be more attractive in this respect, since it should diffuse very much more slowly out of the sample. Carbon monoxide is a radiolysis product of food components (though not of course of water); encouraging preliminary observations by Roberts *et al.* (following Furuta *et al.)* were reported to the previous meeting.[2] Irradiated frozen meat has indeed been successfully identified after chromatographic analysis of the headspace, even after a year in storage.[10] The modification of the detection system used in this study for the sensor-based determination of both hydrogen and carbon monoxide in frozen foods is therefore being investigated.

5 CONCLUSIONS

Previously it was suggested that the potential application of the hydrogen-specific detector for monitoring irradiated food was limited to packaged, solid or frozen foods which retain the characteristic hydrogen marker.[1] It may now be concluded from these initial experiments that the sensor-based determination of hydrogen in 100 g samples of frozen chicken allows a confident classification into (a) irradiated product or (b) non-irradiated product. Sufficient hydrogen is retained in such samples to allow such classification at least 6 months after irradiation at 5 kGy. As a simple screening method, the detection of hydrogen in frozen food complements the detection of photostimulated luminescence in dry food.

Although this demonstrated capability is useful, the approach lends itself to potential improvement. Monitoring the release of hydrogen from the frozen samples by allowing each to thaw over a period of two hours inside the detection system is unnecessary; the samples could be thawed in separate sealed detection vessels, and each connected to the detection system sequentially for less than 5 minutes. If necessary, detection of irradiated chicken after longer storage times would be possible by taking samples over 100 g. The procedure might well be applicable to other irradiated frozen foods. Any interference from the sample matrix could be overcome by purifying the recirculating gases in the detection system, *e.g.* by selective adsorption. This purification would be attractive in the examination of frozen irradiated prawns, since 10 g of non-irradiated prawns gives a signal equivalent to 1.1 µl of hydrogen (unpublished observations); this high blank is doubtless due to volatiles which affect the sensor.

This current simple procedure suffers from several disadvantages. While applicable to frozen foods, it would be possible for fraudulent processors to thwart the analyst by removing the hydrogen after irradiation (*e.g.* by aeration of the unfrozen sample). Its application to unfrozen packaged foods depends on the packaging being impervious to hydrogen; this effectively limits samples to products that are canned and then irradiated, which is not useful in practice. Its potential applicability to solid foods at ambient temperatures must be based on the retention of hydrogen within the solid structures of food components; this remains to be tested, although it has been reported[11] that irradiated pepper (10 kGy) released detectable traces of hydrogen gas when pulverised after 2-4 months storage at ambient temperatures.

When the sensor must be operated at its lower limit of detection where it is very sensitive, electronic drift is significant; it may be possible to minimise this by incorporating an inert reference transistor in the detection system. The hydrogen levels in ambient air (about 0.5 ppm by volume), as well as certain food volatiles, may interfere. The continuous loss of hydrogen from the solid sample at -18°C is unfortunate, limiting the storage time of acceptable samples. It also prevents the hydrogen level being interpreted directly as an accurate measure of the original radiation dose. The modification of the sensor system to accommodate the determination of carbon monoxide therefore merits further investigation.

Overall, this initial investigation has established that monitoring hydrogen levels with this convenient detector complements existing methods for the control of irradiated foods.

Acknowledgements

The author is grateful for useful co-operation and correspondence with Dr. B Birch, Unilever Research Laboratory, Bedford, England; Dr. M. N. Clifford, University of Surrey, Guildford, England; Dr. M. Furuta, University of Osaka, Japan; Dr. P. Roberts, Institute of Geological and Nuclear Sciences, Lower Hutt, New Zealand; Dr. I. Robins, Thorn EMI Central Research Laboratories, Hayes, England; Dr. J. Schwander, Physics Institute, University of Bern, Switzerland; Dr. M. H. Stevenson, Department of Agriculture, Belfast, Northern Ireland; Dr. E. Wolff, British Antarctic Survey, Cambridge, England.

This work was funded in part by the U.K. Ministry of Agriculture, Fisheries and Food (Project No. N2624).

References

1. C. H. S. Hitchcock, *J. Sci. Food Agric.*, 1993, **62**, 301.
2. ADMIT, "Co-ordinated Research Programme on Analytical Detection Methods for Irradiation Treatment of Foods (ADMIT)," Report of the Second Research Co-ordination Meeting, Budapest, Hungary, 15-19 June 1992, Joint FAO/IAEA Division of Nuclear Techniques in Food and Agriculture, Vienna.
3. H. Delincée, "Analytical Detection Methods for Irradiated Foods: a Review of Current Literature," IAEA-Tecdoc-587, ISSN 1011-4289, 1991, IAEA, Vienna.
4. A. Neftel, H. Oeschger, J. Schwandler and B. Stauffer, *J. Phys. Chem.*, 1983, **87**, 4116.
5. B. Stauffer and H. Oeschger, *Annals of Glaciology*, 1985, **7**, 54.
6. P. V. Hobbs, "Ice Physics," Clarendon Press, Oxford 1974.
7. A. Kahane, J. Klinger and M. Philippe, *Solid State Commun.*, 1969, **7**, 1055.
8. J. G. Davy and K. W. Miller, *Solid State Commun.*, 1970, **8**, 1459.
9. J. Haas, B. Bullemer and A. Kahane, *Solid State Commun.*, 1971, **9**, 2033.
10. M. Furuta, T. Dohmaru, T. Katayama, H. Toratani and A. Takeda, *J. Agric. Food Chem.*, 1992, **40**, 1099.
11. T. Dohmaru, M. Furuta, T. Katayama, H. Toratani and A. Takeda, *Radiat. Res.*, 1989, **120**, 552.

DNA Methods

INTRODUCTION TO DNA METHODS FOR IDENTIFICATION OF IRRADIATED FOODS

H. Delincée

Federal Research Centre for Nutrition
Engesserstr. 20
76131 Karlsruhe
Germany

This brief introduction sets the scene with respect to the presentations in this ADMIT meeting dealing with DNA changes as a tool to detect the radiation processing of food. The choice to examine DNA seems obvious, since DNA is a sensitive cellular target to irradiation and the changes in DNA are responsible for many effects observed in irradiated foods, such as the inactivation of microorganisms, elimination of insects, inhibition of sprouting in bulbs and tubers and delay of ripening in several fruits. Therefore, these changes in DNA should be discernible in microbial or insect DNA or in the nucleic acids in the food itself. If DNA changes were specific to irradiation, a detection method could be designed which would have wide applicability, since most foods are derived from living organisms which all contain DNA. Such a method could almost be the universal method for detecting the radiation treatment of foods. Radiation-induced changes in DNA can be analysed by a variety of analytical techniques,[1,2] which have mostly been employed on pure DNA or on DNA in living cells in radiation biology research. Whether or not some of these techniques can be utilised to detect irradiated food has recently been very briefly discussed.[3]

In the foregoing volume "Food Irradiation and the Chemist" Deeble and co-workers[4] described the main effects of irradiation on DNA, leading to base modifications, denaturation of the DNA helix and the appearance of both single and double strand breaks. A number of research groups, particularly in the frame of the Commission Bureau of Reference (BCR) concerted action on new methods for the detection of irradiated food[5-8] but also under the aegis of ADMIT [9,10] have made attempts to use these changes in DNA for the detection of irradiated foods. Various DNA methods have been proposed for this purpose (Table 1). One of the difficulties observed seems to be the efficient isolation of DNA from the food in such a way that subsequent analysis is not interfered with, for instance, due to gross denaturation or disorganisation of the DNA structure during isolation or the presence of other food components. Another problem is the handling of food which also may lead to changes in DNA, e.g. during storage or by other treatments. The applicability of base damage analysis is also hampered by the high background levels of damaged bases also found in control non-irradiated samples.[4] Only very specific base changes, such as the estimation of dihydrothymidine with an

Table 1 *Proposed Methods to Detect Changes in DNA in Irradiated Food*

Detection of base damage
    Chromatographic procedures
        GC/MS
        HPLC
        Post-labelling with ^{32}P and HPLC
    Fluorescence *e.g.* thymine glycol
    Immunoassay
        antibodies to thymine glycol
        antibodies to dihydrothymidine

Detection of DNA denaturation
    Hyperchromicity combined with chemical reactivity

Detection of strand breaks
    Filter elution
    Pulsed field gel electrophoresis
    Agarose electrophoresis
        mt DNA
        native or denatured DNA
    Micro gel electrophoresis of single cells
    Enzymatic analysis of end groups
    Flow cytometry
    Fluorescence
    Immunoassay (mab to ss DNA)

immunoassay possibly show great promise and these will be described by Williams and co-workers. The use of a sensitive immunoassay would greatly simplify detection analysis.

Fragmentation of DNA is analysed by various methods including filter elution, pulsed field gel electrophoresis and other electrophoretic techniques. Filter elution to estimate DNA fragments in frozen Norway lobsters was, however, influenced by a thawing-freezing treatment.[12] Pulsed field gel electrophoresis needs relatively long sample preparation and separation times, and clear differentiation between irradiated and non-irradiated samples is still problematic.[7,8]

Continuous agarose electrophoresis of cellular DNA did not lead to reproducible results for a number of vegetables. When, however, instead of the susceptible cellular DNA, mitochondria were isolated, and mitochondrial DNA (mt DNA) analysed for strand breaks, promising results were obtained using beef liver.[13] The obvious advantage of using mt DNA is that it is protected against enzymatic reactions which generally cleave cellular DNA during storage, and mt DNA also seems to be resistant to freeze-thawing cycles. One drawback of the method as it has hitherto been performed is the long isolation procedure for mt DNA. Marchioni will describe progress in the estimation of strand breaks in mt DNA to detect irradiation of various fresh, chilled and frozen foods.

A further promising technique for detecting DNA fragments in irradiated food is the microgel electrophoresis of single cells or nuclei.[14] Advantages of this method are its simplicity and its speed, the electrophoretic separation only requiring a few minutes. A drawback is the non-specificity of radiation to yield DNA fragments, which also may be produced by heat treatment or certain storage conditions. Nevertheless, for frozen meat such as chicken, beef, pork or veal, a good differentiation was possible between non-irradiated and irradiated samples as will be described by Delincée in this volume. In addition, an interlaboratory blind trial achieved high correct identification rates with samples of frozen chicken and pork.

In summary, DNA methods may contribute successfully to the array of analytical detection methods for the irradiation treatment of foods. However, more work is needed to exploit the full potential of these promising techniques. The influence of radiation parameters and storage variables needs to be studied and the number of foods to which the tests may be applied needs to be extended. Further interlaboratory studies should be encouraged, leading to appropriate standard protocols. Furthermore, due to the very active field of DNA research, possibly some new emerging methods will be able to rapidly identify specific radiation-induced changes, and work along this line should also be encouraged.

Acknowledgement

The author is most grateful for the support of the Joint FAO/IAEA Division of Nuclear Techniques in Food and Agriculture, Food Preservation Section, particularly to L. Ladomery for the opportunity to participate in the ADMIT programme.

References

1. C. von Sonntag, "The Chemical Basis of Radiation Biology," Taylor and Francis, London, 1987.
2. J. Sambrook, E. F. Fritsch and T. Maniatis, "Molecular Cloning - A Laboratory Manual," 2nd Ed., Cold Spring Harbor, New York, 1989.
3. H. Delincée, "Potential New Methods of Detection of Irradiated Food," J. J. Raffi and J.-J. Belliardo, Eds., Commission for European Communities (BCR), Brussels, Luxembourg, 1991, EUR 13331 EN, p 5.
4. D. J. Deeble, A. W. Jabir, B. J. Parsons, C. J. Smith, and P. Wheatley, "Food Irradiation and the Chemist," D. E. Johnston and M. H. Stevenson, Eds., Royal Society of Chemistry, Cambridge, 1991, p 57.
5. J. J. Raffi and J.-J. Belliardo, "Potential New Methods of Detection of Irradiated Food," Commission for European Communities (BCR), Brussels, Luxembourg, 1991, EUR 13331 EN.
6. M. Leonardi, J. J. Raffi and J.-J. Belliardo, "Recent Advances on Detection of Irradiated Food," Commission for European Communities (BCR), Brussels, Luxembourg, 1993, EUR 14315 EN.
7. H. Delincée, E. Marchioni and C. Hasselmann, "Changes in DNA for the Detection of Irradiated Food," Commission for European Communities (BCR), Brussels, Luxembourg, 1993, EUR 15012 EN.

8. J. J. Raffi, H. Delincée, E. Marchioni, C. Hasselmann, A-M. Sjöberg, M. Leonardi, M. Kent, K. W. Bögl, G. Schreiber, H. Stevenson and W. Meier, "Methods of Identification of Irradiated Foods," Commission for European Communities (BCR), Brussels, Luxembourg, 1994, EUR 15261 EN.
9. ADMIT 1992, "Co-ordinated Research Programme on Analytical Detection Methods for Irradiation Treatment of Foods (ADMIT)," Report of Second Research Co-ordination Meeting, Budapest, Hungary, 15-19 June 1992, Joint FAO/IAEA Division of Nuclear Techniques in Food and Agriculture, Vienna, 1992.
10. ADMIT 1994, "Co-ordinated Research Programme on Analytical Detection Methods for Irradiation Treatment of Foods (ADMIT)," Report of Third Research Co-ordination Meeting, Belfast, UK, 20-24 June 1994, Joint FAO/IAEA Division of Nuclear Techniques in Food and Agriculture, Vienna, 1994.
11. M. Mayer, K. W. Bögl, N. Helle, I. Ugi and G. A. Schreiber, "Recent Advances on Detection of Irradiated Food," M. Leonardi, J. J. Raffi and J.-J. Belliardo, Eds., Commission for European Communities (BCR), Brussels, Luxembourg, 1993, EUR 14315 EN, p. 375.
12. M.-P. Copin and C. M. Bourgeois, "Potential New Methods of Detection of Irradiated Food," J. J. Raffi and J.-J. Belliardo, Eds., Commission for European Communities (BCR), Brussels, Luxembourg, 1991, EUR 13331 EN, p. 22.
13. E. Marchioni, M. Tousch, V. Zumsteeg, F. Kuntz and C. Hasselmann. *Radiat. Phys. Chem.*, 1992, 40, 485.
14. H. Cerda, B. V. Hofsten and K. J. Johanson, "Recent Advances on the Detection of Irradiated Food," M. Leonardi, J. J. Raffi and J.-J. Belliardo, Eds., Commission for European Communities (BCR), Brussels, Luxembourg, 1993, EUR 14315 EN, p. 401.

APPLICATION OF THE DNA "COMET ASSAY" TO DETECT IRRADIATION TREATMENT OF FOODS

H. Delincée

Federal Research Centre for Nutrition
Karlsruhe
Germany

1 INTRODUCTION

Since treatment with ionising radiation causes DNA fragmentation, microgel electrophoresis of single cells ("comet assay") offers a simple and rapid tool for identification of irradiated foods. The principle is based on migration of DNA in an agarose gel exposed to an electric field. Single cells or nuclei are embedded in the agarose and after lysis, intact DNA will virtually not move out of the cell upon electrophoresis, whereas if DNA has been fragmented, the fragments are able to migrate and "comets" following the cells become visible after staining.[1]

A drawback of this method is the non-specificity of radiation to fragment DNA. DNA fragmentation also occurs by heat treatment, and repeated freezing/thawing cycles as well as long storage periods may also interfere. Thus for any foods subjected to a heating step, e.g. blanching of shrimps on board the sea vessel, the DNA "comet assay" cannot be applied to detect a radiation treatment.

The advantages of this test, however, for foods not exposed to heat are its speed and simplicity, as the electrophoretic separation only requires a few minutes. DNA is visualised by silver staining, avoiding the need for a fluorescence microscope. Thus the method requires only relatively cheap equipment - in contrast to other methods for identification of irradiated foods such as electron spin resonance (ESR) spectroscopy or gas chromatography/mass spectrometry (GC/MS). The DNA "comet assay", therefore, seems suitable as a pre-screening test to detect whether food has been radiation processed. Suspected samples may subsequently be analysed by established, but more expensive techniques.

2 EXPERIMENTAL

The DNA "comet assay" was performed as described by Cerda and co-workers,[1] using a 0.1% sodium dodecylsulfate (SDS) solution in 40 mM Tris acetate buffer containing 1 mM EDTA (pH 8.0) for lysis and electrophoresis (TAE buffer). Silver staining of DNA was carried out according to the procedure previously published[2] (it should be recognised, however, that the percentages written for staining solution B are ten times too high; the actual correct recipe for mg of chemical to be dissolved should be followed). In recent experiments the lysis and electrophoresis solutions, respectively, were changed to 2.5% SDS in a 45 mM Tris-borate buffer, 1 mM EDTA (pH 8.0), and the TBE electrophoresis buffer being devoid of SDS, as in the protocol of the blind trial,[3] (see Annex). Lysis time was 5-10 min and submarine electrophoresis was carried out at 2 V cm^{-1} for 2 min. The stained slides were observed in a usual transmission microscope and can be documented by photography or image analysis. Frequently just a glance at a slide is enough to decide whether the sample has been irradiated or not. The form of the "comets" varies with radiation dose, and therefore, a rough dose estimation can also be carried out.

3 RESULTS AND DISCUSSION

First experiments using TAE buffer and only 0.1% SDS for lysis were successfully carried out with raw frozen chicken, pork and beef.[2,4] The distance of DNA migration, "comet length", increased with radiation dose, and by counting at least 300 cells (3 slides with 100 cells counted on each slide) frequency histograms with the migration distance (in mm) served as a basis for the decision whether or not the meat had been exposed to ionising radiation. These conditions were also used in a preliminary small intercomparison study[5] with raw frozen chicken meat and gave encouraging results.

However, a problem encountered in this work was the appearance of apparently intact cells with no comets in the irradiated samples. An explanation for this behaviour may be an insufficient lysis of the cell wall. Consequently, the concentration of the lysing agent (SDS) was increased to 2.5%, and also another buffer, *i.e.* TBE buffer, was employed.[3]

In recent experiments with frozen veal, homogeneous patterns of similar "comets" were obtained, and no intact or apparently intact cells could be observed in irradiated samples.[6] The samples could be classified as irradiated or non-irradiated at a glance (Figure 1). In contrast to previous conditions, not so much "comet length" but "comet shape" indicated differences in radiation dose. An experienced experimenter will also be able to roughly estimate the radiation dose just by a glance at the form of the "comets". Using an image analyser an objective measure of the different form of the "comets" may be obtained.

Further experiments with beef and chicken meat were also successful, although the different forms of "comets" were not restricted to just one treatment. Also, in control non-irradiated cells, "comets" of different forms were observed, but always adjacent to intact cells without "comets". The lowest degree of DNA damage observed, therefore, determines the classification of the sample, further factors *e.g.* storage, giving rise to some DNA fragmentation. Similarly, freezing/thawing cycles yield DNA "comets", but still intact cells without comets could still be observed, making a differentiation of irradiated samples possible. On the other hand, DNA "comets" may thereby indicate an interruption of the freezing chain and serve as a quality indicator.

Although this method has its limitations, *e.g.* it could not be successfully applied to shrimps or mushroom spores, it may be applicable to a wide range of foods.[1,7] Experiments with sea-food (oysters, mussels), fruits and vegetables are underway.

Results from a larger intercomparison with frozen chicken and pork meat are described in the Annex.

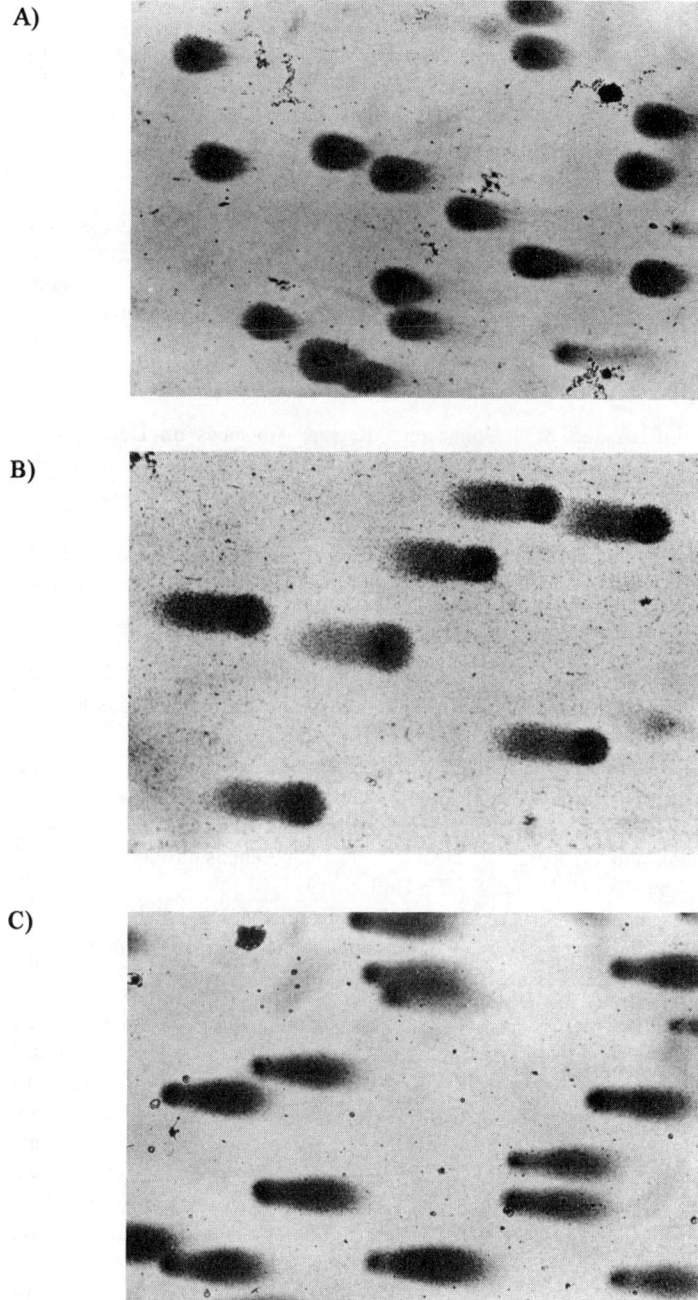

Figure 1 *DNA "comet assay" on frozen veal*
A) non-irradiated, B) irradiated with 1.5 kGy, C) 3 kGy (Original magnification was x 400)

4 CONCLUSION

The DNA "comet assay" offers high potential as a cheap and rapid pre-screening test for detection of the irradiation treatment of foods.

Acknowledgements

The author is most grateful for the support of the Joint FAO/IAEA Division of Nuclear Techniques in Food and Agriculture, Food Preservation Section, particularly to L. Ladomery, for the opportunity to participate in the ADMIT programme. The skilful technical assistance of Mrs. S. Delincée and Mrs. M. Menzler is highly appreciated.

References:

1. H. Cerda, B. v. Hofsten and K. J. Johanson, "Recent Advances on Detection of Irradiated Food," M. Leonardi, J.-J. Belliardo, J. J. Raffi, Eds., Commission of the European Communities (BCR), Brussels, Luxembourg, 1993, EUR 14315 EN, p. 401.
2. H. Delincée, "New Developments in Food, Feed and Waste Irradiation," G. A. Schreiber, N. Helle and K. W. Bögl, Eds., Bundesgesundheitsamt (BGA), Berlin, SozEp-Heft, 16, 1993, p. 112.
3. H. Cerda, 1994, private communication.
4. H. Delincée, "Changes in DNA for the Detection of Irradiated Food," H. Delincée, E. Marchioni and C. Hasselmann, Eds., Commission of the European Communities (BCR), Brussels, Luxembourg, 1993, EUR 15012 EN, p. 7.
5. H. Delincée and E. Marchioni, "Changes in DNA for the Detection of Irradiated Food," H. Delincée, E. Marchioni and C. Hasselmann, Eds., Commission of the European Communities (BCR), Brussels, Luxembourg, 1993, EUR 15012 EN, p. 8.
6. H. Delincée, P. Funk and D. Roig, "4. Deutsche Tagung Lebensmittelbestrahlung," Bundesgesundheitsamt (BGA), Berlin, SozEp Heft, 5, 1994, p. 149.
7. H. Cerda, "Changes in DNA for the Detection of Irradiated Food," H. Delincée, E. Marchioni and C. Hasselmann, Eds., Commission of the European Communities (BCR), Brussels, Luxembourg, 1993, EUR 15012 EN, p. 5.

ANNEX

Inter-laboratory Study Using the DNA "Comet" Assay to Detect Irradiated Chicken and Pork Meat.

Collaboration between the National Food Administration of Sweden (Main organiser H. Cerda) and ADMIT DNA Group (Research co-ordinator H. Delincée).

Nine laboratories each received a total of 18 coded samples, 6 were chicken bone marrow cells, 6 chicken muscle cells (both from chicken legs) and 6 pork chop muscle cells. The foods had been irradiated with various doses in the range of 0 to 5 kGy using a commercial electron accelerator (10 MeV electrons). In addition, each laboratory received a reference set containing chicken bone marrow cells from chicken legs irradiated with 0, 1, 3, and 5 kGy. These cell suspensions were prepared in Uppsala, Sweden according to the accompanying protocol (description prepared by H. Cerda). After adding a freeze protecting agent (DMSO) the suspensions were dispensed into cryotubes and frozen (-70°C). The samples were distributed by air freight in boxes containing dry ice. The participants confirmed that on receipt the samples were still in the frozen state. It was requested that samples be analysed as described in the accompanying protocol. In short, the cells are embedded in agarose on a microscopic slide, lysed with 2.5% sodium dodecyl sulfate (SDS) in a Tris-borate buffer (TBE:45 mM Tris-borate, 1 mM EDTA, pH 8.0) for 10 min, exposed to submarine electrophoresis using the same TBE-buffer (without SDS) at 2 V/cm for 2 min, stained and observed under a microscope.

RESULTS

At present 129 answers have been received (one lab has not yet reported its results, and several samples have been lost), and of these 129 samples, 118 were correctly identified as having been irradiated or not (91.5%). In fact, 93 of the 129 samples had been irradiated, and 85 were identified as such (91.4%). Thus, 8 irradiated samples were classified as false negatives (8.6%). Of the non-irradiated 36 samples, 3 samples were falsely positives being classified as irradiated (8.3%). The reasons for the latter results have yet to be clarified (Figure A1).
In addition to the qualitative questions whether samples had been irradiated or not, a dose estimation was required on comparison with the reference set. Of the total of 85 samples identified correctly as irradiated, the approximate dose of 74 samples was estimated in accordance with the actual applied irradiation dose (87%).

CONCLUSION

This blind trial demonstrates that the DNA "comet assay" used as a rapid screening test offers very high rates of identification (>90%). In addition it must be taken into account that most laboratories were rather inexperienced with this technique. Even a rough dose

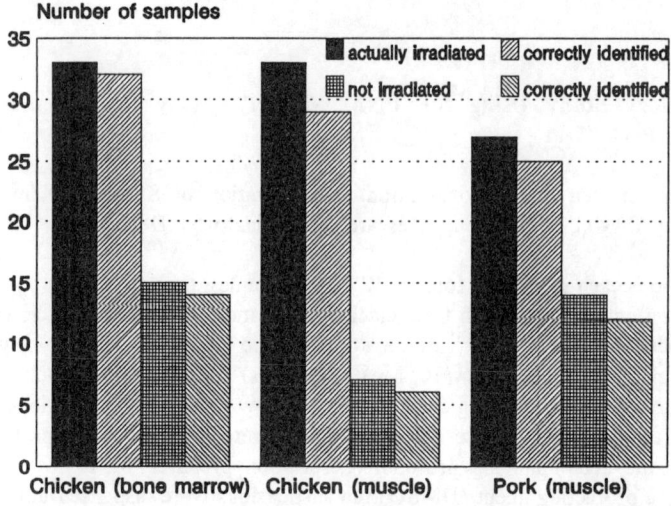

Figure A1 *Interlaboratory study using the DNA "comet assay" to detect irradiated meat*

estimation seems possible using the "comet assay". Since a number of parameters influence this rapid assay, it may be prudent to subject suspected irradiated samples to further analysis, in this case using more demanding techniques, *i.e.* requiring expensive equipment and/or extended sample preparation time, *e.g.* ESR or GC/MS measurements, by which an unequivocal identification can be achieved.

Acknowledgement

The excellent work of all participants and their staff involved should be acknowledged. Thanks are due to the National Food Administration of Sweden for financial support.

DETECTION OF IRRADIATED FRESH, CHILLED, AND FROZEN FOODS BY THE MITOCHONDRIAL DNA METHOD

E. Marchioni,[1] M. Bergaentzle,[1] F. Kuntz,[2] S. Todoriki[1] and C. Hasselmann[1]

[1] Faculté de Pharmacie - Département des Sciences de l'Aliment
74 route du Rhin, 67400 Illkirch-Graffenstaden, France
[2] AERIAL, 19 rue de Saint Junien, BP 23- 67305 Schiltigheim Cedex, France

1 INTRODUCTION

DNA molecules are very sensitive to ionising radiation, even at low doses. Strand breaks are easy to detect despite the generally low DNA content of foods, but such ruptures are not specific to radiation processing. Preliminary experiments showed that cellular DNA in beef underwent strong enzymatic degradation during storage at +4°C and thus radiation effects could not be isolated.[1]

In order to make DNA strand rupture more specific to radiation (other than by deep freezing) it appears necessary to isolate the irradiated DNA from cell enzymes. This is the case for mitochondrial DNA which is protected from enzymatic degradation by the mitochondrial walls but not from radiation. It can, therefore, be assumed that DNA strand breaks in mitochondria will be specific to ionising radiation.

Mitochondrial DNA has a low molecular weight (about 16000 base pairs), therefore, it is much more stable than cellular DNA. It is normally super coiled. With a rupture in one strand, it becomes the open circular relaxed DNA form. Another strand rupture, less than 10 base pairs from the preceding one, will change it to the linear DNA form. These three separate DNA forms are easily distinguishable by agarose gel electrophoresis. Non-irradiated mitochondrial DNA of beef liver is mostly in a super coiled structure (a significant portion is in open circular and linear structures) and irradiation causes an important reduction in super coiled DNA.

A total (or at least notable) disappearance of super coiled DNA could then be proposed as an irradiation test providing, of course, that the disappearance has not been caused by unique storage conditions. The aim of this work is to develop and validate the proposed test on different food samples (meat and fish products) which are already or may be industrially irradiated in the near future.

2 MATERIALS AND METHODS

2.1 Samples

Beef liver was collected directly from the slaughterhouse. Samples (25 g and 4 mm thick) were individually vacuum packed and labelled. Irradiation and storage were carried out either at +4°C or -20°C. The mechanically recovered meats (MRM) were purchased from a specialised French company and all the other foods samples (fishes, prawns, chicken, beef meat and burgers) were directly collected from a local supermarket and stored either at +4°C or -20°C.

2.2 Irradiation Equipment and Dosimetry

A Van de Graaff electron beam accelerator, 2.2 MeV and 70 µA (Vivirad High-Voltage Co, Handschuheim, France), was used in this study. The dose rate used was about 1.5 kGy s.$^{-1}$

Dosimetry was carried out using Far West Technology dosimeters (Goleta, USA), calibrated against a reference alanine dosimeter. Dose uniformity in the sample was ± 10%.

2.3 Mitochondrial DNA Extraction

Samples of about 25 g of beef liver were inserted into a saccharose buffer (5 volumes), pH 7.4 (saccharose 0.25 M, Tris-HCl 10 mM, EDTA-EGTA 1 mM) and ground in a 50 ml Thomas potter. For the other foods, 80 g (chicken or beef meat), 200 g (fish), or 500 g (king prawns) were ground in an electric mixer in a KCl buffer (5 volumes), pH 7.4 (KCl 0.15 M, EDTA 1 mM, Tris HCl 50 mM). The suspension was centrifuged at 2,500 rpm (10 min) to eliminate cellular debris and then at 10,000 rpm (10 min) to recover the mitochondrial pellet (Beckman J2.21 centrifuge and JA.10 rotor). Mitochondria were put in solution in 20 ml NaCl buffer pH 7.4 (Tris-HCl 20 mM, NaCl 140 mM, EDTA-EGTA 1 mM). All these operations were conducted at +4°C. The residual cellular DNA was digested by a DNase (Boehringer 104.159) (20 µg ml^{-1}, 10 min, +30°C, in the presence of $MgCl_2$ 5 mM). The samples were then cooled down to +4°C. EDTA-EGTA (20 mM final) was added in order to stop the enzymatic reaction. After centrifugation at 10,000 rpm (10 min) (Beckman JA.20 rotor), the pellet was recovered in 10 ml NaCl buffer. The mitochondrial walls were broken down by sodium dodecyl sulphate (SDS) (Merck 13.760) in a quantity required to achieve a final concentration of 0.5%. The nucleic acids were then refined through 3 successive phenolic (3 x 10 ml) and chloroformic (3 x 10 ml) extractions and precipitated (-20°C) overnight in 70% ethanol with 0.15 M ammonium acetate. After centrifugation for 10 min at 10,000 rpm (rotor JA.20), the pellet was vacuum dried and dissolved in 1.5 ml Tris-HCl 50 mM, EDTA-EGTA 1 mM (TE 50/1), pH 8. The mitochondrial RNA was digested by a RNase (Sigma R.5503) (50 µg ml^{-1}, 1 h at 37°C). The samples were then cooled down at +4°C. The mitochondrial DNA was refined through 2 phenolic (2 x 1.5 ml) and chloroformic (2 x 1.5 ml) extractions and re-precipitated overnight. After centrifugation (10,000 rpm, 10 min) and drying (under vacuum), the DNA was recovered in 100 µl TE 50/1. The nucleic acid concentration was estimated with a UV spectrophotometer (UVIKON 930 - KONTRON Instruments) set at 260 nm and 280 nm. The ratio of the 2 values should be approximately 1.8 to 1.

2.4 Separation of Mitochondrial DNA by Agarose Gel Electrophoresis

One volume of sample buffer (5 ml glycerol, 3 ml SDS 10%, 1 ml bromophenol blue 1%, 0.5 ml EDTA-EGTA 0.2 M for 10 ml buffer) was added to 10 volumes of the DNA solution. The resulting solution was placed on a 0.8% agarose gel in the electrophoresis buffer (Tris 40 mM, sodium acetate 20 mM, EDTA-EGTA 1 mM, pH 8). The electrophoresis was performed with a continuous electric field of 2.5 $V.cm^{-1}$ during 14 h at 14°C. The gel was then wetted for 10 min, with gentle agitation, in an ethydium bromide solution (1 $\mu g\ ml^{-1}$), rinsed with distilled water, placed on a UV ($\lambda = 312$ nm) transilluminator and photographed. The negatives were analysed with a computerised image analyser (Biocom Compaq Desk pro 286e). They were taken at various exposure times in order to avoid saturated signals. A molecular weight marker (l digested with Hind III) was placed on the same gel in order to have an estimation of the molecular weight of the different fragments of the mitochondrial DNA. The sizes of the weight marker fragments were 23 kb, 9.4 kb, 6.55 kb, 4.36 kb, 2 kb and 0.5 kb respectively.

3. RESULTS AND DISCUSSION

3.1 Plasmidic DNA

In a first experiment, the effects of irradiation (1.5 and 3 kGy) on commercial DNA (plasmidic DNA pBR 322, Boehringer) in aqueous solution (TE 50/1) were studied. This molecule has a lower molecular weight (4300 base pairs) than mitochondrial DNA, but it has a similar super coiled structure. Results previously published have shown that non-irradiated DNA is mostly in a super coiled structure but also that a significant portion is in open circular structure.[3] Irradiation causes an important reduction in super coiled DNA, with a complete disappearance at 3 kGy.

The total disappearance of super coiled DNA could then be used as an irradiation marker providing, of course, that mitochondrial DNA from food behaves in the same way as commercial plasmidic DNA and that the proposed test (super coiled DNA disappearance) is characteristic for radiation. In other words, that the disappearance of super coiled DNA has not been caused by particular storage conditions.

3.2 Beef Liver

The first foodstuff studied was beef liver. In actual fact, the texture of this food allows a relatively easy and efficient extraction of the mitochondrial DNA.
The results previously published confirm those previously obtained.[1-3] In the non-irradiated sample, the mitochondrial DNA did not appear exclusively under a super coiled form (representing only 30% of the total amount of DNA) but, it was also present as an open circular DNA form (59%) and linear DNA form (11%). In the irradiated samples, the super coiled DNA form decreased with the absorbed dose, being exceedingly low (5%) after a 3 kGy irradiation dose and vanishing after irradiation at

5 kGy. The rate of the open circular DNA increased drastically after an irradiation dose of 1 kGy and then decreased when the irradiation dose increased. Meanwhile, the open circular DNA was then transformed into a linear DNA the rate of which increased with the absorbed irradiation dose.

Statistical studies running on 20 different livers, showed that the amount of super coiled DNA was of a striking stability. The average value stated was around 30% with a standard deviation of 3%.[3]

When samples obtained from the same non-irradiated liver were preserved at +4°C during 21 days, no significant changes in super coiled DNA gave evidence of any deterioration. Similar results have been noticed if the preservation is performed at -20°C during 60 days. If during the preservation time, any temperature cycling is done, no noticeable modification of the super coiled DNA grade has been observed, neither at +4°C or -20°C storage. Each measurement has been carried out after such a temperature cycling by leaving the sample for 1 h at +20°C (during this freezing break the sample stored at -20°C was fully defrosted).[3]

Therefore, it is obvious that the super coiled DNA form of a non-irradiated liver sample obtained through this proposed method, is not lower than 25%, independent of the storage temperature (-20°C up to +4°C) and is not modified by any eventual freeze/thaw cycle during storage. Under these conditions, an important decrease of the super coiled DNA form is only attributable to an ionising treatment and not to unsatisfactory storage conditions.

In order to study the influence of an ionising treatment on the super coiled DNA grade of beef liver, 6 series (1 non-irradiated and 5 irradiated at 2, 2.5, 3, 3.5 and 4 kGy respectively) among 20 different liver samples (a total of 120 samples), have been analysed directly after the treatment. When the irradiation dose was higher than 3.5 kGy destruction of the super coiled DNA was found to be total. With doses lower than 3.5 kGy the damage was only partial but sufficiently evident, in the way that no possible mistake could be made between an irradiated and a reference sample.[3]

To achieve a good hygienic quality of the beef liver, an irradiation dose between 3 and 4 kGy should be used.[2] Measurements of the super coiled DNA grade thus allow, without any ambiguity, the differentiation between an irradiated and a non-irradiated sample. Even in the absence of a non-irradiated reference sample it is possible to prove that the sample has or has not been irradiated.

In order to study the influence of the irradiation temperature, different samples of beef liver were irradiated at 8 different doses between 0 and 8 kGy at 4 different temperatures (-70°C, -20°C, +10°C and +30°C). Figure 1 presents the results of the analysis of the negatives from the photographs. It is obvious that the DNA disappeared more rapidly when the sample was not in a frozen state and that the radio-sensitivity of the mitochondrial DNA increased as the irradiation temperature increased. The total disappearance of the super coiled DNA was observed at 3 kGy for the positive irradiation temperatures and was only observed at 5 to 6 kGy for temperatures of about -20°C and -70°C respectively. Nevertheless, to achieve the same bacteriological efficiency, the irradiation treatments of deep frozen foods are usually performed with higher absorbed doses than at chilled temperatures. Therefore, the mitochondrial DNA test remains applicable even with frozen foods.

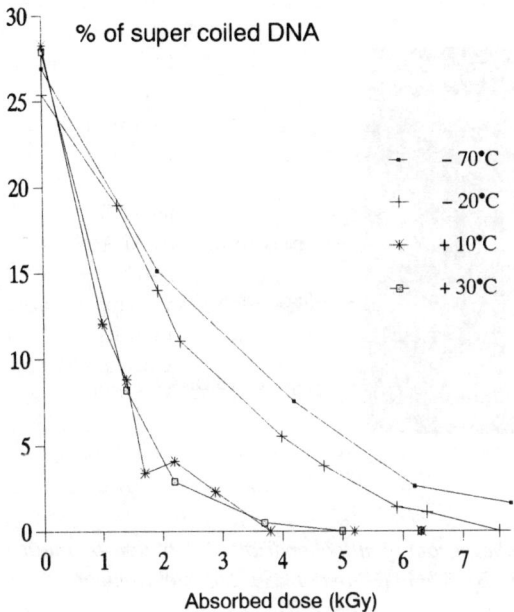

Figure 1 *Percentage of super coiled DNA form from beef liver versus the absorbed dose at different irradiation temperature*

3.3 Other Foods

The other foodstuffs studied presented a more rigid structure which did not allow an easy and gentle extraction as was the case for beef liver. Therefore, this detection test was applied with small modifications to the extraction protocol in order to maximise the mitochondrial DNA extraction rate. In general, and in spite of these modifications, the total quantity of mitochondrial DNA extracted from these foodstuffs was always less than that extracted from beef liver. The electrophoregram presented in Figure 2 shows the migration profile of mitochondrial DNA from chicken meat irradiated at 0, 2, 3 and 4 kGy. The concentration of the mitochondrial DNA decreased as the value of the absorbed dose increased (56% for the reference, 24% for 2 kGy, 9% 3 kGy and less than 6% for 4 kGy) in this electrophoregram. At 4 kGy, there was no significant level of super coiled DNA in the gel.

A statistical study carried out on chicken meat (4 irradiation doses and 9 samples per dose) confirmed the results obtained for beef liver. Super coiled DNA extracted from chicken meat represented 56 ± 10% of the total mitochondrial DNA (this percentage was only 30 ± 3% in beef liver). It was, however, rapidly reduced by irradiation (Table 1). It disappeared completely with a dose of 4 kGy, but the detection of irradiation was possible when the irradiation doses exceeded 2 kGy.

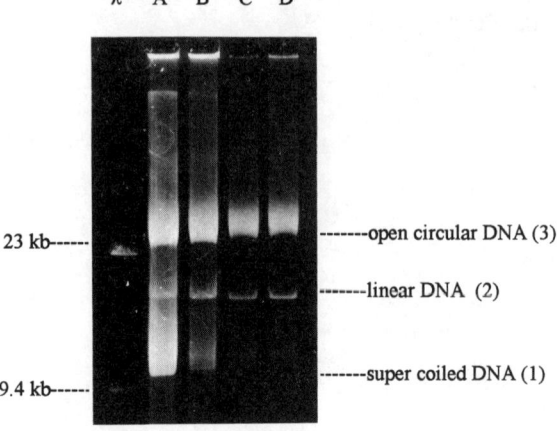

Figure 2 *Electrophoregram of mitochondrial DNA of non-irradiated (A) and irradiated at 2 kGy (B), 3 kGy (C) and 4 kGy (D) chicken meat.*

Table 1 *Average Percentage of the Different Forms of the Mitochondrial DNA of Chicken Meat versus the Absorbed Dose (Replication: 9)*

Absorbed Dose (kGy)	Percentage of Each Geometrical Form		
	Super Coiled	Open Circular	Linear
0	56 ± 10	34 ± 11	10 ± 5
2	24 ± 15	57 ± 16	19 ± 8
3	16 ± 10	65 ± 12	19 ± 8
4	<8 ± 4	75 ± 8	19 ± 8

During storage, the form of the mitochondrial DNA of chicken meat remained very stable as for beef liver (Figure 3). Storage at +4°C during 20 days did not have a significant effect on the amount of super coiled DNA which remained constant during all the storage. The concentration of the super coiled DNA in a reference sample stayed in a much higher concentration than the concentration of super coiled DNA from the sample irradiated at 2 kGy.

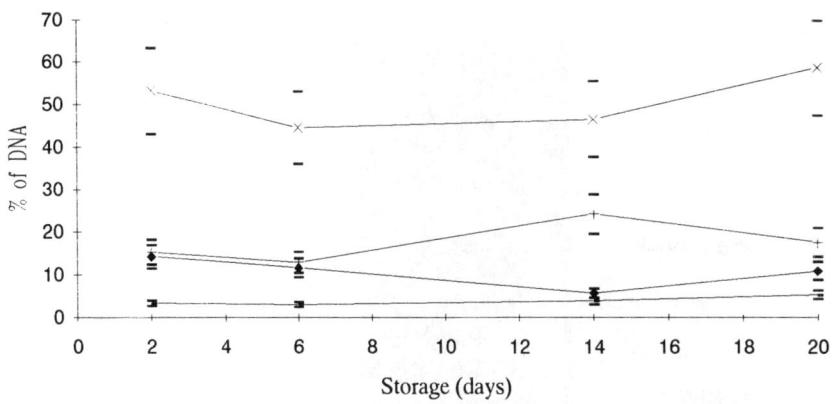

Figure 3 *Stability of the super coiled DNA of chicken meat, during storage (20 days) at +4°C (× reference; + 2 kGy; ♦ 3 kGy; - 4 kGy)*

The detection of the three forms of mitochondrial DNA, in particular the super coiled DNA, was obtained with all the other foodstuffs studied (minced and sliced steaks, king prawns, fish) with the exception of MRM, in which only the circular DNA fragment could be identified, the other fragments being submerged in a continuous background of nucleic acids (smear).

For the minced and sliced beef meat, as well as in the fish and king prawns (Figures 4 to 8), a rapid reduction of the super coiled DNA was observed. The levels in the reference samples were generally in the 30-40% range (Tables 2 to 4). In the minced steaks (burger), detection of a 1 kGy irradiation dose was possible. In the other foodstuffs, the detection limit was 2 kGy. However, for fish, it should be noted that the test could only be carried out on fresh (chilled) fish. When the fish is frozen, it is not possible to extract mitochondrial DNA by the technique used, undoubtedly due to a modification of the fish texture owing to the effects of freezing. For the sliced steaks, the amount of extracted DNA was too low to be quantified, the exploitation of the electrophoregram was only possible *de visu* (by vision).

An interlaboratory blind test showed the efficiency of the proposed test. Ten samples (reference and irradiated at 4 kGy, without specifying the number of each) of 4 different foodstuffs (minced and sliced steaks, chicken, king prawns) supplied by ADRIA, Quimper (France), were analysed in the laboratory. There were, in fact, 4 reference samples and 6 irradiated samples of minced steaks and 6 reference and 4 irradiated samples for all the other foodstuffs. For the sliced steaks and the king prawns, the results were 100% correct. For the minced steaks (burger), one reference sample was considered to be irradiated and for the chicken, one irradiated sample was identified as a reference sample. Overall, the results obtained were very encouraging.

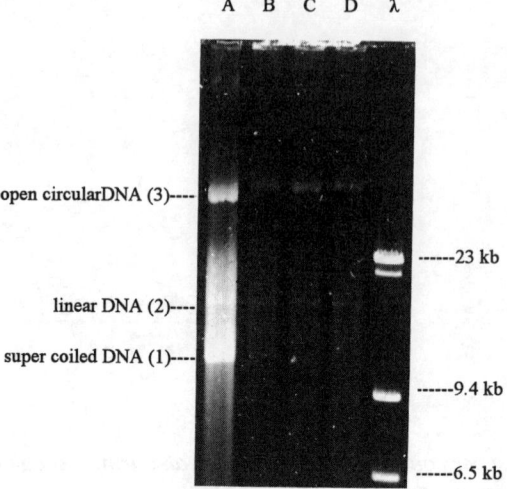

Figure 4 *Electrophoregram of mitochondrial DNA of non-irradiated (A) and irradiated at 1 kGy (B), 2 kGy (C) and 3 kGy (D) chilled trout fillet*

Table 2 *Average Percentage of the Different Forms of the Mitochondrial DNA of Fresh Trout Fillet versus the Absorbed Dose*

Absorbed Dose (kGy)	Percentage of Each Geometrical Form		
	Super Coiled	Open Circular	Linear
0	29	58	13
1	12	74	14
2	4	76	20
3	0	75	25

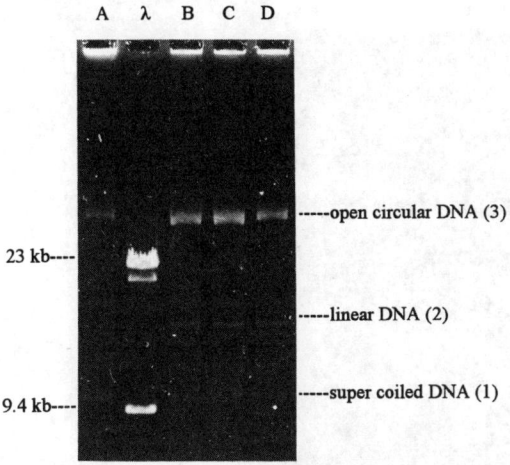

Figure 5 *Electrophoregram of mitochondrial DNA of non-irradiated (A) and irradiated at 1 kGy (B), 2 kGy (C) and 3 kGy (D) king prawns*

Table 3 *Average Percentage of the Different Forms of the Mitochondrial DNA of King Prawns versus the Absorbed Dose*

Absorbed Dose (kGy)	Percentage of Each Geometrical Form		
	Super Coiled	Open Circular	Linear
0	30	51	19
1	17	52	31
2	0	82	18
3	0	78	22

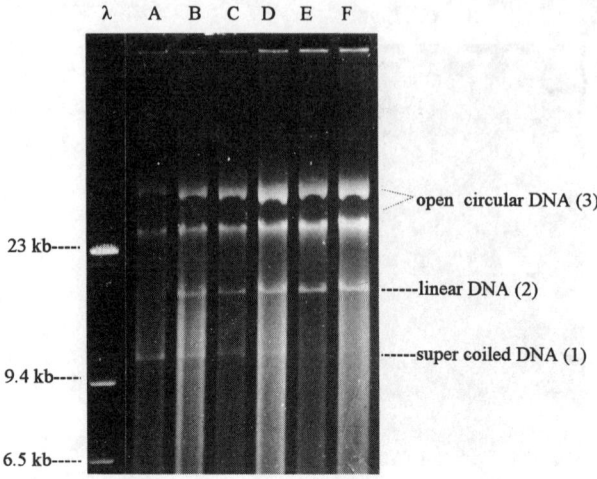

Figure 6 *Electrophoregram of mitochondrial DNA of non-irradiated (A) and irradiated at 1 kGy (B), 2 kGy (C) 3 kGy, (D), 4 kGy (E) and 5 kGy (F) beef burger*

Table 4 *Average Percentage of the Different Forms of the Mitochondrial DNA of Beef Burger versus the Absorbed Dose*

Absorbed Dose (kGy)	Percentage of Each Geometrical Form		
	Super Coiled	Open Circular	Linear
0	34	58	8
1	7	86	7
2	7	84	9
3	7	77	16
4	2	84	16
5	0	92	8

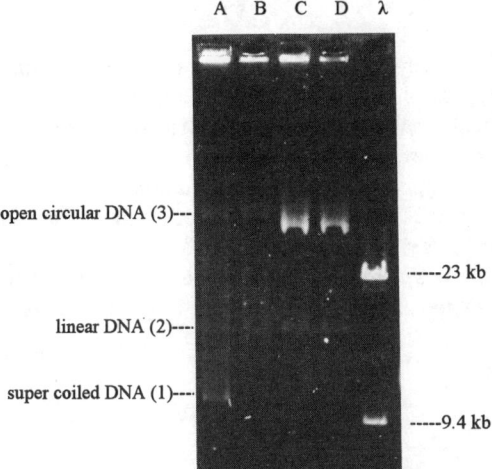

Figure 7 *Electrophoregram of mitochondrial DNA of non-irradiated (A) and irradiated at 2 kGy (B), 3 kGy (C) and 4 kGy (D) beef meat*

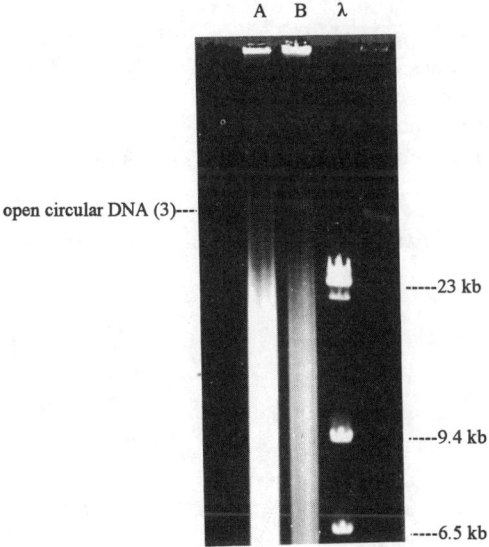

Figure 8 *Electrophoregram of mitochondrial DNA of non-irradiated (A) and irradiated at 5 kGy (B) MRM*

4 CONCLUSION

The appraisal of the mitochondrial super coiled DNA form constitutes an unambiguous detection test for irradiation treatment in a variety of fresh and frozen foods. This fact has been confirmed by the good results obtained in the blind test. Nevertheless, this method still presents one major defect *i.e.* the extraction process of the mitochondrial DNA is complicated and time consuming, thus the test is not routine.

Now, research is evolving towards simplifying the method of the extraction protocol and increasing the detection sensitivity.

The present aim is to apply this detection test to vegetables. Indeed, it must be remembered that the main advantage of the test is to be able to detect *a priori*, any fresh or frozen irradiated foodstuffs.

Acknowledgements

This work was supported by two grants: No. 51.88.02, Ministère de l'Economie et des Finances (France) and No. 90.H.0616, Ministère de la Recherche et de la Technologie (France).

References

1. C. Hasselmann and E. Marchioni, *Ann. Fals. Exp. Chim.*, 1989, **82 (876)**, 169.
2. M. H. Desmonts, "L'ionisation des Produits Carnés d'origine Bovine: Conséquences Microbiologiques, Nutritionelles et Organoleptiques," Thèse de Doctorat de l'Université Louis Pasteur de Strasbourg, 1991.
3. E. Marchioni and C. Hasselmann, *Radiat. Phys. Chem.*, 1992, **40 (6)**, 485.

IMMUNOLOGICAL DETECTION OF MODIFIED DNA BASES IN IRRADIATED FOOD

J. H. H. Williams, A. L. Tyreman, D. J. Deeble, M. Jones, C. J. Smith,[1] J. F. Christiansen and P. C. Beaumont.

Multi Disciplinary Research and Innovation Centre (MRIC)
Athrofa Gogledd Ddwyrain Cymru (NEWI), Glannau Dyfrdwy, Ffordd Celstryn
Cei Connah, Clwyd, North Wales, CH5 4BR
[1]Current Address: Cortecs Diagnostics Ltd., Techbase 1, Newtech Square, Deeside, Clwyd, North Wales, CH5 2NU.

1 INTRODUCTION

Ionising radiation is fatal to all known life forms given sufficient exposure in terms of dose and duration. This property has been used beneficially to sterilise a range of materials, particularly medical products where the removal of all contaminating organisms is deemed essential. Irradiation has long been used to sterilise food for consumption by certain categories of patients. The method is attractive because all potentially contaminating organisms can be removed by one simple treatment. Irradiation also slows down or stops certain processes such as sprouting. There are, however, disadvantages to irradiating food. It is not only DNA that is affected as alterations in lipid and protein components of the food may lead to a loss of quality. Although irradiation will kill bacteria it will probably not affect any toxin produced by those bacteria prior to treatment.

These disadvantages are compounded by the consumer who may perceive dangers where they may not exist. There is, therefore, considerable debate on the suitablity of irradiation for use in foodstuffs. Irradiation is banned in many countries and allowed within certain guidelines in others. The worldwide trade in foodstuffs means that there is a need for detection systems to monitor how foods are treated.

Irradiation achieves its effects by damaging molecules, particularly nucleic acids. Consequently, if any of the damage to nucleic acids could be shown to be by a process unique to irradiation, and the products of this unique process could be measured, then there is the basis for a detection system. Furthermore, if the damage could be shown to be proportional to the dose of radiation received then the dose could also be quantified.

Legislation, therefore, requires that assays be developed for use in different countries, those which totally ban irradiated food, those which require irradiated food to be labelled and those which have selective laws relating to specific foods and specific levels of irradiation. The constraints are that any test must be:
    a) rapid;
    b) simple to use;
    c) flexible in terms of product that can be tested;
    d) highly specific for the products of ionising radiation.

Ionising radiation alters DNA by producing strand breaks (single and double) and modified bases.[1] Strand breaks as the basis for a routine test are covered elsewhere in this volume. Although there are many different base modifications which occur on irradiation of nucleic acids[1,2] it was decided to concentrate on two modifications of thymidine, viz thymidine glycol and dihydrothymidine[1] (Figure 1). The mechanisms by which these two modified bases are formed on irradiation have previously been reviewed.[1,2] The differences between thymidine and the modified bases are small (Figure 1) and, therefore, any analytical technique designed to distinguish between these compounds must be highly specific.

2 IMMUNOASSAYS

A group of compounds capable of distinguishing small differences between molecules are antibodies. Antibodies bind with a varying degree of specificity to antigens and those which bind with high affinity and extreme specificity can be selected from an heterogenous population. Immunoassays can be simple, robust, very rapid and can be carried out in non-specialised laboratories (*e.g.* most home pregnancy testing kits are immunoassays). For these reasons antibody based assays are widely used in medicine and food analysis. Several groups have demonstrated that immunqassays are suitable for the detection of normal or modified nucleic acid bases.[3-5] Immunoassays could fulfil the criteria for a routine assay, being rapid, simple to use and potentially having the required specificity.

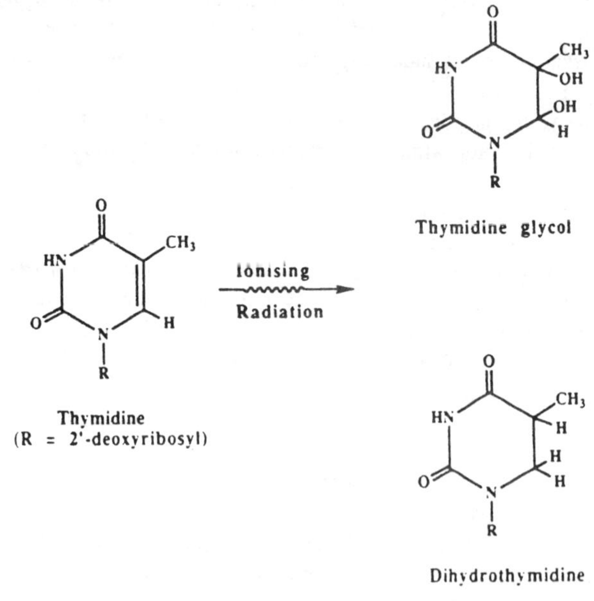

Figure 1 *Structures of thymidine and two of the products formed following exposure to ionising irradiation*

This paper will first briefly describe the process of producing a suitable antibody with the required characteristics and then the development of a suitable immunoassay, with reference to the detection of thymidine glycol and dihydrothymidine.

Often a higher degree of specificity can be obtained by using monoclonal antibodies. These were produced by standard procedures.[6] Compounds of molecular weight below 1000, such as thymidine glycol and dihydrothymidine, do not elicit an immune response. Therefore, an immune response can only be obtained by conjugating the antigen to a large molecule such as a protein. The carrier molecule will also produce an immune response, therefore, when testing the antibodies a different conjugate is used. The two modified bases were conjugated to a number of different molecules (BSA, thyroglobin or poly-L-alanine: poly-L-lysine).

3 ENZYME LINKED IMMUNOSORBENT ASSAY (ELISA)

There are several different types of immunoassay, including radioimmunoassay and enzyme linked immunosorbent assay (ELISA). The use of radioactive isotopes is not suitable for routine analyses because of time constraints and the need for special precautions. ELISA methods have proved to be appropriate for different types of food analysis, therefore, the aim of this work was to develop ELISA tests for thymidine glycol and dihydrothymidine.

Two types of ELISA were used: one for detection of antibody in serum and culture fluid samples, and the other for testing specificity of the antibodies and development of a test ELISA. The 2 methods are non-competitive (or direct) ELISA and competitive ELISA (Figure 2).

In a non-competitive ELISA the first antibody concentration is unknown, and in the competitive ELISA the first antibody is added at a known concentration with the test solution containing an unknown amount of the antigen.

Both non-competitive and competitive ELISAs were carried out on 96 well plates. ELISA plates were coated with conjugates of the antigen: thymidine glycol-BSA or dihydrothymidine-BSA (BSA conjugates were not used for immunisation) (Figure 2). Serum was collected from mice immunised with the thymidine glycol and the dihydrothymidine conjugates and added to the appropriate plate (Figure 2). After washing the second antibody, rabbit anti-mouse IgG conjugated to horse radish peroxidase (HRP), was added (Figure 2). Unbound second antibody was then washed off the plate. The amount of antibody in the serum was proportional to the HRP activity bound to the plate. HRP activity was measured by the addition of 2, 2' azino-bis 3-ethylbenzthiazoline-6-sulphonic acid (ABTS) and hydrogen peroxide, resulting in oxidation of ABTS, which absorbs strongly at 405 nm.

Culture supernatants from hybridoma cultures were tested using a competitive ELISA. The plates were coated as with the non-competitive ELISA (Figure 2). The test samples were applied to the plate in the presence of the antibody containing cell supernatant. There is competition between the antigen coated onto the plate and the antigen in the sample. The larger the amount of antigen in the sample the lower amount of antibody that binds to the plate. The second antibody and assay were carried out as for non-competitive ELISA. In a competitive ELISA a low reading at 405 nm represents a high amount of antigen in the sample.

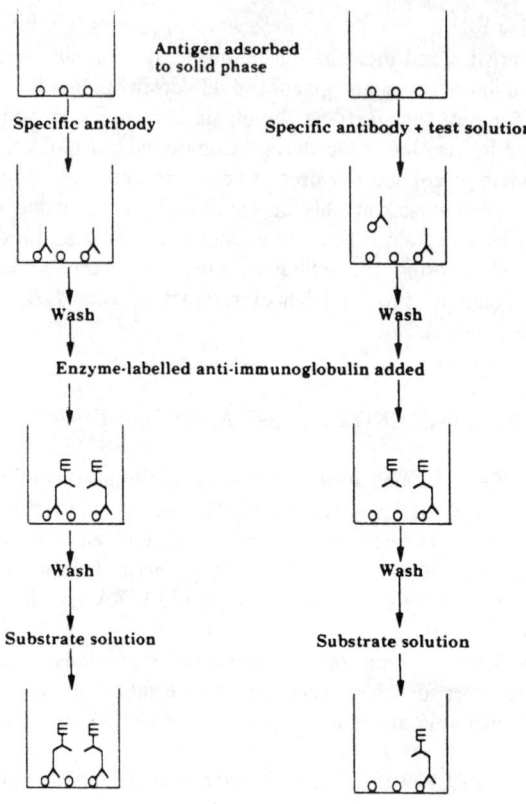

Figure 2 *Non-competitive (left) and competitive (right) ELISA for detection of antibody*

4 THYMIDINE GLYCOL AND DIHYDROTHYMIDINE ANTIBODIES.

Two sets of mice were immunised with the modified DNA base conjugated to a polypeptide carrier molecule. The first group were immunised with thymidine glycol-BSA and the second with dihydrothymidine-(poly-L-alanine : poly-L-lysine). Antibody was detected within 1 month of immunisation with the conjugates. Although the titre was very low 1 month after immunisation the amount of antibody increased in both sets of mice with time after the initial immunisation (Table 1).

Sufficient antibody was found in the sera of both groups of mice after 4 months to give an A_{405} of over 1.5 when diluted 1 in 1000. The mice were sacrificed and the spleens removed. Standard procedures were used for the production of hybridomas. Several cell lines were produced, grown on and then tested for antibody production using a non-competitive ELISA with different plate coatings. Cell lines that gave responses specific for irradiated DNA were selected. Two cell lines gave such a response for the thymidine glycol and 4 cell lines for dihydrothymidine.

Table 1 *ELISA for Serum Samples from Mice Immunised with Thymidine Glycol or Dihydrothymidine (Serum was diluted 1:1000 prior to the assay)*

Time after Initial Immunisation (Months)	Absorbance at 405 nm	
	Thymidine glycol	Dihydrothymidine
1	0.10	0.02
2	0.52	0.51
3	0.87	0.63
4	1.63	1.80

Supernatants were then tested using different plate coatings: thymidine, dihydrothymidine, thymidine glycol, denatured calf thymus DNA, irradiated denatured calf thymus DNA. At this stage the thymidine glycol antibodies were found to give no response to irradiated calf thymus DNA or wheat DNA.[7] Efforts were therefore subsequently concentrated on producing antibodies to dihydrothymidine.

Four dihydrothymidine cell lines were selected on the basis of a high response on irradiated DNA coated plates in a non-competitive ELISA. However, when the best of these cell lines (4B3) was tested using the competitive ELISA, non-irradiated DNA proved to be inhibiting better than irradiated DNA. Cell lines at this stage were still polyclonal. The 4B3 cell line supernatant was fractionated by FPLC. Five main fractions were separated. Two of these fractions contained IgM antibodies, which could explain the lack of specificity. Of the other 3 fractions one gave very good differentiation between non-irradiated and irradiated DNA. Inhibition curves for this fraction show much larger inhibition by irradiated DNA than non-irradiated DNA (Figure 3). Dihydrothymidine is not formed in the presence of O_2.[1] Therefore, there was no difference between denatured calf thymus DNA and calf thymus DNA irradiated in the presence of O_2 (Figure 3).

Calf thymus DNA was irradiated under nitrogen at different doses. The DNA was then used in a competitive ELISA with fraction 2 from the Mono Q purification. There is a clear dose response (Figure 4). Similarly, when wheat was irradiated there was a dose response. Dihydrothymidine was detected in wheat DNA when irradiated in the presence of O_2. The O_2 concentration in the nucleus has been calculated to be very low (Swallow, personal communication) and is rapidly depleted during irradiation. Therefore, dihydrothymidine is produced during irradiation of whole wheat germ. Within the 4B3 cell line there is at least 1 antibody which can differentiate between irradiated and non-irradiated DNA. Currently the 4B3 cell line is being subcloned to produce a monoclonal line with these properties.

It is likely that the antibodies from the thymidine glycol cell lines could also be fractionated to give a specific antibody. However, as thymidine glycol can be produced by other treatments apart from ionising irradiation, such as UV radiation, efforts are currently being concentrated on producing a monoclonal anti-dihydrothymidine antibody.

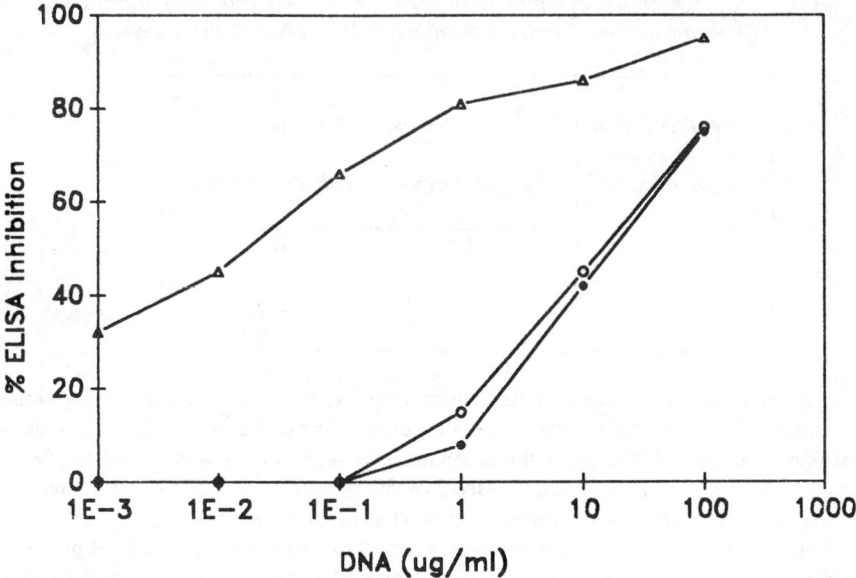

Figure 3 *Inhibition assay using fraction 2 from the Mono Q purified 4B3 supernatant comparing the inhibitory properties of denatured calf thymus DNA before (△) and after irradiation (120 Gy) in N_2 (○) and in O_2 (●)*

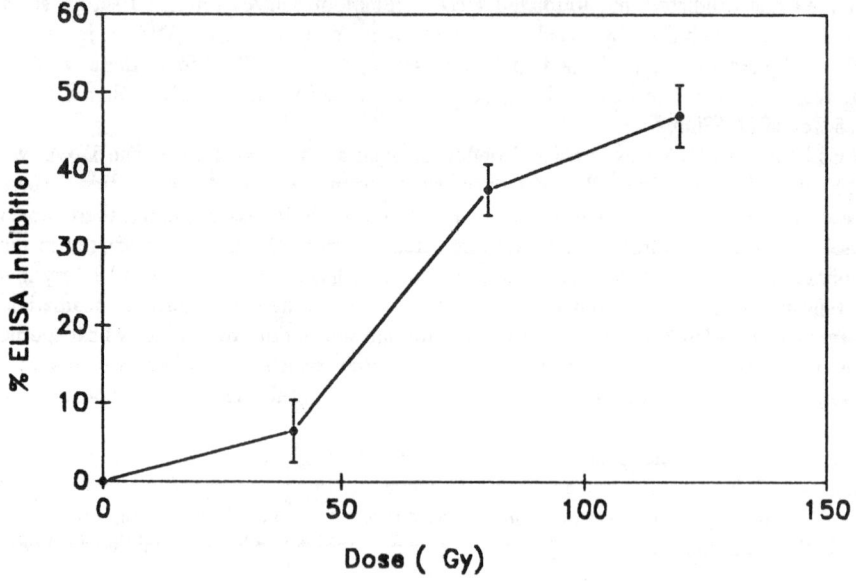

Figure 4 *Dose response curve of percentage ELISA inhibition against degree of irradiation for denatured calf thymus DNA (100 ng ml^{-1}) using fraction 2 from Mono Q purified 4B3 supernatant*

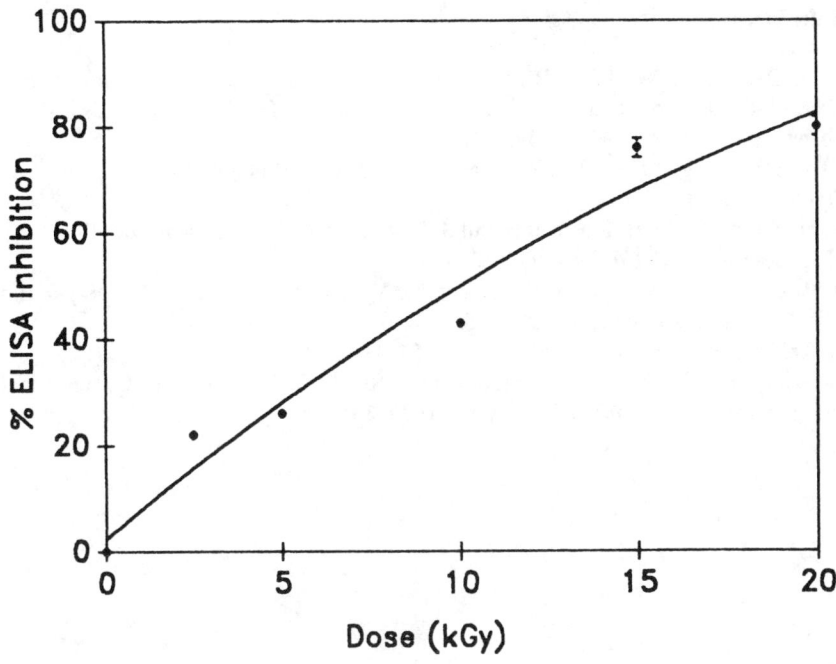

Figure 5 *Dose response curve of percentage ELISA inhibition against degree of irradiation for denatured wheat DNA (100 ng ml^{-1}) using fraction 2 from Mono Q purified 4B3 supernatant*

5 CONCLUSIONS

Highly specific antibodies for dihydrothymidine, a unique product of ionising irradiation, have been produced. It has been shown that the antibodies produced can detect irradiated calf thymus DNA and wheat DNA. Work is now concentrating on subcloning to obtain a monoclonal antibody. A simple ELISA test will then be developed.

ELISA tests are rapid, simple to use. The type of product that can be tested will range from fresh to cooked to frozen food, providing the foodstuff contains DNA. The only proviso with fresh food is that it is not known how much, if any, repair can occur following the types of doses used in the food industry. It has been shown that these are specific antibodies for dihydrothymidine. Dihydrothymidine is only produced when DNA is irradiated with ionising radiation. Therefore, the ELISA test based on dihydrothymidine should fulfil all the criteria for a routine test for irradiated food.

Acknowledgements

We are grateful to the Ministry of Agriculture, Fisheries and Food for a grant to support this work

References

1. D. J. Deeble, A. W. Jabir, B. J. Parsons, C. J. Smith and P. Wheatley, "Food Irradiation and the Chemist," D. E. Johnson and M. H. Stevenson, Eds., Royal Society of Chemistry, 1990, p. 57.
2. C. von Sonntag, "The Chemical Basis of Radiation Biology," 1987, Taylor and Francis, London.
3. K. Hubbard, H. Ide, B. F. Erlanger and S. S. Wallace, 1987, *Biochem.*, **28**, 4382.
4. S. A. Leadon, *Br. J. Cancer*, 1987, **55**, 113.
5. A. Raza, I. Mehdi, W. J. Guo, N. Yousuf, M. Masterson, S. Mirto, L. E. Motyka and G. L Mayers, *Leukemia Res.*, 1991, **15**, 9.
6. G. Kohler and C. Milstein, *Nature*, 1975, **256**, 495.
7. W. Jabir, D. J. Deeble, P. A. Wheatley, C. J. Smith, B. J. Parsons, P. C. Beaumont and A. J. Swallow, *Radiat. Phys. Chem.*, 1989, **34**, 935.

Biological Methods

BIOLOGICAL METHODS FOR THE DETECTION OF IRRADIATED FOODS

K. M. Hammerton

Australian Nuclear Science and Technology Organisation
PMB 1
Menai NSW 2234
Australia

1 INTRODUCTION

The main purpose of most applications of food irradiation is to obtain a change in a biological system associated with or within the food commodity. The biological effects in systems associated with the food include the reduction or elimination of spoilage and pathogenic microorganisms, insect disinfestation and parasite disinfection. For fresh plant foods, the effects are due to changes in biological systems in the food such as inhibition of sprouting of tubers, bulbs and roots, loss of seed viability and delay of maturation and senescence of fruits and vegetables. Ultimately, the efficacy of a food irradiation process is determined by reference to whether the desired effect can be observed.

These biological effects have formed the basis for investigating various methods for the detection of irradiated foods. A review of the methods developed up to 1990 was carried out within the framework of the FAO/IAEA Coordinated Research Programme on Analytical Detection Methods for Irradiation Treatment of Foods (ADMIT),[1] and a state-of-the-art presented in 1992.[2] A summary of the current status of biological methods is given in Table 1. Changes in the microbial flora and inhibition of root or shoot formation of plant products appear to be reliable indicators of irradiation treatment. The main drawback of these methods is the lack of specificity since other processes, such as heat or chemical treatment, can induce similar changes. Other disadvantages may be the time required to measure the change and the inherent variability of biological systems necessitating the availability of untreated control samples. However, some of these methods could be used in the routine screening of products for irradiation treatment in food control laboratories.

2 MICROBIOLOGICAL METHODS

Treatment of food with substerilising doses of ionising radiation can be used to eliminate certain types of pathogenic microorganisms, such as Salmonellae in poultry meat, and to extend the shelf-life of food by reducing food spoilage microorganisms. Therefore, a

Table 1 *Biological Methods for Detection of Irradiated Foods*

Biological Change	Applicable Food	Status*	References
Changes in microflora			
Reduced viability [DEFT/APC method]	spices	A	4-8
	meat	B	9-12
Reduced viable gram-negative bacteria [LAL/GNB method]	poultry	A	13,14
Shift in microbial profile	strawberry	B	15-17
	fish	C	18
Radiosensitivity	poultry	C	19
Change in susceptibility to bacterial spoilage	fish, meat	C	21,22
Inhibition of seed germination			
Whole seeds	cereals, legumes	B	22-25
Half-embryo test	citrus fruit	B	23,26,27
	cherry, apple	C	28
Histological/morphological changes			
Buds or tissues of potatoes		C	29,30
Roots of onions		C	31
Hyphae, spores or tissues of mushroom		C	32,33
Morphological change in fruit flies		B	34-36

* A - method investigated in several laboratories; interlaboratory trial undertaken
  B - method investigated in more than one laboratory
  C - research undertaken requiring further evaluation

change in the microbial flora with a loss of radiation-sensitive types of microorganisms and an increase in the proportion of radiation-resistant types should be indicative of irradiation treatment. These characteristic shifts in the microbial populations of irradiated foods have been reviewed and the feasibility of using these changes for developing detection methods for irradiated foods discussed.[3]

The gram-negative rod-shaped bacteria, belonging to the Enterobacteriaceae group and Pseudomonads, are among the most radiation-sensitive microorganisms. The very low numbers of these bacteria in irradiated strawberries compared with untreated strawberries was found to be indicative of irradiation treatment for strawberries grown outdoors but not for those grown in green houses.[15,16] However, only the absence of Enterobacteriaceae appeared to be suitable for distinguishing irradiated from untreated strawberries grown outdoors in Australia.[17] In irradiated fish and shellfish (not preserved with benzoic acid) the absence of Pseudomonads and the predominance of *Moxarella-type* bacteria was indicative of irradiation treatment.[18] Further evaluation of these methods is warranted.

An alternative microbiological method has been proposed for identifying irradiated seafood and meat which was based on the lowered susceptibility of these irradiated foods to bacterial spoilage. It was found that inoculated bacteria proliferated to the same extent in both untreated and irradiated seafood but the formation of total volatile acids (TVA) and bases (TVBN) was lower in irradiated than untreated counterparts.[20] A further study showed that the method could be used for identifying irradiated fresh and frozen meat samples (chicken, beef, mutton, pork) by incubating the meat samples at 37°C for 6-7 h with bacteria followed by measurement of the TVA and TVBN.[21] Further testing of this method should be carried out.

Another approach to the use of microbiological methods for detecting irradiated foods is to estimate the number of microorganisms rendered non-viable by the process. Betts *et al*.[9] proposed that a method based on the combined use of the direct epifluorescent filter technique (DEFT) and the conventional aerobic plate count (APC) could be used for detecting an irradiated food sample. The DEFT is used to count the total number of microorganisms, irrespective of viability, in the food sample; the microorganisms are captured on a filter and stained with a fluorochrome, acridine orange. The APC enumerates the viable microorganisms capable of forming colonies on an agar plate. Viable organisms which were fluorescing orange before irradiation continued to fluoresce orange after irradiation.[9] For irradiated samples of minced beef, steak, bacon, pate, gammon and ham, the difference between the DEFT and APC counts estimated the number of microorganisms rendered non-viable by the process. It was concluded that the DEFT/APC method would give an indication of whether or not a food item has been irradiated and in the case of irradiated food would provide information on the microbiological quality of the food before irradiation.

The DEFT/APC method has been used to distinguish a range of untreated and irradiated spices.[4-8] A collaborative trial of the method has been conducted with 8 spices.[6] In these studies it was found that the DEFT count was 4 log units higher than the APC for spices irradiated with either 5 or 10 kGy. The method is not specific for irradiation treatment since spices treated with ethylene oxide give results similar to those for spices treated with 5 kGy.[5] However, recent studies have suggested that heat or ethylene oxide treated spices could be distinguished from irradiated spices on the basis of an aerobic spore count together with the DEFT and APC counts.[7] Besides spices, the DEFT/APC method has been used for identifying irradiated frozen food products including poultry meat, liquid egg white and parsley.[10-12] For frozen poultry meat irradiated with 3 kGy or more, the DEFT count was more than 2 log units higher than the APC for storage periods up to 12 months.[11,12]

Another approach similar to the DEFT/APC method uses the *Limulus* amoebocyte lysate (LAL) test in combination with the enumeration of viable gram-negative bacteria (GNB). The LAL test provides a measure of the endotoxins derived from viable and non-viable GNB. The endotoxin levels were found to be similar in untreated and irradiated chicken meat but the GNB count was lower in irradiated samples.[13] With fresh chicken meat, comparison of the LAL test with the GNB count provided an index for interpreting the GNB rendered non-viable by irradiation treatment. Interlaboratory studies indicated that the method was suitable for screening fresh poultry meat for irradiation treatment.[14]

In summary, these microbiological methods can be used for screening food products for the possibility of irradiation treatment. The methods also provide information on the microbiological status of the food samples. In particular the DEFT/APC method gives an estimation of microbial numbers pre- and post-irradiation in a sample. This information could be used to indicate that the foodstuff was microbiologically unacceptable prior to irradiation. The methods could be easily incorporated into the routine microbiological regimes already used in food control laboratories.

3 GERMINATION METHODS

The irradiation of plant foods that are seeds (cereal grains and legumes) or contain seeds (fruits) affects the viability of the embryo or germ even at the relatively low doses used in these applications. Inhibition of seed germination has been used for identifying irradiated cereal grains and legumes.[23,24] The elongation of roots or shoots from germinating seeds was inhibited in husked rice[22] and wheat[25] irradiated with doses exceeding 300 Gy. The germination of seeds removed from grapefruits was affected by doses as low as 50 Gy,[37] but the test required 6 to 14 days of incubation. The time required for this testing was considerably improved (2 to 4 days) by removing the half-embryos from the seeds.[26] The half-embryo test has been used for identifying citrus fruits (grapefruits, oranges and lemons),[23-27] cherries and apples,[28] irradiated with 150 Gy or more.

4 HISTOLOGICAL/MORPHOLOGICAL METHODS

Various methods for identifying irradiated plant foods based on changes in histological or morphological characteristics have been proposed.[29-33] A review[1] of these methods concluded that, even though some of the methods such as the rooting of onions were promising, further evaluation was required which is still the current status (Table 1).

Treatment of food with ionising radiation for insect disinfestation can be used for preventing insect damage to the produce or for quarantine control purposes. Since the radiation doses needed for immediate mortality of insects (> 1 kGy) would be phytotoxic to most fruits and vegetables, lower doses can be used that would either inhibit the feeding of the insect or result in the inability of the insect to reproduce. For quarantine control doses of radiation from 75 to 300 Gy are effective. Therefore, radiation-induced changes in insects would be useful for indicating that the produce with which they are associated had been irradiated or for aiding quarantine security. One such morphological change has been observed for fruit flies. The original work found that the supraoesophageal ganglion decreased in size in irradiated eggs and larvae of the Mediterranean fruit fly but the size of the proventriculus remained unchanged.[34] Similar observations have been reported for melon fly, oriental fruit fly[35] and Queensland fruit fly.[36] Further testing under conditions pertaining to the practical implementation of the method should be undertaken.

5 IMMUNOCHEMICAL METHOD

The radiolysis of macromolecules can result in degradation or aggregation reactions. The possibility of detecting these radiolytic products of proteins in foods with immunochemical methods was recognised.[1] It has now been shown that low-molecular weight fragments of egg-white proteins can be detected with specific antibodies in uncooked and cooked eggs irradiated with 2 kGy.[38] The technique may be applicable to other irradiated foods.

6 CONCLUSION

A range of biological methods has been established for indicating irradiation treatment of foods. With these methods, the absence of the biological effect, *e.g.* very high numbers of viable microorganisms in spice samples or normal seed germination, would indicate that the food product had not been irradiated or inappropriately treated. On the other hand, most of these tests do not allow the presumption to be made that the sample has been irradiated due to their lack of specificity for a radiation-induced change. However, the advantages of most of these methods are that they are relatively simple, inexpensive and uncomplicated requiring standard laboratory resources. The tests can be used for screening large numbers of samples as part of routine process control in food laboratories. This screening would reveal food samples that do not comply with labelling requirements; *i.e.* unlabelled food products that appear to be irradiated or conversely, labelled irradiated products that appear to be non-irradiated. Further testing with more definitive physical or chemical methods, as described in these proceedings, would confirm that the food sample had been irradiated. Indeed confirmation by independent tests would be important for forensic purposes.

References

1. H. Delincée, "Analytical Detection Methods for Irradiated Foods. A Review of the Current Literature," IAEA-TECDOC-587, IAEA, Vienna, 1991.
2. H. Delincée, *Radiat. Phys. Chem.*, 1993, **42**, 351.
3. P. A. Gibbs and V. M. Wilkinson, "Feasibility of Detecting Irradiated Foods by Reference to the Endogenous Microflora: A Literature Review," Scientific and Technical Surveys No. 149, Leatherhead Food R. A., 1985.
4. A-M. Sjoberg, M. Manninen, P. Harmala and S. Pinnioja, *Z. Lebensm. Unters. Forsch.*, 1990, **190**, 99.
5. M. Manninen and A-M. Sjoberg, *Z. Lebensm. Unters. Forsch.*, 1991, **192**, 226.
6. G. Wirtanen, A-M. Sjoberg, F. Boisen and T. Alanko, *AOAC Int.*, 1993, **76**, 674.
7. K. M. Hammerton and C. Banos, ADMIT see these Proceedings.
8. T. Leth, ADMIT see these Proceedings.
9. R. P. Betts, L. Farr, P. Bankes and M. F. Stringer, *J. Appl. Bacteriol.*, 1988, **64**, 329.
10. M-P. Copin and C. M. Bourgeois, *Sci. Aliments*, 1992, **12**, 533.
11. M-P. Copin, D. Jehanno and C. M. Bourgeois, *J. Appl. Bacteriol.*, 1993, **75**, 254.

12. G. Wirtanen, S. Salo, M. Karwoski and A-M. Sjoberg, ADMIT see these Proceedings.
13. S. L. Scotter, R. Wood and D. J. McWeeny, *Radiat. Phys. Chem.*, 1990, **36**, 629.
14. S. L. Scotter, K. Beardwood and R. Wood, *Food Sci. Technol. Today*, 1994, **8**, 106.
15. S. K. Tamminga, R. R. Beumer, J. G. van Kooij and E. H. Kampelmacher, *Eur. J. Appl. Microbiol.* 1975, **1**, 79.
16. S. K. Tamminga, R. R. Beumer and E. H. Kampelmacher, "Food Preservation by Irradiation," Vol. II, Proceedings of a Symposium, Wageningen, IAEA, Vienna, 1978, p. 171.
17. R. E. O'Connor and G. E. Mitchell, *Int. J. Food Microbiol.*, 1991, **12**, 247.
18. K. J. A. van Spreekens and L. Toepoel, "Food Preservation by Irradiation," Vol. II, Proceedings of a Symposium, Wageningen, IAEA, Vienna, 1978, p. 157.
19. M-P. Copin and C. M. Bourgeois, "Recent Advances on Detection of Irradiated Food," M. Leonardi, J. J. Raffi and J.-J. Belliardo, Eds., Commission of the European Communities (BCR), Brussels, Luxembourg, EUR 14315 EN, 1992, p. 78.
20. M. D. Alur, V. Venugopal, D. P. Nerkar and P. M. Nair, *J. Food Sci.*, 1991, **56**, 332.
21. M. D. Alur, S. P. Chawla and P. M. Nair, *J. Food Sci.*, 1992, **57**, 593.
22. Y. Kawamura, N. Suzuki, S. Uchiyama and Y. Saito, *Radiat. Phys. Chem.*, 1992, **39**, 203.
23. A. Leffke, N. Helle, B. Linke, K. W. Bögl and G. A. Schreiber, "Recent Advances on Detection of Irradiated Food," M. Leonardi, J. J. Raffi and J.-J. Belliardo, Eds., Commission of the European Communities (BCR), EUR 14315 EN, 1992, p. 110.
24. L. Qiongying, K. Yanhua and Z. Yuemei, *Radiat. Phys. Chem.*, 1993, **42**, 387.
25. S. Zhu, T. Kume and I. Ishigaki, *Radiat. Phys. Chem.*, 1993, **42**, 421.
26. Y. Kawamura, S. Uchiyama and Y. Saito, *J. Food Sci.*, 1989, **54**, 379.
27. Y. Kawamura, S. Uchiyama and Y. Saito, *J. Food Sci.*, 1989, **54**, 1501.
28. Y. Kawamura, M. Murayama, Y. Saito and S. Uchiyama, ADMIT see these Proceedings.
29. H. Sparenberg, "The Identification of Irradiated Foodstuffs," Proc. Int. Colloq., Commission of the European Communities, Luxembourg, 1974, p. 325.
30. R. Jona and A. Fronda, *Radiat. Phys. Chem.*, 1990, **35**, 317.
31. H. J. Zehnder, *Alimenta*, 1984, **23**, 114.
32. H. J. Zehnder, *Mitt. Gebiete Lebensm. Hyg.* 1988, **79**, 362.
33. E. Kovacs, Z. Voros and J. Farkas, *Acta Aliment*, 1981, **10**, 379.
34. R. Rahman, R. Rigney and E. Busch-Petersen, *J. Econ. Entomol.*, 1990, **83**, 1449.
35. R. Rahman, A. D. Bhuiya, S. M. S. Huda, R. M. Shahjahan, G. Nahar and M. A. Wadud, "Use of Irradiation as a Quarantine Treatment of Food and Agricultural Commodities," Proc. Final Res. Co-ord. Meet., Malaysia, IAEA, Vienna, 1992, p. 133.
36. A. J. Jessup, C. J. Rigney, A. Millar, R. F. Sloggett and N. M. Quinn, "Use of Irradiation as a Quarantine Treatment of Food and Agricultural Commodities," Proc. Final Res. Co-ord. Meet., Malaysia, IAEA, Vienna, 1992, p. 13.
37. S. Uchiyama, S. Konno, T. Toyooka, Y. Kawamura and Y. Saito, *J. Food Hyg Soc. Japan*, 1989, **30**, 152.
38. T. Kume and T. Matsuda, ADMIT see these Proceedings.

DEVELOPMENT OF HALF-EMBRYO TEST AND GERMINATION TEST FOR DETECTION OF IRRADIATED FRUITS AND GRAINS

Y. Kawamura, M. Murayama, S. Uchiyama* and Y. Saito

National Institute of Health Sciences, Setagaya, Tokyo 158, Japan
(*Present address: Food and Drug Safety Center, Hatano Research Institute, Hatano, Kanagawa 257, Japan)

1 INTRODUCTION

Irradiation treatment of fruit and grains is allowed in many countries for purposes such as insect disinfestation and the extension of shelf-life; the irradiation doses used being mainly below 1 kGy. The detection of such low doses is difficult for most techniques. On the other hand, biological systems are sensitive to low doses of γ-irradiation. Therefore, a detection method based on biological changes would be applicable. It is well known that germination, or sprouting, is inhibited by γ-irradiation and it has been observed that the seed germination of grapefruit is affected by irradiation and shoot growth is totally inhibited by a 0.3 kGy dose.[1,2] However, the immediate use of this inhibition for the detection of irradiated grapefruit was prevented by variability caused by such factors as variety, harvest date and storage conditions. The time required to conduct the detection was also too lengthy. Accordingly, the germination test was improved and the half-embryo test was established for the detection of irradiated grapefruit and was applied to other citrus fruits, e.g. apples and cherries.[3-5] Moreover, the test was also applied as a screening method for irradiated rice and wheat.[6,7]

In this report, the development of a standardized half-embryo test and germination test for the detection of irradiated fruit and grains is described.

2 MATERIALS AND METHODS

2.1 Samples and Irradiation

Citrus fruit and imported wheat were obtained from importers in Tokyo while apples, cherries, rice and domestic wheat were obtained on the market. Citrus fruit was irradiated with a dose of 1 kGy, apples, cherries and rice with 2 kGy and wheat with 10 kGy in a 130 TBq ^{60}Co γ-irradiator.

2.2 Half-Embryo Test

Seeds were removed from the fruit and washed with distilled water. The outer and inner seed coats were removed with forceps to reveal the embryos. Only the outer seed coats of cherries were cracked using e.g. pliers. In the case of citrus fruits, since in some cases their seeds contained several embryos, these were separated and the largest embryo was used. From the embryo, one cotyledon was cut off and the remaining seed parts, one cotyledon and embryo axis, were defined as the "half-embryo". Any half-embryo with an extremely small cotyledon or damaged axis was discarded.

At least 10 half-embryos were placed on distilled water moistened filter paper in a covered Petri dish and incubated at 35°C in the case of citrus fruits and at 30°C in the case of apples. The half-embryos of cherries were incubated at 25°C on filter paper moistened with 10 µM benzyladenine solution instead of water. Shoot growth was then observed. Shooting was defined as the elongation of the shoot to the extent of at least a 0.5 mm length in citrus fruits and 1 mm length in apples and cherries.

2.3 Germination Test

The germination test was carried out with at least 10 seeds each of rice and wheat. These were placed on distilled water moistened filter paper in a covered Petri dish and incubated at 30°C. Shoot and root growth were then observed.

3 RESULTS AND DISCUSSION

3.1 Development of Half-Embryo Test for Grapefruit

The germination rate was enhanced by dissection of the seed. Root elongation started at 23°C after 14 days for the whole seed, 5 days for a seed without outer seed coat, 4 days for the naked embryo and 2 days for the half-embryo of grapefruit (Figure 1). Shoot elongation started at 23°C after 20, 14, 4 and 4 days, respectively, and accelerated in the same order. The growth of half-embryos was the quickest while the variation of shooting became smaller.

There was little difference in the germination and growth parameters for the half-embryo when the 3 lots of grapefruit were analysed, although the lots differed in variety, growing province, harvest time and ripeness (Figure 2). The effect of incubation temperature was examined and the optimum temperature of 35°C for growth was determined.

The effect of irradiation dose on the growth of the half-embryo was studied (Figure 3). Irradiation caused obvious effects on root and shoot growth of the half-embryos. The germination of half-embryos irradiated with 0.05 kGy was slower than that for non-irradiated samples, but shooting percentage did reach 100% after 3 days similar to the control. A dose of 0.15 kGy or more, on the other hand, almost totally retarded elongation of both the root and shoot and the shooting percentage was zero. Storage periods of up to 8 weeks had little effect. There could not be other factors that totally inhibit shooting of seeds in fruit, since the seed is protected by the fruit and the fruit is stored at good conditions.

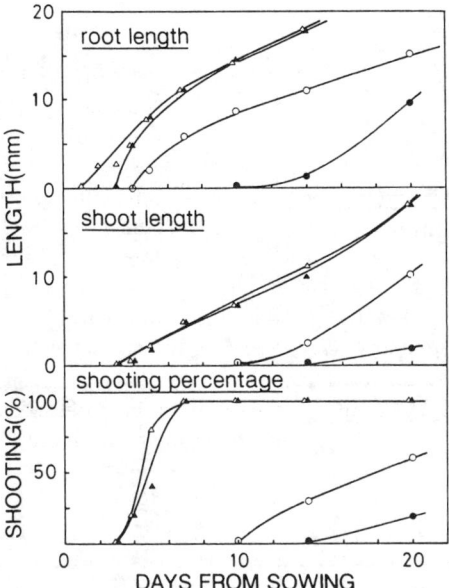

Figure 1 *Comparison of growth with intact seed (●), seed without outer seed coat (○), embryo (▲) and half-embryo (△) of grapefruit* [3]

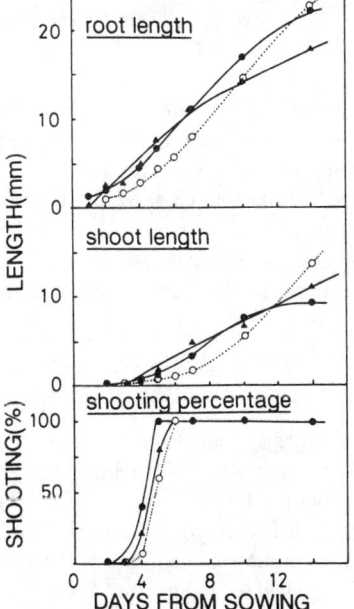

Figure 2 *Comparison of growth curve with 3 lots of non-irradiated grapefruit[3]*
● : *Marsh White from California, ripe*
○ : *Marsh White from Florida, unripe*
▲ : *Ruby Red from California, ripe*

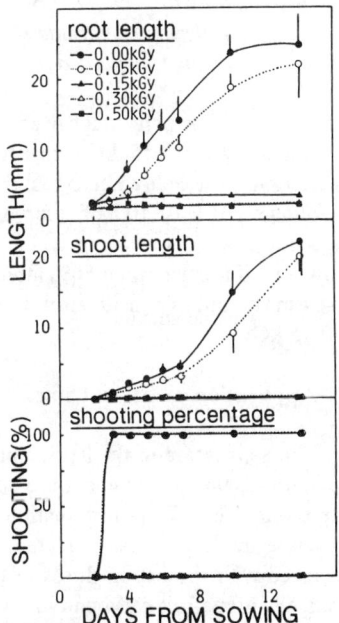

Figure 3 *Effect of γ-irradiation on growth curve of half-embryos extracted from grapefruit var. Marsh White from Florida incubated at 35°C*[4]

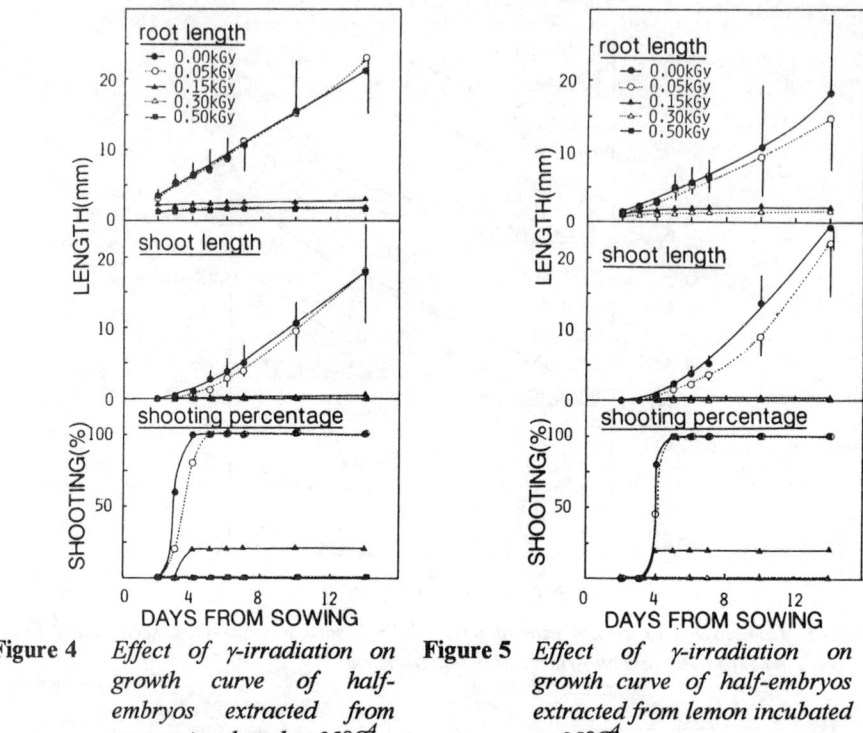

Figure 4 *Effect of γ-irradiation on growth curve of half-embryos extracted from orange incubated at 35°C*[4]

Figure 5 *Effect of γ-irradiation on growth curve of half-embryos extracted from lemon incubated at 35°C*[4]

The shooting percentage was the best parameter to discriminate between grapefruit irradiated with a 0.15 kGy dose or more and non-irradiated grapefruit 8 weeks after irradiation. The half-embryo test for the detection of irradiated grapefruit was developed as follows. At least 10 half embryos are prepared and incubated at 35°C. If the shooting percentage is 50% or more within 4 days, the grapefruit is identified as non-irradiated. However, if it is less than 50% after 4 days, the grapefruit is identified as irradiated. This assessment could be made after 3 to 4 days and the detection limit of the irradiation dose is 0.15 kGy.

3.2 Application to Other Citrus Fruits

The application of the half-embryo test to oranges and lemons was also studied. The optimum incubation temperature of their half-embryos was 35°C which was similar to grapefruit. The effect of irradiation is shown in Figures 4 and 5, respectively. The results are comparable to those obtained with grapefruit half-embryos. Limited shoot growth occurred with irradiation doses of 0.15 kGy and shooting percentages were under 20%. Thus, the grapefruit half-embryo test can also be applied to oranges and lemons.

3.3 Application to Apples

The application of the half-embryo test to apples was also studied. The optimum incubation temperature was 30°C. Apples were then irradiated at 0.05, 0.15, 0.3, 0.5, 1 and 2 kGy and their half-embryos were incubated. The results were similar to those of citrus fruits (Figure 6). The shoot of non-irradiated half-embryos grew well. Shooting reached almost 100% after 2 to 3 days and it was faster than that of citrus fruit. Those samples irradiated with 0.05 kGy were somewhat reduced and those irradiated over 0.15 kGy were almost totally retarded. The shooting percentage was also a useful parameter to distinguish between irradiated and non-irradiated apples.

The half-embryo test for the detection of irradiated apples is almost the same as that for citrus fruits except that the incubation temperature and period were as follows. At least 10 half-embryos are prepared and incubated at 30°C. If the shooting percentage is 50% or more within 3 days, the apples are identified as non-irradiated. However, if it is less than 50% after 3 days, the apples are identified as irradiated. This assessment could be made after 1 to 3 days and the detection limit of the irradiation dose is 0.15 kGy.

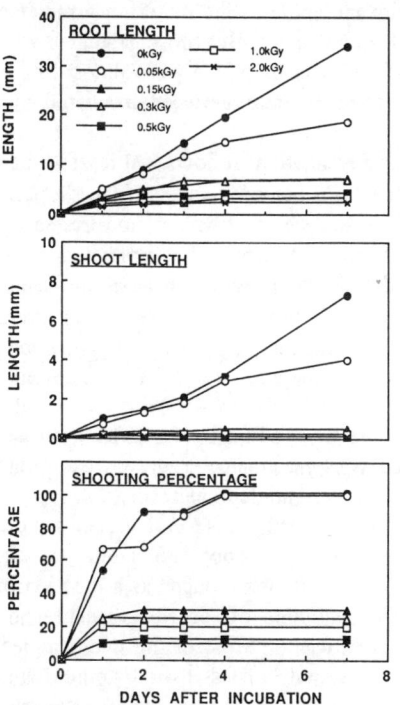
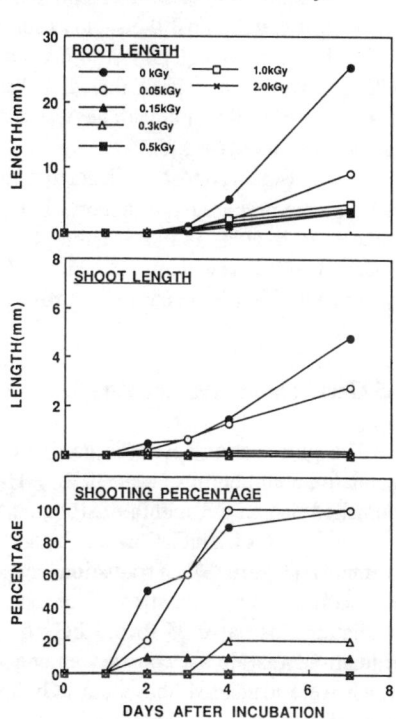

Figure 6 Effect of γ-irradiation on growth curve of half-embryos extracted from apples incubated at 30°C[5]

Figure 7 Effect of γ-irradiation on growth curve of half-embryos extracted from cherries incubated at 25°C in 10 μM benzyladenine solution[5]

3.4 Application to Cherries

The application of the half-embryo test to cherries was studied. The optimum incubation temperature was 25°C. However, it took 7 days or more for the shooting percentage of non-irradiated half-embryos to be over 50%. This is too long for a detection method, so plant hormones for the stimulation of shooting were tested.

Four kinds of auxins, 3-indoleacetic acid, 3-indolebutylic acid, 2,4-dichlophenoxyacetic acid and α-naphthylacetic acid, and three kinds of cytokinins, kinetin, gibberellin and benzyladenine were used in the incubation solutions. The concentration used was 5 μM. Only the benzyladenine stimulated shoot growth and the shooting percentage reached 60% after 4 days. The concentration of benzyladenine was then changed to 0, 0.1, 0.5, 1, 5 and 10 μM. The most stimulative concentration was found to be 10 μM. At this concentration, shooting started after 3 days and reached 90% after 4 days. Thus, it was decided to use 10 μM benzyladenine for the incubation solution of cherries.

The cherries were then irradiated at dose levels of 0.05, 0.15, 0.3, 0.5, 1 and 2 kGy and their half-embryos were incubated at 25°C with 10 μM benzyladenine (Figure 7). The results were almost the same as for citrus fruits and apples. Shoots of non-irradiated half-embryos grew well and shooting reached 90% after 4 days. But those irradiated with 0.05 kGy were somewhat reduced, and those irradiated over 0.15 kGy were almost totally retarded. Shooting percentage was also useful to distinguish between irradiated and non-irradiated cherries.

The half-embryo test for irradiated cherries was developed as follows. At least 10 half embryos are prepared and incubated at 25°C with 10 μM benzyladenine. If the shooting percentage is 50% or more within 4 days, the cherries are identified as non-irradiated. However, if it is less than 50% after 4 days, the cherries are identified as irradiated. This assessment could be made after 3 to 4 days and the detection limit of irradiation dose is 0.15 kGy.

3.5 Germination Test for Rice

The germination test was used for the detection of irradiated rice. The optimum incubation temperature was 30°C. Husked rice was used since it grew faster than unhusked rice after 1 month of storage and is the most common commercially.

The effect of irradiation on the germination of rice seeds of several varieties were examined (Figure 8). Irradiation caused obvious effects on root and shoot growth, especially on root elongation. The root length of rice seeds was reduced as a function of irradiation dose at 0.15 to 0.3 kGy or more. The reduction of root length caused by irradiation was almost constant among the various varieties. Most of the rice samples, which were irradiated above 0.5 kGy, exhibited root growth of less than 10 mm after 3 days of incubation. The shooting and rooting percentages of the irradiated rice were 80-100% and there were no differences between non-irradiated and irradiated rice at the irradiation doses studied. Storage periods of up to 12 months had little effect.

Therefore, the germination test for the detection of irradiated rice was developed as follows. At least 10 seeds are prepared and incubated at 30°C. If the root length is 10 mm or longer within 3 days, the rice is identified as non-irradiated. However, if it is shorter than 10 mm after 3 days, the rice is identified as irradiated. This assessment could

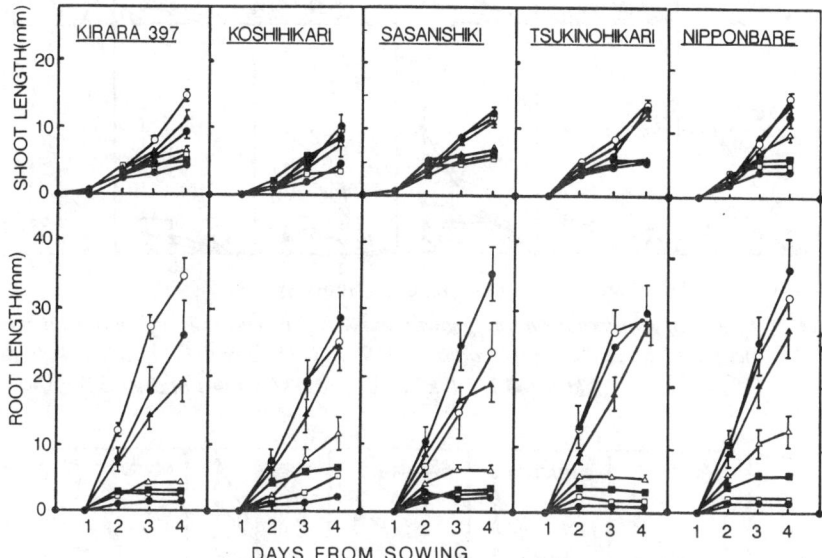

Figure 8 *Effect of γ-irradiation on growth curve of 5 rice varieties without husks incubated at 30°C : non-irradiated (●), 0.05 kGy (○), 0.15 kGy (▲), 0.3kGy (Δ), 0.5 kGy (■), 1.0 kGy (□) and 2.0 kGy (●)* [6]

be made after 2 to 3 days and the detection limit of the irradiation dose is 0.15 kGy. This test can thus discriminate between irradiated and non-irradiated rice stored for 12 months or more after irradiation.

3.6 Germination Test for Wheats

The germination test was applied to the detection of irradiated wheat. The optimum incubation temperature was also 30°C. The effect of irradiation on the germination of wheat seeds from several varieties were examined (Figures 9 and 10). Irradiation caused an obvious depression in both shoot and root growth and the effects were dependent upon the irradiation dose. The critical dose that inhibited root elongation varied from 0.15 to 0.5 kGy for imported varieties and from 0.3 to 1.0 kGy for domestic varieties. Imported wheat was more sensitive to γ-irradiation than domestic wheat. The germination percentage was reduced above 5.0 kGy. Storage periods of up to 12 months had little effect on the reduction in root length.

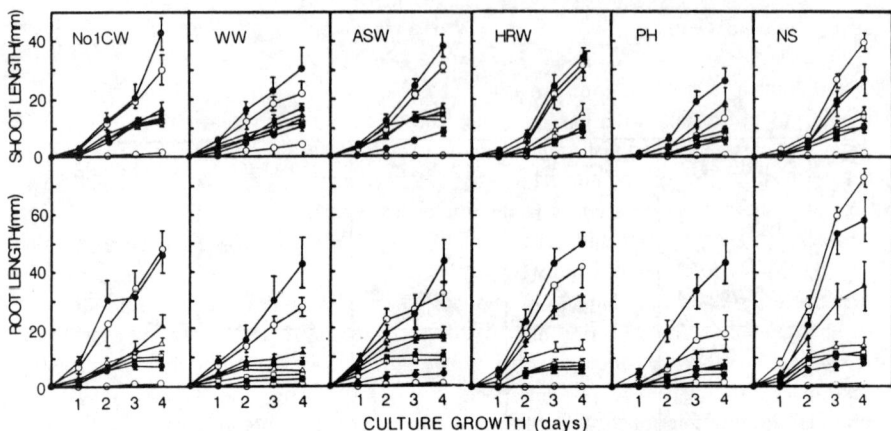

Figure 9 *Effect of γ-irradiation on growth curve of 6 varieties of imported wheat incubated at 30°C: non-irradiated (●), 0.15 kGy (○), 0.3 kGy (▲), 0.5 kGy (△), 1.0 kGy (■), 2.0 kGy (□), 5.0 kGy (●) and 10 kGy (○)*[7]

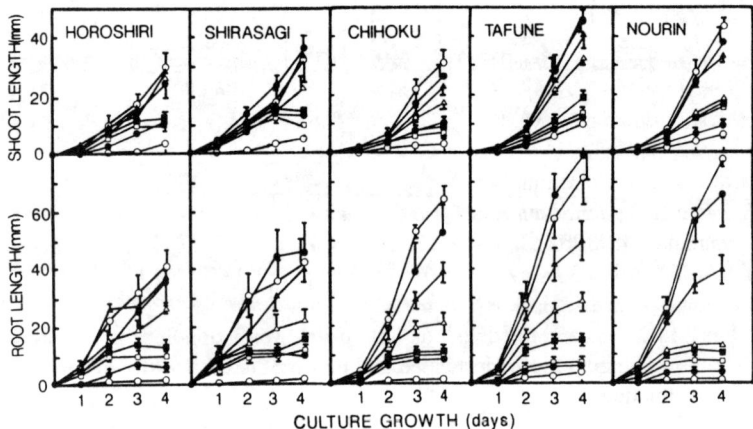

Figure 10 *Effect of γ-irradiation on growth curve of 6 varieties of domestic wheat incubated at 30°C: non-irradiated (●), 0.15 kGy (○), 0.3 kGy (▲), 0.5 kGy (△), 1.0 kGy (■), 2.0 kGy (□), 5.0 kGy (●) and 10 kGy (○)*[7]

Therefore, the germination test for the detection of irradiated wheat was developed as follows. At least 10 seeds are prepared and incubated at 30°C. If the root length is 20 mm or longer within 4 days, the wheat is identified as non-irradiated. However, if it is shorter than 20 mm after 4 days, the wheat is identified as irradiated. The assessment could be made after 2 to 4 days and the detection limit of irradiation dose is 0.5 kGy for imported wheat and 1 kGy for domestic wheat. This test can thus discriminate between irradiated and non-irradiated rice for 12 months or more after irradiation.

4 CONCLUSIONS

The inhibition of shoot and root growth is sensitive to γ-irradiation. Shoot inhibition of half-embryos is the definitive parameter for the detection of irradiated fruit. The half-embryo test can be applied not only to citrus fruits but also to apples and cherries as the detection method for irradiation. The detection limit of the irradiation dose is 0.15 kGy or less and it coincides with citrus fruits, apples and cherries. The half-embryo test is a very sensitive, easy, cheap and reliable method. Thus, it can be used as a practical technique for the detection of irradiated fruits.

On the other hand, the damage of rice and wheat to root growth by γ-irradiation is a useful parameter for the detection of radiation treatment. However, root elongation can be retarded by other factors, *e.g.* high temperature, high moisture, long storage and disease, since they are not protected and not stored under the suitable conditions. Therefore, the germination test for rice and wheat would be better used as a screening test. It is also a sensitive, easy and cheap method and can be used as a screening technique for the detection of irradiated rice, wheat and other cereals.

References

1. S. Uchiyama, K. Nagashima, Y. Kawamura, M. Toyooka and Y. Saito, *J. Food Hyg. Soc. Japan*, 1988, **29**, 395.
2. S. Uchiyama, S. Konno, M. Toyooka, Y. Kawamura and Y. Saito, *J. Food Hyg. Soc. Japan*, 1989, **30**, 152.
3. Y. Kawamura, S. Uchiyama and Y. Saito, *J. Food Sci.*, 1989, **54**, 379.
4. Y. Kawamura, S. Uchiyama and Y. Saito, *J. Food Sci.*, 1989, **54**, 1501.
5. Y. Kawamura, A. Miura, T. Sugita, T. Yamada and Y. Saito, *Radiat. Phys. Chem.*, 1995, **46**, 371.
6. Y. Kawamura, N. Suzuki, S. Uchiyama and Y. Saito, *Radiat. Phys. Chem.*, 1992, **39**, 203.
7. Y. Kawamura, N. Suzuki, S. Uchiyama and Y. Saito, *Radiat. Phys. Chem.*, 1992, **40**, 17.

DETECTION OF IRRADIATED SPICES WITH A MICROBIOLOGICAL METHOD - DEFT/APC METHOD

K. M. Hammerton and C. Banos

Australian Nuclear Science and Technology Organisation
PMB 1
Menai NSW 2234
Australia

1 INTRODUCTION

The decontamination of spices that are to be used as ingredients in processed foods is necessary in order to prevent the introduction of spoilage microorganisms and more rarely disease causing organisms. Spices can be contaminated with bacteria and moulds in concentrations from 10^3 to 10^8 microorganisms per gram so that, even when used in small amounts, they can contaminate food with large numbers of microorganisms. The most effective means of decontaminating spices is irradiation treatment with an absorbed radiation dose from 5 to 10 kGy. Several countries are commercially using radiation processing of spices. A method for monitoring the treatment of spices whilst at the same time indicating the hygienic status of the sample would be very useful for food control authorities.

A microbiological screening method based on the use of the direct epifluorescent filter technique (DEFT) and the conventional aerobic plate count (APC) has been established for the detection of irradiated spices.[1] The DEFT count enumerates the total number of contaminating microorganisms, irrespective of viability, in an untreated or treated spice sample. The APC indicates the number of viable microorganisms capable of forming colonies on an agar plate and is expressed as colony forming units (cfu). The quotient of the two counts [log DEFT/APC = $\log_{10}$(DEFT count/g) - $\log_{10}$(cfu/g)] can be used for assessing whether the spice sample has been subjected to a decontamination treatment. It has been found that for untreated spices the log DEFT/APC was $\leq$ 4.0 and for spices treated with a radiation dose $\geq$ 5 kGy the log DEFT/APC was > 4.0. However, with this method, conclusive evidence of irradiation treatment relies on the knowledge that the sample has not been fumigated or heat treated. This paper reports recent investigations on the possibility that the inclusion of a mesophilic aerobic spore count will enable irradiated spices to be distinguished from ethylene oxide (EtO) or heat treated spices.

2 METHODS

Untreated spice samples were obtained from spice suppliers in Australia and food research laboratories in Finland, Germany and New Zealand. Where multiple samples of a particular spice were used, each sample had been obtained from a different source.

The samples were irradiated in ^{60}Co research facilities at the Australian Nuclear Science and Technology Organisation (ANSTO) with absorbed doses of either 5 or 10 kGy (±1.5%). EtO or steam treated spice samples and their untreated counterparts were obtained from commercial processors.

Prior to the APC and DEFT counts, spice samples were homogenised, centrifuged and filtered through a cellulose filter and 5μm nylon mesh. This procedure was found to yield good recovery of the microorganisms and was effective in removing particulate materials that interfere with the DEFT count. The mesophilic aerobic spore count (ASC) was determined after heating the filtered spice sample at 80°C for 1 min. The APC and ASC agar plates were incubated at 30°C for 2 to 6 days. For the DEFT count, the microorganisms were collected on a 0.4 μm polycarbonate filter, stained with acridine orange and counted with an epifluorescent microscope.

3 RESULTS AND DISCUSSION

3.1 DEFT/APC Method

The DEFT/APC method was carried out with 60 samples from 30 different spices. The log DEFT/APC [log DEFT/APC = $\log_{10}$(DEFT count/g) - $\log_{10}$(cfu/g)] was determined for each sample, untreated or irradiated with either 5 or 10 kGy (Figure 1).

The log DEFT/APC was ≤ 4.0 for the untreated spice samples except for one sample of cloves (Figure 1). Thermoluminescence analysis confirmed that this sample had not been irradiated. Indeed many of the spices with known anti-microbial activity such as cloves, cinnamon, garlic and mustard had a relatively higher log DEFT/APC, probably due to the inhibitory effect of the anti-microbial constituent on the APC.

For samples irradiated with a dose 2 5 kGy, the log DEFT/APC was > 4.0 except for 3 samples of basil (Figure 1). The 4 samples of basil, obtained from different sources and irradiated with 5 kGy had a relatively low log DEFT/APC (3.5, 3.6, 3.8, 4.5). However, these samples could be easily distinguished from their untreated counterparts (log DEFT/APC = 1.4, 1.6, 1.6, 1.7). Comparison of the results obtained for basil with those obtained for the other spices suggest that the microflora associated with basil may be more resistant to the effects of irradiation.

The log DEFT/APC for untreated and irradiated spice samples that had been stored for up to two years remained unaltered. Even though the DEFT/APC method cannot always be used for specifically identifying irradiation treatment, the method would be very useful for screening spices in food control laboratories and for indicating the hygienic status of spice samples.

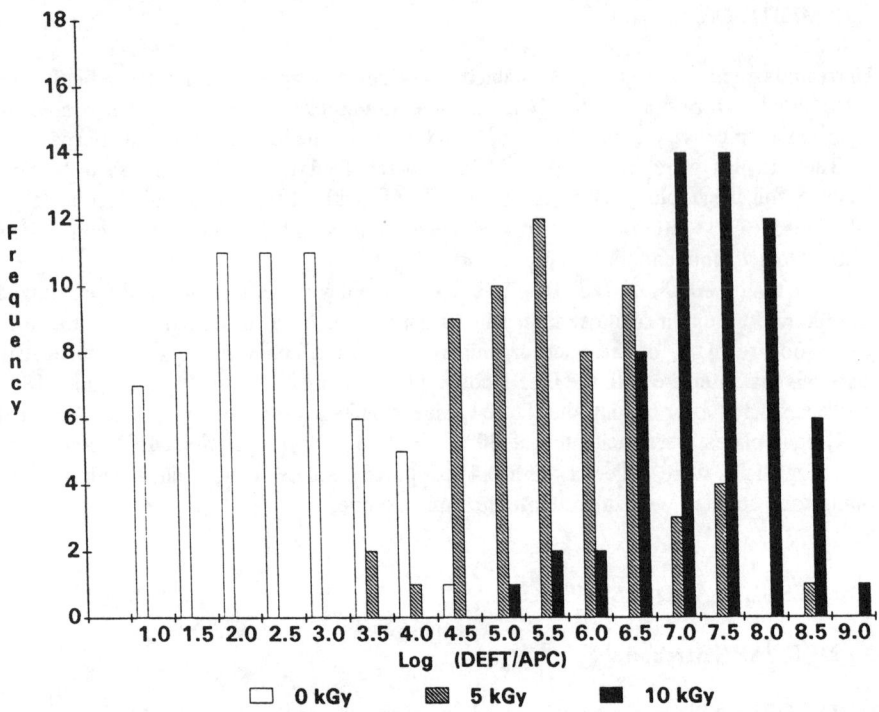

Figure 1 *Histogram of DEFT/APC ratios (log DEFT/APC) for 60 samples of 30 different spices, untreated or irradiated with 5 kGy or 10 kGy*

3.2 DEFT/APC/ASC Method

During the course of this work the effects of irradiation, EtO and steam treatment on reducing the microbial flora of spices using the DEFT/APC method have been compared. Since aerobic bacterial spores frequently form more than 50% of the mesophilic total viable cell count of spices, a mesophilic ASC was included in the assay system. The ASC was determined by heating the filtered spice sample at 80°C for 1 min prior to plating onto agar and was expressed as a percentage of the aerobic plate count (% APC). The results obtained for the APC, log DEFT/APC, and ASC for paprika and black pepper, untreated and subjected to treatment with irradiation, steam or EtO, are summarised in Figure 2. Several observations, pertinent to the assessment of the microbiological quality, the effectiveness of decontamination treatments and the identification of irradiation treatment, were made in view of the results obtained:

(1) the high levels of viable microorganisms (log APC, Figure 2) found in untreated samples of paprika and black pepper, from 10^5 to 10^8 cfu/g, would necessitate a decontamination treatment;
(2) irradiation treatment with absorbed doses greater than 5 kGy would be effective in reducing the microbial contamination below the acceptable level of 10^4 cfu/g;

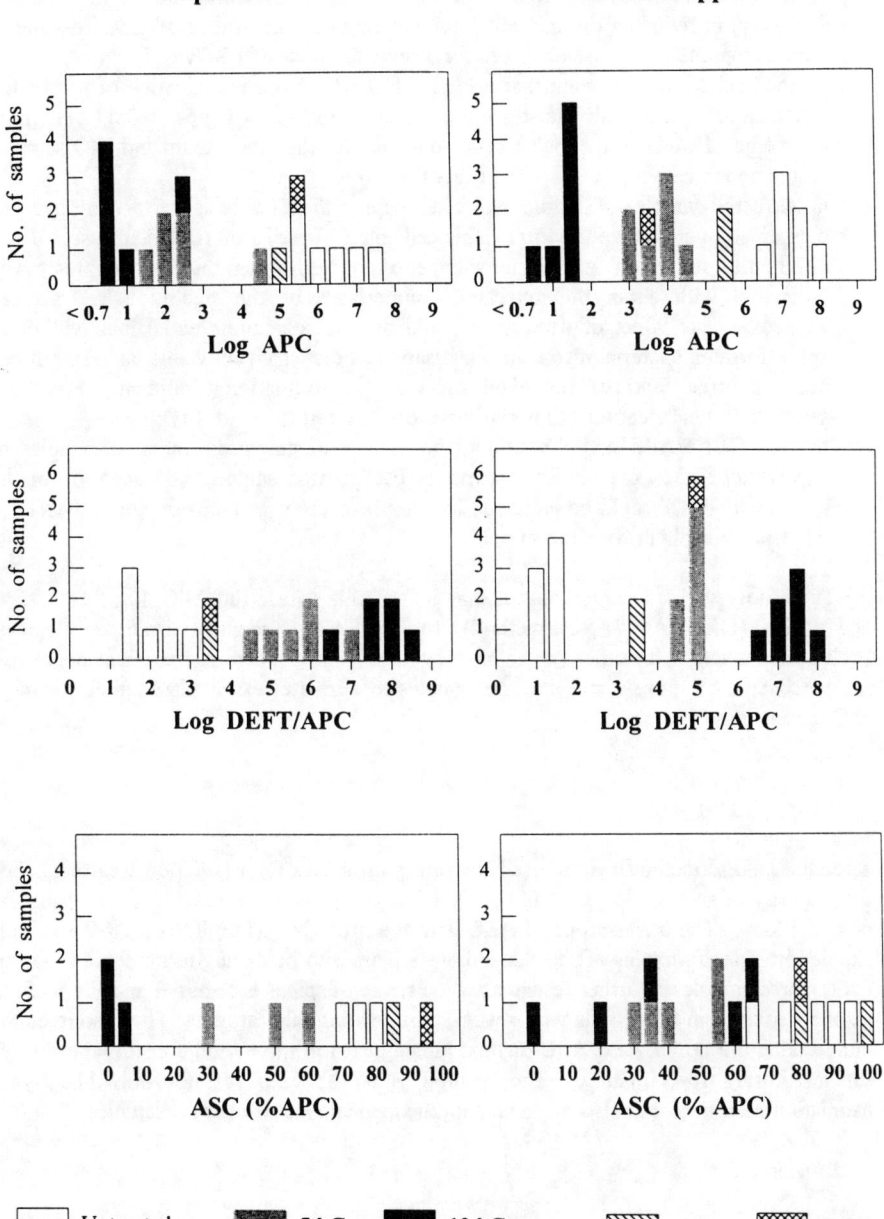

Figure 2 *Histogram of the effects of irradiation, steam or ethylene oxide treatment on the microbial flora of 13 samples of paprika and black pepper*

(3) steam or EtO treatment would not be as effective as high doses of radiation (> 5 kGy) in reducing the microbial contamination. The effects of these treatments were comparable to those observed for a radiation dose of 5 kGy;
(4) on the basis of the criterion that a log DEFT/APC > 4.0 is indicative of irradiation treatment of spices, only the steam treated sample of black pepper would fall within this range. However, the higher ratios obtained for the other steam and EtO samples would be indicative of a treatment other than irradiation;
(5) in untreated samples of paprika and black pepper the aerobic spore forming bacteria comprised over 60% of the total viable cell count. Irradiation treatment resulted in a differential reduction in the percentage of spore-formers in the samples when compared with either the untreated counterparts or the steam or EtO treated samples. This effect of irradiation could be due to a preferential inactivation of spore-forming bacteria or to heat sensitisation of the surviving viable bacteria. It has been reported that the microbial survivors of irradiation treatment were more sensitive to heat treatment than the survivors of treatment with EtO;[2]
(6) the log DEFT/APC and ASC could be used to distinguish irradiated samples of paprika and black pepper from steam or EtO treated samples. Fumigation of the spices with EtO would be indicated by the presence of residues or derivatives (*e.g.* ethylene chlorohydrin) in the sample.

These observations suggest that it may be possible to use the APC, log DEFT/APC and ASC (DEFT/APC/ASC method) to specifically identify irradiated spices. Preliminary results with other spices have indicated that irradiation treatment decreased the percentage of spore-formers in the samples to a greater extent than either steam or EtO in most cases.

4 CONCLUSIONS

The DEFT/APC method can be used for screening spices for irradiation treatment. For the majority of spices a log DEFT/APC > 4.0 is indicative of treatment with a radiation dose 2.5 kGy. The inclusion of an ASC with the procedure (DEFT/APC/ASC method) enabled irradiated samples of paprika and black pepper to be distinguished from steam or EtO treated samples. Further research with a range of spices is required and the method should be tested in blind trials with several participating laboratories. The resources for undertaking the DEFT/APC/ASC method already exist in most food control laboratories for monitoring food quality. The method could be used for not only identifying irradiation treatment, but also for indicating the hygienic status of spice samples.

References

1. G. Wirtanen, A-M. Sjoberg, F. Boisen and T. Alanko, *AOAC Int.*, 1993, **76**, 674.
2. J. Farkas and E. Andrassy, "Proceedings of IUMS-ICFMH 12th International Symposium: Microbial Associations and Interactions in Food," I. Kiss, T. Deak and K. Incze, Eds., Akademiai Kiado, Budapest, 1984, p. 393.

Poster Presentations

CONTROL OF FOOD IRRADIATION IN DENMARK

T. Leth

Institute of Food Chemistry and Nutrition
The National Food Agency of Denmark
DK-2860 Søbora
Denmark

1 INTRODUCTION

In Denmark food irradiation is not allowed. However, with the considerations within EEC about directives regulating food irradiation and the development of methods for detection, it has been found necessary to ascertain that illegally irradiated foods are not found on the Danish market.

Irradiation of spices is allowed in many countries even without being declared in the foods. It thus seemed logical to begin the control of food irradiation by screening a number of spices. This resulted in 1992 in the assessment of 105 samples and in 1993 of 48 samples.

2 MATERIALS AND METHODS

In 1992, 100 g samples of spices were taken from big and small importers throughout Denmark. The most used spices were sampled with the highest frequency. In 1993 the local food control units were invited to send samples under suspicion of irradiation treatment to the Odense Regional Laboratory.

The methods used were DEFT/APC (direct epifluorescent filter technique/aerobic plate count)[1,2] which measures the difference between the total number of microbes and viable microbes. A log difference greater than 4 is taken as an indication of irradiation, heat treatment, gassing or other operations capable of killing bacteria.

If a log difference > 4 was found, the sample was sent to the Risø Research Station for thermoluminescence (TL) analysis, where the glow curve was measured before and after re-irradiation with a dose of 1 kGy. Coincidental curves are taken as proof of irradiation.[3]

3 RESULTS

Of the 153 samples examined 34 samples had a higher log difference than 3, and 11 samples had a higher difference than 4, which is indicative of irradiation (Figure 1).

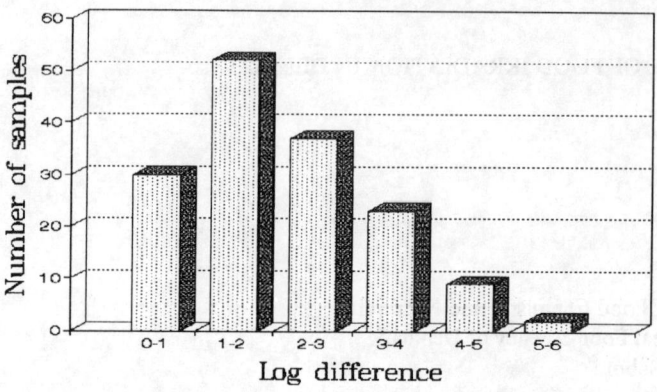

Figure 1 *Log difference DEFT/APC*

One sample of pepper with a log difference > 4 was shown to have been irradiated; this was clearly seen from the glow curves when compared with a sample of a non-irradiated spice. However, the pepper sample had been legally irradiated for export use.

For vanilla, ginger, cinnamon and spice mixtures most of the samples had a log difference of > 3, perhaps indicating a rather harsh treatment during processing. Curry and pepper, the most used spices, almost always had a log difference < 3 in contrast to spice mixtures. All samples with log difference > 3 were checked by the TL method.

5 CONCLUSION

During the period 1992-93, 153 samples of spices, with the greatest importance being attached to the most used spices, taken country-wide from big and small importers have been controlled for food irradiation.

The spices were screened with the DEFT/APC method and for samples with a log difference > 3, the use of irradiation was confirmed or disproved with the TL method. Eleven samples had a log difference > 4 and 34 out of the 153 samples had a log difference > 3. No illegally irradiated spices were found in this investigation. However, one legally irradiated sample of pepper for export use was detected.

References

1. F. Boisen, "Detection of irradiated spices using a combined DEFT/APC method."
2. G. Wirtanen and A.-M. Sjoberg, "Recent Advances on Detection of Irradiated Food," Commission of the European Communities (BCR), Brussels, Luxembourg, 1992, EUR 14315 EN.
3. T. Autio and S. Pinnioja, "Recent Advances on Detection of Irradiated Food," Commission of the European Communities (BCR), Brussels, Luxembourg, 1992, EUR 14315 EN.

INFLUENCING FACTORS ON ESR DOSE ASSESSMENT IN IRRADIATED CHICKEN LEGS

S. Baccaro,* P. Fuochi,+ S. Onori° and M. Pantaloni°

* ENEA-INN/TEC - CRE Casaccia
Via Anguillarese 301, 00060 S.Maria di Galeria (Rome), Italy.
+ Istituto FRAE, Area delta ricerca - CNR
Via P. Gobetti 101, 40129 Bologna, Italy.
° Istituto Superiore di Sanita, Physics Laboratory
Viale Regina Elena 299, 00161 Rome, Italy.

Electron spin resonance (ESR) dosimetry of irradiated chicken legs is based on the additive dose or the calibration curve methods. In both cases the practical assumption is made that the behaviour of the chicken bone does not depend on factors such as temperature during irradiation, storage conditions and dose rate. So the aim of the present work was to investigate to what extent the above mentioned factors could influence the post-irradiation dose assessment using the ESR technique.

The results show that dose rate and temperature during irradiation have a significant effect on the yield of the radiation induced radicals and should, therefore, be taken into account for dose assessment (Figure 1a). A 20% difference in signal amplitude was found between gamma and electron irradiation and a 40% difference between +20°C and -20°C gamma irradiation. Dose rate had a slight, but significant, influence on the stability of the radiation-induced radicals, while a more marked effect was found with temperature during irradiation (Figure 1b).

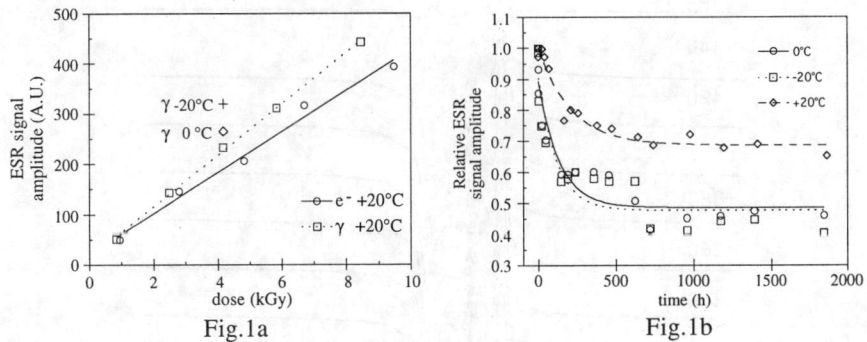

Figure 1 *Influence of dose-rate (0.6 Gy s^{-1} for ^{60}Co source and 6 x 10^6 Gy s^{-1} for electron-beam) and temperature on ESR signal amplitude of irradiated chicken legs (Figure 1a). Influence of irradiation temperature on room temperature fading (Figure 1b).*

THE NATURE AND ORIGIN OF THE EPR SPECTRA FROM IRRADIATED BONE AND HYDROXYAPATITE.

S. M. Glidewell and B. A. Goodman

Scottish Crop Research Institute
Invergowrie
Dundee DD2 5DA
U.K.

Although the principal constituents of bone are hydroxyapatite and collagen, the characteristic EPR spectrum of irradiated bone, which is used as an indicator of exposure to ionising radiation, corresponds to the CO_2^- radical. When trapped in material such as bone, this is an extremely stable radical and permits the identification of irradiated bone even in cooked samples. Hydroxyapatite has the idealised formula $Ca_5(OH)(PO_4)_3$, but isomorphous replacement of some PO_4^{3-} by CO_3^{2-} ions is commonly observed, and is particularly associated with the mineral surfaces. This CO_3^{2-} is generally regarded as the origin of the CO_2^- radical in irradiated bone.

EPR spectra of hydroxyapatite irradiated at ambient temperatures are dominated by the CO_3^- radical which shows a dramatic decrease in intensity on annealing to ~460 K. In contrast, the CO_2^- radical, the intensity of which represents only a minor component in irradiated hydroxyapatite at 300 K, remains virtually unaltered on annealing to 460 K over 90 min (Figure 1).

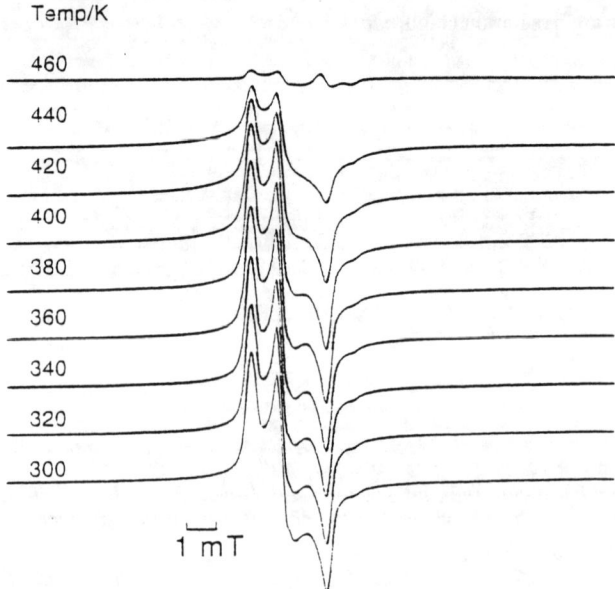

Figure 1 *EPR spectrum of hydroxyapatite at increasing temperature*

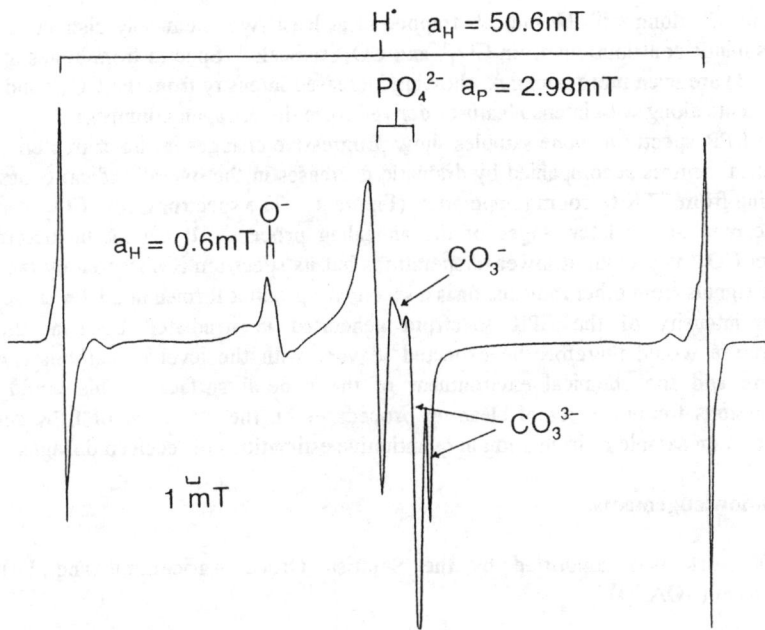

Figure 2 *EPR spectrum of hydroxyapatite at 77 K*

Measurements on samples irradiated and maintained at low temperatures show that the CO^{2-} radical[1] is only a minor product of the radiation-induced chemistry of bone. At 77 K, the hydroxyapatite spectrum (Figure 2) shows features from the unstable radicals

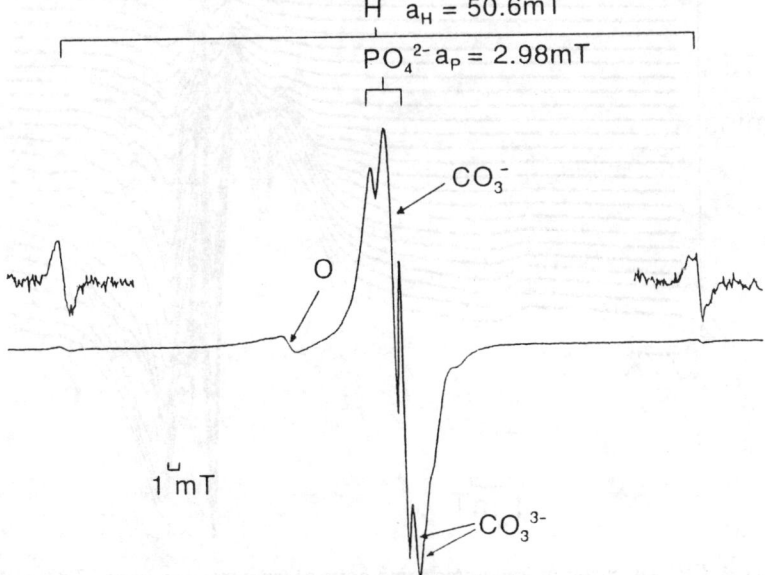

Figure 3 *EPR spectrum of lamb bone at 77 K*

PO_4^{2-} and O^- along with H· radicals trapped in at least two chemically distinct sites[2] as well as minor contributions from CO_3^{3-} and CO_3^- radicals.[3] Spectra from bones at 77 K (Figure 3) are even more complex, showing increased intensity from the CO_3^{3-} and CO_3^- components along with intense features derived from the collagen component.

The EPR spectra of bone samples show progressive changes in the characteristics of the radical centres accompanied by dramatic decreases in the overall radical content on annealing from 77 K to room temperature (Figure 4). The spectrum from CO_2^- can only be discerned at the later stages of the annealing process. It cannot be determined whether CO_2^- is present at lower temperatures but its spectrum is obscured by the more intense signals from other radicals, or is a secondary product formed at a later stage.

The intensity of the EPR spectrum generated in irradiated bone at ambient temperature would therefore be expected to vary with the level of carbonate in the structure and the chemical environment of the mineral surface. This could have repercussions for the effect of clean-up procedures on the efficiency of EPR spectral generation on sample re-irradiation in quantitative estimations of received dosages.

Acknowledgements

This work was supported by the Scottish Office Agricultural and Fisheries Department (SOAFD).

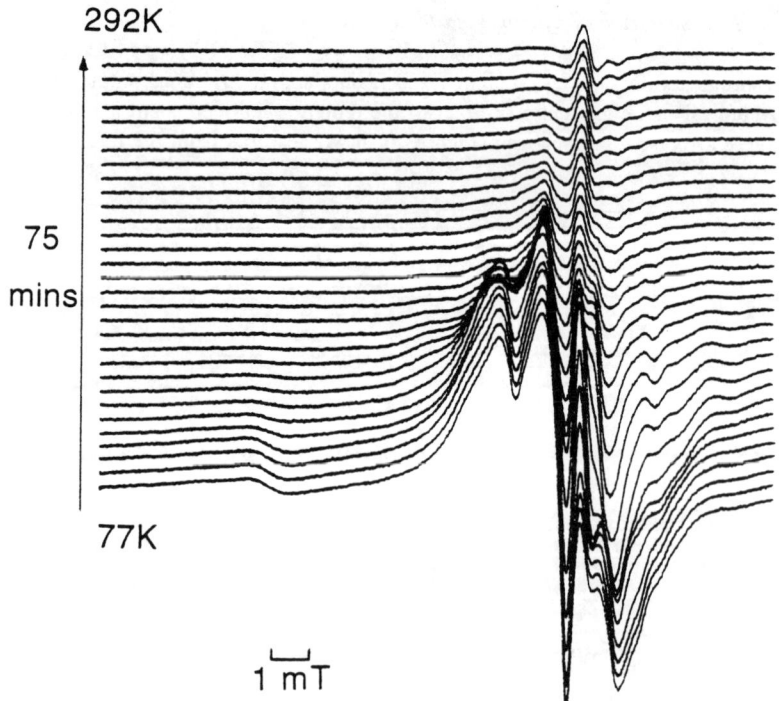

Figure 4 *EPR spectrum of irradiated lamb bone at temperatures from 77 K to 292 K*

References

1. M. Geoffroy and H. J. Tochon-Danguy, *Calcif. Tissue Int.*, 1982, **34**, S99.
2. B. V. Fisher, R. E. Morgan. G. O. Phillips and H. W. Wardle, *Radiat. Res.*, 1971, **46**, 229.
3. L. G. Gilinskaya, M. Ya. Shcherbakova and Yu. N. Zanin, *Sov. Phys. Cryst.*, 1971, **15**, 1016.

PRELIMINARY STUDIES ON THE DETECTION OF IRRADIATED PRAWNS USING 2-ALKYLCYCLOBUTANONES

B. T. McMurray,[1] W. C. McRoberts,[2] J. T. G. Hamilton[2] and M. H. Stevenson[1,2]

[1]Department of Food Science, The Queen's University of Belfast and
[2]Department of Agriculture for Northern Ireland
Newforge Lane, Belfast BT9 5PX
U.K.

1 INTRODUCTION

The use of ionising radiation for the preservation of food has been under investigation for many years but has yet to receive worldwide acceptance. Although irradiation can be carefully controlled, it is generally accepted that the development of a test or tests for the detection of irradiated food would enhance consumer confidence and might help to enforce labelling regulations.

The 2-alkylcyclobutanones are reported to be the only cyclic compounds formed as the products of the radiolysis of saturated and unsaturated triglycerides.[1] The synthesis of 2-dodecylcyclobutanone (DCB) and 2-tetradecylcyclobutanone (TCB), which are formed from palmitic and stearic acid respectively following irradiation, has been carried out and using irradiated chicken meat as the model for a high-lipid containing food, both cyclobutanones have been extracted with hexane and then identified using gas chromatography/mass spectrometry (GC/MS). DCB and TCB have been detected in chicken meat irradiated at doses well below 10 kGy and were not detectable in non-irradiated chicken. Systematic studies have been undertaken to confirm the specificity of these compounds as markers for irradiation treatment.[2,3] The presence of both DCB and TCB in other irradiated meats and irradiated liquid whole egg has also been confirmed.

Prawn meat also contains these precursor fatty acids but the total lipid content is much lower (normally < 1%). The corresponding 2-alkylcyclobutanones should, however, be formed in irradiated prawn meat. The main objective of the project was to modify or change the methodology presently used for the detection of irradiated high-lipid containing foods using 2-alkylcyclobutanones so that it can be applied to low-lipid containing foods, in particular, prawns.

2 MATERIALS AND METHODS

Samples of Norway Lobster (*Nephrops norvegicus*) were either irradiated at 5 kGy or non-irradiated and the lipid extracted using a number of solvent systems. The Soxhlet extraction, using hexane, of irradiated prawn meat was investigated; however, this system only afforded < 20 mg of lipid 100 g^{-1} of prawn meat which was much lower than the 1% fat content expected. One explanation for obtaining such a low percentage of fat from the prawn meat could be that hexane is not a polar enough solvent to extract all the lipid. Subsequently, the polarity of the extraction solvent was increased by adding diethyl ether to the hexane in a 1:1 ratio. The percentage fat obtained was still considerably lower than the value quoted in literature, although it is approximately 10 times higher than that found using hexane. Literature procedures for total lipid extraction of prawn meat use chloroform/methanol (2:1) as the solvent system which is a considerably more polar system than hexane/ether and this was also used to extract the lipid from prawn meat.

Various drying procedures were also investigated due to the highly wet nature of prawn meat and these included freeze-drying, microwaving and air-drying.

3 RESULTS

The lipid extracts were purified using deactivated Florisil[2] and both DCB and TCB were qualitatively detectable in irradiated prawn meat using GC/MS. Selected ion monitoring of the extracts from the irradiated prawn meat for ions m/z 98 and m/z 112 produced two peaks with retention times and ion ratios corresponding to that of standard DCB and TCB. The non-irradiated control sample showed no detectable peaks at these retention times under the same analytical conditions. The concentrations of DCB and TCB reflected also the ratio of their precursor fatty acids.

References

1. W. W. Nawar, *Food Rev. Int.*, 1986, **2**, 45.
2. D. R. Boyd, A. V. J. Crone, J. T. G. Hamilton, M. V. Hand, M. H. Stevenson and P. J. Stevenson, *J. Agric. Food Chem.*, 1991, **39**, 789.
3. A. V. J. Crone, J. T. G. Hamilton and M. H. Stevenson, *J. Sci. Food Agric.*, 1992, **58**, 249.

IDENTIFICATION OF IRRADIATED FRUIT FROM THE PECTIN-DERIVED EPR SIGNAL

N. Deighton, S. M. Glidewell, B. A. Goodman, G. P. McMillan and M. C. M. Perombelon

Scottish Crop Research Institute
Invergowrie
Dundee DD2 SDA
U.K.

An EPR method based upon the observation of an unidentified multiline spectrum has been reported (Raffi, 1992) for the detection of irradiated fruit (dried papaya, banana, dates). A similar long-lived EPR spectrum is observed after irradiation of dried amorphous apple pectin (BDH, 6% methyl ester) (spectrum 1) suggesting that pectin might be the origin of the signal in the dried fruit. Compositions of pectins vary greatly according to source and also with the age of the fruit itself. Therefore, a series of pectins of citrus fruit and potato tuber origin have been investigated to examine the effects of different levels of esterification on the irradiation-induced EPR signal.

EPR spectra such as that presented in Figure 1 were not obtained from purified pectins (9.8-93% esterification), instead only a singlet was observed (spectrum 2). A sample of polygalacturonic acid *(i.e.* a 0% esterified pectin) produced a spectrum similar to the low field component of that observed in apple pectin, but at a higher field (lower g-value) possibly indicating the loss of a large hydrogen splitting (spectrum 3). The spectrum from apple pectin closely resembles that from irradiated α-methyl glucoside (spectrum 4). This resemblance is also apparent from the respective ENDOR spectra (not shown).

The EPR signal from the irradiated pectin component of fruit is dependent upon the composition of the pectin and quite likely the other components with which it is associated. Because of these complications, it is considered unlikely that EPR of pectin-derived free radicals can be used to generally determine an irradiation history of dried fruits. However, as pointed out by Raffi,[1] there will be occasions where observation of a multi-line component within the EPR spectrum will provide a positive indication of an irradiation history

Acknowledgement

We thank the Scottish Office Agriculture and Fisheries Department (SOAFD) for funding this work.

Reference

1. J. Raffi, "Electron Spin Resonance Intercomparison Studies on Irradiated Foodstuffs," Commission of the European Communities (BCR), Brussels, Luxembourg, EUR 13630, 1992.

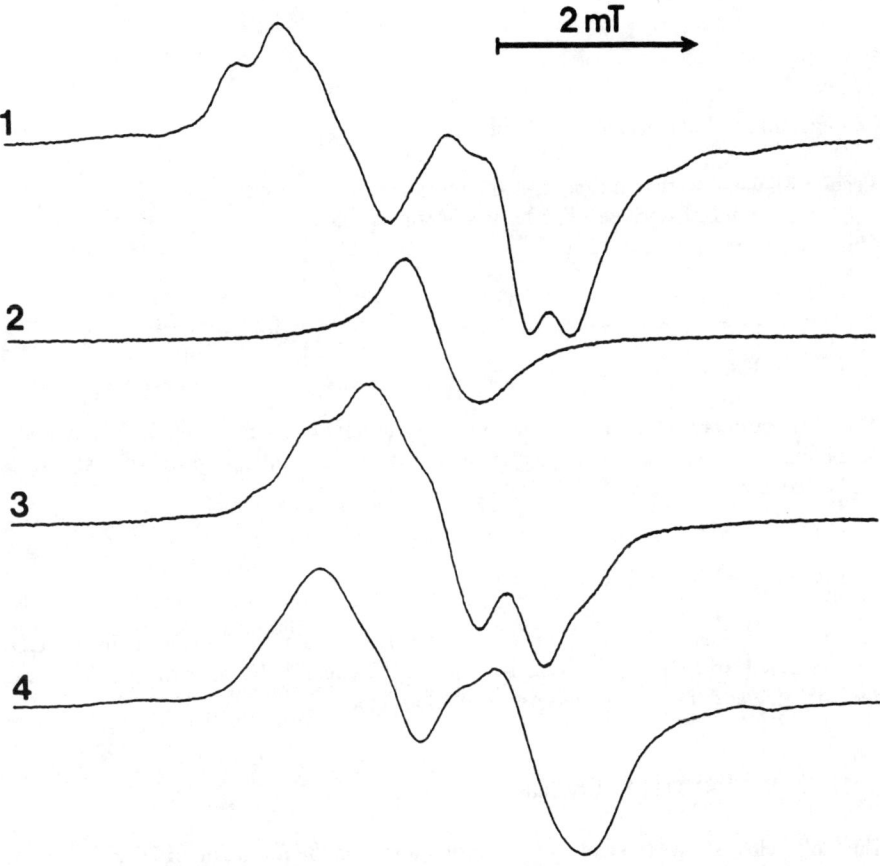

Figure 1 EPR spectra of γ-irradiated (1) apple pectin, (2) potato tuber pectin, (3) polygalacturonic acid, and (4) α-methyl glucoside

PRACTICAL METHOD FOR DETECTING IRRADIATED CHICKEN AND TURKEY

D. Schwartz, L. Lakritz and K. Kohout

Eastern Regional Research Center, Agricultural Research Service
U.S. Department of Agriculture, 600 East Mermaid Lane
Philadelphia, Pennsylvania 19118
U.S.A.

1 INTRODUCTION

This abstract details a test applicable to poultry breasts exposed to relatively low doses of gamma rays. The screen is simple, inexpensive, uses little organic solvent and is applicable to a large number of samples.

2 EXPERIMENTAL PROCEDURE

The method is based on the observation that small amounts of formaldehyde (HCHO) are generated when poultry tissue is irradiated. The HCHO is extracted following derivatisation and the derivative is estimated fluorimetrically.

3 RESULTS AND DISCUSSION

The table below shows the effect of irradiation doses on the formation of HCHO in some chicken and turkey samples. Following irradiation, the samples were stored at -18°C for approximately 2 months and analysed at 0, 1, 4, 8 and 10 weeks.

4 CONCLUSIONS

Gamma irradiation of poultry breast tissue generates HCHO in concentrations estimated to be < 10 ppm (parts per million) when doses up to 6 kGy have been applied. Under the conditions used in the experiment it was possible to distinguish irradiated from control tissue even after 2 months storage at -18°C.

The work thus far should be considered as preliminary as a number of problems must still be investigated.

Table 1 *Effect of Storage and Irradiation Dose on Sample Fluorescence*

Species	Storage (Weeks)	Fluorescence Reading					
		0 kGy	1 kGy	2 kGy	3 kGy	6 kGy	10 kGy
Chicken	0	20	41	103	146	396	532
	1	13	39	62	169	311	446
	4	17	33	78	107	191	318
	8	8	22	55	92	142	205
Turkey	0	14	26	98	162	327	547
	1	12	38	71	110	326	468
	4	16	37	58	100	232	371
	8	9	28	54	80	143	265
	10	10	42	111	167	316	476

APPLICATION OF A MICROBIOLOGICAL SCREENING METHOD FOR THE INDICATION OF IRRADIATION OF POULTRY MEAT

G. Wirtanen,[1] S. Salo,[2] M. Karwoski[1] and A-M. Sjoberg[1]

[1]VTT Biotechnology and Food Research, Tietotie 2,
[2]University of Technology, Department of Chemical Engineering, Otakaari 1
Espoo
Finland

1 INTRODUCTION

The FDA (Food and Drug Administration of the Department of Health and Human Services, USA) ruling of May 1990 permits the use of irradiation of fresh or frozen poultry and poultry parts, including ground and mechanically separated poultry products, at absorbed doses of 1.5 to 3 kGy to control foodborne pathogens and bacteria.[1] The aim of this study was to apply a microbiological method (DEFT/APC) to assess the possible irradiation treatment of samples of frozen poultry meat.

2 SAMPLE PREPARATION

The samples used in this assessment were received from the University of Bologna in March (preliminary samples), April (batches I and II) and May (batches III and IV) 1993. The 40 samples were labelled according to the microbial load and the doses used. These samples were non-irradiated or irradiated with doses of 3, 5 and 7 kGy. The irradiation was performed using a linear electron beam accelerator.

3 METHOD

The principle of the method applied to poultry meat samples is based on comparison of the aerobic plate count (APC) results obtained by cultivation on nutrient agar and by the direct epifluorescent filter technique (DEFT).[2,3] The APC count indicates the number of microorganisms present in the sample at the time of determination that are capable of growth under the culturing conditions used. The DEFT count is the total number of microorganisms, both viable and non-viable, that have ever been present in the sample.

4 RESULTS

The irradiation doses used in this study (3-7 kGy) were sufficient to lower the viable count by at least 2 log units, as shown in Figure 1.

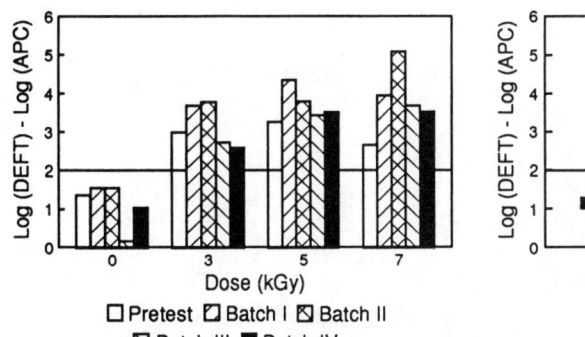
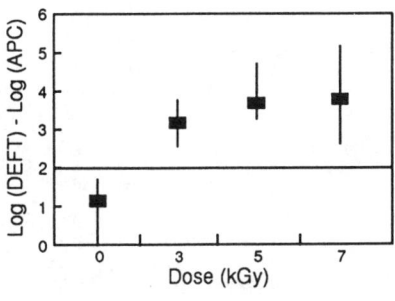

Figure 1 *The difference between the manually counted DEFT units and the aerobic plate counts DEFT/APC (given in log units/g) for both non-irradiated and irradiated samples in the 5 different batches (the pretest, batches I, II, III and IV)*

Figure 2 *Basic statistics (min., max. and mean values) of the microbial load in poultry meat samples (10 from each dose level) both non-irradiated and irradiated with doses of 3, 5 and 7 kGy using the difference between manually counted DEFT units and conventional plating DEFT/APC (given in log units/g).*

The greatest difference between the DEFT and APC values was 1.7 log units for non-irradiated samples and the smallest difference for irradiated poultry meat samples, independent of the irradiation dose used (3-7 kGy), was 2.6 log units (Figure 2). The mean values of the differences between DEFT and APC counts in the determinations performed were 1.14 log units for non-irradiated samples and 3.16, 3.68 and 3.79 log units for samples irradiated with doses of 3, 5 and 7 kGy respectively (Figure 2).

5 DISCUSSION

The effects of gamma irradiation on microbial loads of chicken carcasses were reported by Katta et al.[4] The results indicated that irradiation of chicken at 2 kGy, or more, substantially reduced bacterial loads. The results from the study outlined in this paper are in accordance with those of Katta et al. A difference of 2 logarithmic units can be considered as a limit value in cases where the use of irradiation must be confirmed.

References

1. Food and Drug Administration, *Fed. Reg.*, 1990, **55**, 18538.
2. G. Wirtanen, A-M. Sjoberg, F. Boisen and T. Alanko, *J. AOAC Int.*, 1993, **76**, 674.
3. J. Raffi, H. Delincée, E. Marchioni, C. Hasselmann, A-M. Sjöberg, M. Leonardi, M. Kent, K. W. Bögl, G. Schreiber, H. Stevenson and W. Meier, "Concerted Action of the Community Bureau of Reference on Methods of Identification of Irradiated Foods," European Commission (BCR), Brussels, Luxembourg, 1994, EUR 15261 EN.
4. S. Katta, D. Rao, G. Sunki and C. Chawan, *J. Food Sci.*, 1991, **56**, 371.

Participants

Dr Sándor BARABÁSSY
University of Horticulture and Food Industry
Pf 53, Ménesi út 45
H.-1502 Budapest
Hungary

Mr Carl BLACKBURN
Ministry of Agriculture, Fisheries and Food
Nobel House
17 Smith Square
London SW1P 3JR
U.K.

Dr Lorna A CARMICHAEL
Scottish Universities Research and Reactor Centre
East Kilbride
Scotland G75 OQU
U.K.

Ms Dominique DEFRISE
Ion Beam Applications/EB Division
Chemin Du Cyclotron/Av. J. Lenoir 6
1348 Louvain-la-Neuve
Belgium

Dr Henry DELINCÉE
Bundesforschungsanstalt für Ernährung
Engesserstrasse 20
D-76131 Karlsruhe
Germany

Dr Marc F DESROSIERS
Building 245, Room C229
National Institute of Standards and Technology
Gaithersburg, MD 20899
U.S.A.

Professor Dieter AE EHLERMANN
Institute of Food Engineering
Federal Research Centre for Nutrition
Engesserstrasse 20
D-76131 Karlsruhe
Germany

Dr Chris ELLIOTT
Veterinary Sciences Division
Department of Agriculture for Northern Ireland
Stoney Road
Upper Newtownards Road
Belfast BT4 3SD
U.K.

Professor Luciano de FRANCESCHI
Cattedra Di Fisica
Facoltà di Agraria
Università Degli Studi
Vla del Borghetto 80
56124 Pisa
Italy

Dr René FUHLENDORFF
DTI Chemistry
Danish Technological Institute
Teknologiparken
DK-8000 AARHUS C
Denmark

Dr Sheila M GLIDEWELL
Scottish Crop Research Institute
Invergowrie
Dundee DD2 SDA
Scotland
U.K.

Mr William D GRAHAM
Food Science Division
Department of Agriculture for Northern Ireland
Newforge Lane
Belfast BT9 5PX
U.K.

Dr Richard GRAY
Food Science Division
Department of Agriculture for Northern Ireland
Newforge Lane
Belfast BT9 5PX
U.K.

Dr Lynne HAMILTON
Department of Food Science
The Queen's University of Belfast
Newforge Lane
Belfast BT9 5PX
U.K.

Dr Kerie M HAMMERTON
Radiation Technology
Australian Nuclear Science and Technology Organisation
PMB 1, Menai, NSW 2234
Australia

Professor A K Muzhar M HAQUE
School of Applied Plant Science
South Bank University
London SE1
U.K.

Dr Toru HAYASHI
Radiation Technology Laboratory
National Food Research Institute
Ministry of Agriculture, Forestry and Fisheries
Kannondai, Tsukuba, Ibaraki 305
Japan

Professor Christopher H S HITCHCOCK
(University of Surrey)
2 Chudleigh Close
Bedford MK40 3AW
U.K.

Dr Yoko KAWAMURA
National Institute of Health Sciences
1-18-1 Kamiyoga, Setagaya
Tokyo 158
Japan

Professor József KISPÉTER
College of Food Industry, Szeged
Mars tér 20
H-6722 Szeged
Hungary

Dr Tamikazu KUME
Japan Atomic Energy Research Institute
Tatasaki Radiation Chemistry Research Establishment
1233 Watanuki, Takasaki 370-12, Gunma
Japan

Dr Leslie G LADOMERY
ADAC Associate Referee
Irradiated Food
Herman Otto U 7
4400 Nyiregyhaza
Hungary

Mr Torben LETH
The National Food Agency of Denmark
Institute of Food Chemistry and Nutrition
Morkoj Bygade 19, DK 2860 Soborg
Denmark

Dr Eric MARCHIONI
Faculté de Pharmacie
74 Route du Rhin
Laboratoire de Chimie Analytique 2
67400 Illkirch
France

Dr Joe McKEOWN
AECL Accelerators
436B Hazeldean Road
Kanata, Ontario K2L IT9
Canada

Dr Cecil H McMURRAY
Department of Agriculture for Northern Ireland
Room 647
Dundonald House, Upper Newtownards Road
Belfast BT4 3SB
U.K.

Dr Brian T McMURRAY
Department of Food Science
The Queen's University of Belfast
Newforge Lane
Belfast BT9 5PX
U.K.

Dr Werner MEIER
Kantonales Laboratorium Zürich
Postfach
8030 Zürich
Switzerland

Dr Wojciech MIGDAL
Institute of Nuclear Chemistry and Technology
Dorodna 16, 03-195 Warsaw
Poland

Dr Kim M MOREHOUSE
US Food and Drug Administration
Centre for Food Safety and Applied Nutrition
Chemistry Methods Branch, HFS-248
200 C Street, SW
Washington DC 20204
U.S.A.

Professor Wassef W NAWAR
Department of Food Science
University of Massachusetts at Amherst
Chenoweth Laboratory
Amherst MA 01003
U.S.A.

Mr Håkan NILSON
The Swedish National Food Administration
Chemical Division 2
Box 622
751 26 Uppsala
Sweden

Mr Hans NOOTENBOOM
Food Inspection Service
PO Box 260
6500 AG Nijmegen
Netherlands

Dr Astrid NORDBOTTEN
Norwegian Food Control Authority
PO Box 8187 DEP
N-0034 Oslo
Norway

Dr Sandro ONORI
Istituto Superiore Ai Sanita
Physics Laboratory
Viale Regina Elena 299
00161 Rome
Italy

Dr Marco PANTALONI
Istituto Superiore Ai Sanita
Physics Laboratory
Viale Regina Elena 299
00161 Rome
Italy

Dr Jürgen PFORDT
Staatliches Lebensmitteluntersuchungsamt
Postfach 2462
26014 Oldenburg
Germany

Professor Jacques RAFFI
LARQUA
Faculteé de Sainte-Jérôme,
Avenue Escadrille Normandie Niémen
F-13397 Marseille cedex 20
France

Ms Ruksana RAHMAN
School of Applied Science
Food Research Centre
South Bank University
London SEI OAA
U.K.

Dr Peter ROBERTS
Nuclear Sciences Group
Institute of Geological and Nuclear Sciences
PO Box 31-312
Lower Hutt
New Zealand

Dr David C W SANDERSON
Scottish Universities Research and Reactor Centre
East Kilbride
Scotland G75 OQU
U.K.

Dr Georg A SCHREIBER
Bundesinstitut für gesundheitlichen Verbraucherschutz und
Veterinarmedizin
BgVV, General-Pape-Strasse 62
D-12101 Berlin
Germany

Dr Grit SCHULZKI
Bundesinstitut für gesundheitlichen Verbraucherschutz und
Veterinarmedizin
BgVV, General-Pape-Str. 62
D-12101 Berlin
Germany

Mr Mohamed Mansour SHERIF
University of Horticulture and Food Industry Budapest
Ménesi út. 43-45
1118 Budapest
Hungary

Dr Daniel P SCHWARTZ
USDA
600 E Mermaid Lane
Philadelphia, PA 19118
U.S.A.

Ms M Paula SEQUEIRA ESTEVES
Physics Department
National Institute for Industrial Engineering and Technology
Estrada Nacional 10
2685 Sacavém
Portugal

Professor Waclaw STÀCHOWICZ
Institute of Nuclear Chemistry and Technology
Department of Radiation Chemistry and Technology
Dorodna 16
03-195 Warsaw
Poland

Dr M Hilary STEVENSON
Food Science Division
Department of Agriculture for Northern Ireland, and
The Queen's University of Belfast
Newforge Lane
Belfast BT9 5PX
U.K.

Dr Eileen M STEWART
Department of Food Science
The Queen's University of Belfast
Newforge Lane
Belfast BT9 5PX
U.K.

Dr Sam SUMAR
School of Applied Science
South Bank University
London SEI OAA
U.K.

Dr Setsuko TODORIKI
Radiation Technology Laboratory
National Food Research Institute
Ministry of Agriculture, Forestry and Fisheries
Kannondai, Tsukuba, Ibaraki 305
Japan

Mrs Maria J TRIGO
Departamento de Tecniologia des Produtos
de Origem Animal (DTPOA)
Instituto Nacional de Investigacàs Agralia
Quinta do Marquês - 2780 Leiias
Portugal

Professor Gordon J TROUP
Department of Physics
Monash University
Clayton, Victoria 3168
Australia

Dr Sadao UCHIYAMA
Hatano Research Institute
Food and Drug Safety Centre
729-5, Ochiai
Hatano City
Kanagawa, 257
Japan

Dr Pierre VIDAL
Association of International Industrial Irradiation (AIII)
59 Route de Paris
6920 Charbonnières les Bains
France

Dr John H H WILLIAMS
MRIC
Athrofa Gogledd Ddwyrain Cymru (NEWI)
Cei Connah, Glannau DyErdwy
CLwyd, North Wales CH5 4BR
U.K.

Professor Jilan WU
Department of Technical Physics
University of Beijing
Beijing 100871
China

Subject Index

Absorbed dose, 15
  maximum, 15-18
  minimum, 15-18
  overall average, 16
Agarose gel electrophoresis, 346, 355, 357
  (*see* also Mitochondrial DNA)
Alanine dosimeter, *see* Dosimeters
Alfalfa, 72, 74
Alkylcyclobutanones, *see* Cyclobutanones
Allspice, 85-90, 151, 187-191, 193, 196-197
Ammonia, *see* Gas evolution
Antibody, generation to
  cyclobutanones, 288
    and cross reactivity evaluation, 288-291
  thymidine glycol, 370-373
  dihydrothymidine, 370-373
Apples, 24, 378, 384, 387, 408
Apricot, 94
Autoxidation, 300, 317
Avocado, 73, 98, 100-103, 247, 294-295

Bacillus spp.
  pumulis, 9-10
  stearothermophilus, 11
Banana, 94, 408
Basil, 153, 155, 169, 242, 393
Barley, 46-47, 49-51, 72-73
Beef burgers, 53-59, 356, 364
Beef, 24-25, 271-350, 356-359, 361, 364-365
Biallylic radicals, 322-324
Bilberries, 94-95
Black tiger prawns, *see* Tiger prawns
Black pepper, *see* Peppers

Blackberries, 24
Blackcurrants, 24, 46-49
Blind trials, *see* Interlaboratory blind trials
Brown shrimp, 36, 40, 99, 101, 134-135, 141-142, 146

Calcite, 131-132, 140-144, 146
Camembert cheese, 101, 103, 247, 260-262
Caraway, 169, 187-189, 199
Carbon monoxide, *see* Gas evolution
Cardamom, 76-78
Carp, 24, 26-27
Celery leaf, 91, 150-151
Cellulose, 28-29, 46, 50-51, 93, 95
Charlock, 28
Cherries, 378, 384, 387-388
Chicken
  bone, 24-27, 53, 62-69, 75-94, 401-402
  meat, 245-247, 270-272, 277, 291, 293, 295, 300, 306, 310, 327-328, 332-333, 336-339, 350, 353, 356, 378, 410-413
Chilli, 169, 242
Chromatography
  column, 99, 251
  gas, 99, 251, 260-261, 270, 293-294, 336
  high pressure liquid, 248, 259-260, 303-304, 311, 346
Cinnamon, 72, 76-79, 85-90, 151, 153, 169-171, 178-180, 187-191, 193, 196-197, 199, 400
Clams, 94, 163, 250, 253, 256
Clostridium sporogenes, 10
Cloves, 76, 78, 91, 168-171

Cobalt-60 irradiator, 4
  (*see also* Irradiation facilities)
Cockles, 164-165
Cocoa, 72
Cod, 24-26, 250, 254, 260, 263-264
Coffee, 72
'Comet' assay, *see* DNA
Common mussel, *see* Mussels
Computer-simulation, of ESR signals, 85, 89, 91
Coriander, 75-78, 91, 151, 153
Crevette, Mediterranean, 34-40, 140-142, 145
Crabs, 141
Criteria, for an irradiation detection method, 33, 249, 269, 367
Crustacea, 33, 79-81, 104, 139, 256
Cumin, 24, 29, 76-78, 153, 155, 242, 244
Curry powder, 72, 76-78, 169, 400
Cyclobutanones, 269, 285, 406
  detection in
    irradiated liquid whole egg, 271, 278-279, 281-282
    irradiated meats, 270-282, 291, 406-407
  effect of irradiation dose, 270
  formation mechanism, 270
  parent fatty acids, 271, 286
  stability, 270-271, 273-277

Dates, 24, 27-28, 94-95, 408
DCI system, 293
Differential scanning calorimetry, 187
Dihydrothymidine, structure, 368
Dill, 153, 155, 171, 168-171
Direct Epifluorescent Filter Technique/ Aerobic Plate Count (DEFT/ APC), 378, 392
  to detect irradiated chicken, 412-413
  to detect irradiated spices, 393-396
DNA
  alterations by irradiation
  and methods of detection, 346-347
  'comet assay', to detect, 349
    irradiated beef, 350
    irradiated chicken, 350, 353-354
    irradiated pork, 350, 353-354

DNA (continued)
    irradiated veal, 350-351
  (*see also* mitochondrial DNA and ELISA)
Dodecylcyclobutanone, 270, 271, 286, 406
  (*see also* Cyclobutanones)
Dose estimation, of irradiated chicken bone
  by ESR/calibration curve, 64, 401
  by ESR/dose-additive method, 26, 64, 401
  and influencing factors, 65-69, 401
Dosimeters
  alanine, 11-12, 70-72, 272
  dose rate dependence, 12
  Far West, 11-12, 54, 356
  perspex, 272
  plastic, 6-7
  radiochromic film, 216
  sucrose/silicone, 11
Dried vegetables, 24, 29
Duck, 24, 26

Egg
  liquid whole, 271, 277-279, 281, 327
  powder, 245-247
  shell, 24, 27, 64-69, 80, 93-94, 108-115, 117-118
  white, 306, 310-315
  (*see also* Proteins)
Electron-beam irradiation
  dose distribution, 4-7
  effect of dose rate, 5-7, 9-12
  and ESR dosimetry, 11-12
  facilities, 7-9
  and microorganism sensitivity, 9-11
ELISA
  competitive, 369-370
  development methodology, 285-288, 369-373
  non-competitive, 369-370
  to detect cyclobutanones 285
  to detect dihydrothymidine, 368
  to detect thymidine glycol, 368
EPR spectroscopy, *see* ESR spectroscopy
Echerichia coli, 10
ESR signal
  effect of animal age, 25, 64

Subject Index 427

ESR signal (continued)
  effect of animal origin, 25
  effect of animal sex, 42, 64
  effect of animal size, 42
  effect of animal species, 33-36, 25, 101
  effect of bone extraction methods, 58-59
  effect of cooking
    post-irradiation, 37, 41
    pre-irradiation, 36, 41
  effect of dose rate, 11-12, 68-69
  effect of sample preparation, 38
  effect of temperature, 40, 68
  effect of irradiation dose, 25-26, 38-39, 65-66, 86-88, 256
  effect of relative humidity, 89-90
  effect of sample site used, 25, 37, 41
  effect of sample water content, 36, 47, 50-51
  origin, 23-24, 27-28, 46, 89, 94-95, 402-403, 408
  reproducibility, 65
  specificity, 36
  stability, 25, 27-31, 36-37, 40-41, 47-48, 51, 66-67, 73-75, 77, 79, 81-83, 86, 88, 94, 104, 113-114, 256
ESR spectroscopy, to detect
  irradiated cereals, 72-73, 46
  irradiated Crustacea/shellfish, 33, 79, 94, 99, 252
  irradiated eggshell, 24, 80, 94, 108
  irradiated fish, 24, 252
  irradiated fruits, 24, 46, 93, 408
  irradiated herbs and spices, 24, 70, 75, 85, 93
  irradiated meats, 24, 53, 62, 80, 96
  irradiated mechanically recovered meat (MRM), 53
  irradiated mushrooms, 24, 29-30
  irradiated Pistachio nuts, 24, 80, 93, 108
  irradiated processed foods, 30-31, 53-54
  irradiated vegetables, 24, 409
Exoskeleton
  components, 34, 36-37, 41

Falling number, 187, 192
  (*see also* Viscosity measurement)

Far West Technology dosimeters, *see* Dosimeters
Fat extraction methods, 99, 250-251, 260, 294
Fatty acid methyl esters, 245
  preparation, 251, 260
Fennel, 75-78, 94, 96, 160, 162-163, 242-245
Fenugreek, 75-76
Figs, 27, 94
Fish, 26-27, 94, 96, 250, 259, 294, 303, 306, 356, 378
Fluorescence, 346, 410-411
Florisil column chromatography, 99, 248, 251, 260, 263
Formaldehyde, 410
Food irradiation, advantages and disadvantages, 357
Frogs legs, 94, 253, 303, 306
Fruits, 129-130, 247, 293, 378, 383, 408
  dried, 94, 96
  fresh, 27-28, 45, 73, 94, 98

Gamma-radiation, 3-4, 9-12
Garlic, 75-78, 169, 179-180, 188-189
Gas chromatography, *see* Chromatography
Gas evolution
  ammonia, 326, 328
  carbon monoxide, 300, 326-328, 331, 335
  hydrogen, 300, 326-328, 331, 335
  hydrogen sulphide, 326, 328
  measurement by gas sensors, 327, 337
    effect of electronic drift, 339, 341
    effect of irradiation dose, 332-333
    effect of storage, 328, 332-333, 337-339
    effect of thawing, 337, 341
    from ice, 336-337, 339-340
    from irradiated foods, 327-329, 331-334, 337-341
Gelatin, 24, 30
Germination tests, 380
  for irradiated rice, 384, 388-389
  for irradiated wheat, 384, 389-390
  (*see also* Half-embryo test)
Ginger, 29, 72, 77-78, 169, 178-179, 188-189, 193, 400

Glass prawns, 36
Glow curves, *see* Thermoluminescence
Good Irradiation Practice, 14-19
Good Manufacturing Practice, 14
Goose, 24, 26
Gooseberries, 24
Grapefruit, 24, 27, 384-386
Grapes, 94, 96
Grooved carpet shell, 164-165

Haddock, 250
Half-embryo test, development
  effect of irradiation dose, 384
  effect of seed dissection, 384
  for irradiated apples, 387
  for irradiated cherries, 388
    and use of plant hormones, 388
  for irradiated grapefruit, 386
Herbs, 24, 28, 121, 125, 159, 178
High performance liquid chromatography, *see* Chromatography
Histological/Morphological detection methods, 378, 380
Hapten, 285
  (*see also* ELISA)
Hydrocarbons
  detection in
    irradiated cheese, 260, 98-99, 101-103
    irradiated eggs, 245-247
    irradiated fish, 243, 249, 259
    irradiated fruits, 247, 98-99, 101-103, 295
    irradiated meats, 245, 247, 295
    irradiated oils, 263, 295
    irradiated shellfish, 249, 259
    irradiated spices, 242-244
  effect of co-extractants, 253
  effect of irradiation dose, 103, 246, 262
  effect of lipid content, 253
  precursor fatty acids, 102, 245, 252, 261, 265
Hydrogen, *see* Gas evolution
Hydroperoxides, 300, 318, 322
  (*see also* Peroxides)
Hydroxyapatite, 23-24, 68, 94-95, 402-404

Ice, 336-337, 339-340

Immunoassays, *see* ELISA
Immunoblotting, 313-315
Immunochemical detection, 346
  of irradiated eggs, 310
  of modified DNA bases, 367
  (*see also* ELISA)
Impedance measurement
  electrode system, 188, 203, 198
  of irradiated potatoes, 188, 202
    dose dependency, 209, 197-199
    effect of cultivar, 197, 210-211
    effect of electric current, 204-206
    effect of electrode type, 204-207
    effect of measuring region, 209
    effect of measuring temperature, 204
    effect of sample pre-incubation, 207-208
Interlaboratory blind trials
  cyclobutanones, 269, 272-283
  DNA 'comet' assay, 353-354
  ESR, 37, 79-80, 59-61, 99-100, 108-113
  hydrocarbons, 98-99, 242, 244-247
  Mitochondrial DNA, 361
  peroxides, 318, 321
  thermoluminescence, 100-101, 178-181
  o-tyrosine, 300
  viscosity measurement, 229-234
Intestinal grits, *see* Thermoluminescence
Irradiation facilities, 7-9

Joint FAO/IAEA/WHO Expert Committee on Food Irradiation (JECFI), 16, 269

King prawn, 34-40, 79
King scallop, 140-141, 145-146

Labelling of irradiated food, 14, 17, 125
Lactone, 290-291
Lamb, 94, 271
Laurel, 151, 153
Legumes, 378
Lemons, 24, 380, 386
Limulus amoebocyte lysate (LAL) test, 378-380
Linoleic acid, 99, 252, 271, 285-286, 293
Linear accelerator, 4-9, 53

Lipid extraction, *see* Fat extraction
Liquid whole egg, *see* Eggs
Liquor, 317, 322
Listeria monocytogenes, 10, 260
Lobster, 141-142, 146
  (*see also* Norway lobster)
Luminescence detection techniques, 121, 124, 139
Lysozyme, 311-313

α-Methyl glucoside, 408-409
Macaroni, 30-31
Mackerel, 243, 250, 253
Malondialdehyde, 323-324
Manganese, 24, 34, 77, 79, 93-95, 113, 117, 132, 144
Mango, 98-103, 247
Marjoram, 24, 29, 153, 169, 179-180, 188-189, 199
Micro gel electrophosesis, 346-347
Mass spectrometry, 252, 255, 270, 285, 406
Meat, 25-26, 94, 96, 271, 326, 331, 365, 378-379
Mechanically recovered meat (MRM), 53, 271, 328-329, 356
  methods of bone extraction, 55
Mediterranean crevette, *see* Crevette
Mesophilic aerobic spore count (ASC), 392-394
Microorganisms
  irradiation sensitivity, 9-11, 378
Milk protein concentrate powder, 173-175
Mitochondrial DNA (mtDNA), 346
  different forms, 355, 357
  effect of irradiation dose, 357-361
  effect of irradiation temperature, 358, 359
  effect of storage, 358, 360, 361
  extraction, 356, 357
  separation, 357
    from irradiated beef, 364-365
    from irradiated beef liver, 357-359
    from irradiated chicken meat, 359-361
    from irradiated fish, 362
    from irradiated MRM, 361, 365
    from irradiated prawns, 363

Moisture/water content, 47, 51
Morphological changes, 378
Multiple gas sensors, *see* Gas evolution
Mushrooms, 29-30, 122, 350
Mussels, 94, 141-142, 145-146, 164, 306
Mustard, 151, 153, 155, 169, 171

Nephrops norvegicus, *see* Norway lobster
Near infra-red spectrometry, 185, 187-188, 193-197
Norway lobster, 33, 36-43, 99, 101, 104, 141-142, 146-147, 346, 406
Nucleic acids, 368
Nutmeg, 72, 76-78, 151, 153, 155, 188-189, 242, 244
Nuts, *see* Pistachio nuts

o-Tyrosine
  analysis of, 299-300, 303
  production, 299
Oats, 72
Oils, 241, 293-296
Oleic acid, 99, 241, 252, 261, 271, 286, 293
Onion, 94, 122, 169, 171, 179-180, 188-189, 199
Optical bleaching, 130
Optical exposure, 129
Oranges, 24, 27, 380, 386
Oregano, 73, 94, 96, 159, 162-163, 169, 171, 179-180, 242
Ovalbumin, 310-314
Overall average dose, 16, 18
Ovomucoid, 310-314
Ovotransferrin, 310-314
Oysters, 141-142, 250, 253, 256-257

Palmitic acid, 99, 241, 252, 261, 286
Papaya, 93-96, 98-104, 247, 271, 286, 408
Paprika, 24, 29, 80, 98, 104, 108-110, 112-117, 169, 179-180, 187, 193-195, 394-395
Paramagnetic centres, 23, 25, 27, 116
Parsley, 73, 94, 96, 133, 160, 162-163, 379
Pear, 24, 27
Peas, 72-74
Pectin, 185, 408-409

Peppers, 24, 29, 76-78, 85-91, 108-110
  112-117, 150-151, 153, 169, 179-
  180, 185-193, 215-226, 229-237,
  241, 244, 328, 341, 394-395, 400
Perch, 250
Peroxides, 299
  analysis, 300, 318
  of irradiated docosahexaenoic acid,
    322-324
  of irradiated liquor, 322
  of irradiated pork, 318-321
  effect of irradiation dose, 319-323
  effect of pH, 322-324
  effect of storage, 318-321
Phenylalanine, 299
Photo-transfer luminescence (PTTL), 139
Photostimulated luminescence (PSL)
  of irradiated herbs and spices, 133-135
  of irradiated shellfish, 134-13, 143-147
  research spectrometer, 132-134
  using infra-red stimulation, 132, 144
Pike, 94
Pink shrimp, 34-39
Pistachio nut, 28, 80, 82-83, 93-94, 96,
  108-109, 111-117
Plum, 24
Polar ice, 339-340
Polygalacturonic acid, 408-409
Polyacrylamide gel electrophoresis, see
  SDS-PAGE
Pork, 24-25, 82, 94, 271, 277-279, 282,
  317-321, 338, 350, 353-354, 379
Potato, 186, 197-198, 202-213, 296, 408
Prawns, 34-35, 79-81, 271, 356, 363, 406
Pre-cooked meals, 58-61
Processed food, 30
Protein blotting, 311, 313-315
Proteins
  carrier, 287, 311
  from egg white, 310-315, 381
    changes by irradiation, 311-312
Prunes, 94
Pulsed field gel electrophosesis, 346
Pulsed photostimulated luminescence, see
  PSL

Quahaug, 164-165

Queen scallops, 141

Raisins, 96
Raspberry, 46, 95
Re-irradiation, 26, 62, 125, 135-136, 149,
  168, 171, 404
Red currants, 95
Relative humidity
  influence on ESR signals, 89-90
  influence on viscosity, 187-190
Rice, 47, 49, 380, 388-389
Roach, 26
Rosemary, 73, 94, 96, 151, 153, 155, 160,
  162-163

Sage, 94, 96, 151, 153, 160, 162-163, 168,
  179-180
Sago, 72-73
Saithe, 263-265
Salmon, 250-256
Sardine, 94
Satay, 75-76, 78
Savory, 94, 96, 160, 162-163
Scallop, 94, 141, 145-146, 163-165
School prawns, 79
Screening methods, 113, 147, 162, 234,
  301, 326, 331, 341, 349, 381, 391,
  396, 410
SDS-PAGE, 310-315
Shaetfish, 26
Shrimp, 34-38, 100-101, 104-105, 122,
  131, 135, 141-142, 242, 250, 260-
  266, 304, 327, 331, 333, 392, 399
Spices, 28-29, 73, 75-78, 85-92, 116-117,
  121, 125, 133-135, 139, 150-156,
  168-170, 178, 185, 215, 229, 242,
Starch, 47, 49, 185, 187, 217-220, 232-233
  indices of damage, 187-188
Stearic acid, 241, 252-261, 271, 286, 406
Strawberries, 28, 45-50, 73, 95, 122, 378
Sugars, 27, 93, 95

Tea, 72-73
Tertiary food products, 53
Tetradecylcylcobutanone, 271, 286, 406
  (see also Cyclobutanones)

Thermoluminescenc, 95, 100, 121, 168-172
  and blending, 127-128, 136
  and optical bleaching, 130
  glow curves, 125, 127-128, 140-141, 143, 159, 174-175
  glow ratios, 126, 128-131, 144
  measuring conditions, 125-126, 132, 141, 143, 168, 173
  of fruits and vegetables, 129-131
  of herbs and spices, 124, 149, 159,
  of irradiated shellfish, 100-101, 131-132, 139-143, 163-165
    using intestinal grits, 101, 104, 131, 142-143
    using shells, 140-142
  of milk protein concentrate powder, 173-175
  quality assurance parameters, 126
  concordance tests, 126-127
  and optical exposure, 129
  sample preparation, 125, 131, 141-143, 161, 168
  signal stability, 96, 132, 141-142, 164
Thyme, 94-95, 150, 151, 153, 160, 162-163, 169, 179-180
Thymine glycol, structure, 368
Tiger prawn, 34-41, 142-142, 146
Trout, 26, 94, 362
True shrimp, 36
Turkey, 26, 54, 410-411
Turmeric, 76, 78, 91, 150, 151, 153, 155, 169, 179-180

Vanilla, 400
Veal, 25, 350-351
Vegetable oils, 241, 293

Viscosity measurement
  effect of heat gelatinisation, 187, 190-193, 223, 224, 226
  effect of irradiation dose, 188, 218-220
  effect of relative humidity, 187, 189-190
  effect of storage, 187, 189, 220-223, 226
  effect of temperature, 187, 189-191
  falling number, 187, 192
  of irradiated spices, 187, 189, 215, 229
  effect of decontamination treatments, 220, 223, 233
  effect of measuring conditions, 223, 224, 232
  parameter values, 220, 224
  effect of planting locality, 225-226
  effect of pH adjustment, 232-233

Wheat, 47-49, 72-74, 371, 373, 380, 389-390
White pepper, *see* Peppers

Xe-excitation, 174

Yeast, 72-74